CAMBRIDGE LIBRARY COLLECTION

Books of enduring scholarly value

Technology

The focus of this series is engineering, broadly construed. It covers technological innovation from a range of periods and cultures, but centres on the technological achievements of the industrial era in the West, particularly in the nineteenth century, as understood by their contemporaries. Infrastructure is one major focus, covering the building of railways and canals, bridges and tunnels, land drainage, the laying of submarine cables, and the construction of docks and lighthouses. Other key topics include developments in industrial and manufacturing fields such as mining technology, the production of iron and steel, the use of steam power, and chemical processes such as photography and textile dyes.

Papers, Literary, Scientific, Etc.

An electrical engineer, university teacher and wide-ranging writer, Fleeming Jenkin (1833–85) filed thirty-five British patents in the course of his career. Edited by Sidney Colvin (1845–1927) and J.A. Ewing (1855–1935) and first published in 1887, this two-volume work brings together a selection of Jenkin's varied and engaging papers. The collection ranges from notes on his voyages as a marine telegraph engineer, to a critical review of Darwin's *On the Origin of Species*, essays on literature, and thoughts on technical education. A memoir written by Robert Louis Stevenson, his former student, provides biographical context and attests to Jenkin's many interests and talents across the arts and sciences. Volume 1 begins with Stevenson's memoir, incorporating Jenkin's records of his voyages. This is followed by writings on literature and drama. Three pieces on scientific subjects, including the review of Darwin, conclude the volume.

Cambridge University Press has long been a pioneer in the reissuing of out-of-print titles from its own backlist, producing digital reprints of books that are still sought after by scholars and students but could not be reprinted economically using traditional technology. The Cambridge Library Collection extends this activity to a wider range of books which are still of importance to researchers and professionals, either for the source material they contain, or as landmarks in the history of their academic discipline.

Drawing from the world-renowned collections in the Cambridge University Library and other partner libraries, and guided by the advice of experts in each subject area, Cambridge University Press is using state-of-the-art scanning machines in its own Printing House to capture the content of each book selected for inclusion. The files are processed to give a consistently clear, crisp image, and the books finished to the high quality standard for which the Press is recognised around the world. The latest print-on-demand technology ensures that the books will remain available indefinitely, and that orders for single or multiple copies can quickly be supplied.

The Cambridge Library Collection brings back to life books of enduring scholarly value (including out-of-copyright works originally issued by other publishers) across a wide range of disciplines in the humanities and social sciences and in science and technology.

Papers, Literary, Scientific, Etc.

With a memoir by Robert Louis Stevenson

VOLUME 1

FLEEMING JENKIN
EDITED BY SIDNEY COLVIN
AND J.A. EWING

CAMBRIDGE
UNIVERSITY PRESS

University Printing House, Cambridge, CB2 8BS, United Kingdom

Published in the United States of America by Cambridge University Press, New York

Cambridge University Press is part of the University of Cambridge.
It furthers the University's mission by disseminating knowledge in the pursuit of
education, learning and research at the highest international levels of excellence.

www.cambridge.org
Information on this title: www.cambridge.org/9781108068031

© in this compilation Cambridge University Press 2014

This edition first published 1887
This digitally printed version 2014

ISBN 978-1-108-06803-1 Paperback

PAPERS AND MEMOIR OF

FLEEMING JENKIN

VOL. I.

PRINTED BY
SPOTTISWOODE AND CO., NEW-STREET SQUARE
LONDON

FLEEMING JENKIN

AGED 51

FROM A PHOTOGRAPH BY MR. JOHN MOFFAT, EDINBURGH

PAPERS

LITERARY, SCIENTIFIC, &c.

BY THE LATE

FLEEMING JENKIN, F.R.S., LL.D.

PROFESSOR OF ENGINEERING IN THE UNIVERSITY OF EDINBURGH

EDITED BY

SIDNEY COLVIN, M.A. AND J. A. EWING, F.R.S.

WITH A MEMOIR BY ROBERT LOUIS STEVENSON

IN TWO VOLUMES

VOL. I.

LONDON

LONGMANS, GREEN, AND CO.

AND NEW YORK: 15 EAST 16th STREET

1887

CONTENTS

OF

THE FIRST VOLUME.

───•◇•───

MEMOIR BY ROBERT LOUIS STEVENSON.

PAPERS BY FLEEMING JENKIN.

ILLUSTRATIONS.

MEMOIR

OF

FLEEMING JENKIN

BY

ROBERT LOUIS STEVENSON

MEMOIR.

CHAPTER I.

The Jenkins of Stowting—Fleeming's grandfather—Mrs. Buckner's fortune—
Fleeming's father; goes to sea; at St. Helena; meets King Tom; service
in the West Indies; end of his career—The Campbell-Jacksons—
Fleeming's mother—Fleeming's uncle John.

IN the reign of Henry VIII., a family of the name of Jenkin,
claiming to come from York, and bearing the arms of Jenkin ap
Philip of St. Melans, are found reputably settled in the county
of Kent. Persons of strong genealogical pinion pass from
William Jenkin, Mayor of Folkestone in 1555, to his contemporary 'John Jenkin, of the Citie of York, Receiver General of
the County,' and thence, by way of Jenkin ap Philip, to the proper
summit of any Cambrian pedigree—a prince; 'Guaith Voeth,
Lord of Cardigan,' the name and style of him. It may suffice,
however, for the present, that these Kentish Jenkins must have
undoubtedly derived from Wales, and being a stock of some
efficiency, they struck root and grew to wealth and consequence
in their new home.

Of their consequence we have proof enough in the fact that The Jen-
not only was William Jenkin (as already mentioned) Mayor of kins of
Folkestone in 1555, but no less than twenty-three times in the Stowting.
succeeding century and a half, a Jenkin (William, Thomas,
Henry, or Robert) sat in the same place of humble honour. Of
their wealth we know that, in the reign of Charles I., Thomas
Jenkin of Eythorne was more than once in the market buying
land, and notably, in 1633, acquired the manor of Stowting Court.
This was an estate of some 320 acres, six miles from Hythe,

in the Bailiwick and Hundred of Stowting, and the Lathe of Shipway, held of the Crown *in capite* by the service of six men and a constable to defend the passage of the sea at Sandgate. It had a chequered history before it fell into the hands of Thomas of Eythorne, having been sold and given from one to another— to the Archbishop, to Heringods, to the Burghershes, to Pavelys, Trivets, Cliffords, Wenlocks, Beauchamps, Nevilles, Kempes, and Clarkes : a piece of Kentish ground condemned to see new faces and to be no man's home. But from 1633 onward it became the anchor of the Jenkin family in Kent ; and though passed on from brother to brother, held in shares between uncle and nephew, burthened by debts and jointures, and at least once sold and bought in again, it remains to this day in the hands of the direct line. It is not my design, nor have I the necessary know- ledge, to give a history of this obscure family. But this is an age when genealogy has taken a new lease of life, and become for the first time a human science ; so that we no longer study it in quest of the Guaith Voeths, but to trace out some of the secrets of descent and destiny ; and as we study, we think less of Sir Bernard Burke and more of Mr. Galton. Not only do our character and talents lie upon the anvil and receive their temper during generations; but the very plot of our life's story unfolds itself on a scale of centuries, and the biography of the man is only an episode in the epic of the family. From this point of view I ask the reader's leave to begin this notice of a remark- able man who was my friend, with the accession of his great- grandfather, John Jenkin.

This John Jenkin, a grandson of Damaris Kingsley, of the family of ' Westward Ho!' was born in 1727, and married Eliza- beth, daughter of Thomas Frewen, of Church House, Northiam. The Jenkins had now been long enough intermarrying with their Kentish neighbours to be Kentish folk themselves in all but name ; and with the Frewens in particular their connection is singularly involved. John and his wife were each descended in the third degree from another Thomas Frewen, Vicar of Northiam, and brother to Accepted Frewen, Archbishop of York. John's mother had married a Frewen for a second

husband. And the last complication was to be added by the
Bishop of Chichester's brother, Charles Buckner, Vice-Admiral
of the White, who was twice married, first to a paternal cousin
of Squire John, and second to Anne, only sister of the Squire's
wife, and already the widow of another Frewen. The reader
must bear Mrs. Buckner in mind; it was by means of that lady
that Fleeming Jenkin began life as a poor man. Meanwhile,
the relationship of any Frewen to any Jenkin at the end of these
evolutions presents a problem almost insoluble; and we need
not wonder if Mrs. John, thus exercised in her immediate circle,
was in her old age ' a great genealogist of all Sussex families,
and much consulted.' The names Frewen and Jenkin may
almost seem to have been interchangeable at will; and yet
Fate proceeds with such particularity that it was perhaps on the
point of name the family was ruined.

The John Jenkins had a family of one daughter and five
extravagant and unpractical sons. The eldest, Stephen, entered
the Church and held the living of Salehurst, where he offered,
we may hope, an extreme example of the clergy of the age. He
was a handsome figure of a man; jovial and jocular; fond of his
garden, which produced under his care the finest fruits of the
neighbourhood; and like all the family, very choice in horses.
He drove tandem; like Jehu, furiously. His saddle horse,
Captain (for the names of horses are piously preserved in the
family chronicle which I follow) was trained to break into a
gallop as soon as the vicar's foot was thrown across its back; nor
would the rein be drawn in the nine miles between Northiam
and the Vicarage door. Debt was the man's proper element; he
used to skulk from arrest in the chancel of his church; and the
speed of Captain may have come sometimes handy. At an early
age this unconventional parson married his cook, and by her he
had two daughters and one son. One of the daughters died
unmarried; the other imitated her father, and married ' imprudently.' The son, still more gallantly continuing the tradition,
entered the army, loaded himself with debt, was forced to sell
out, took refuge in the Marines, and was lost on the Dogger
Bank in the war-ship *Minotaur*. If he did not marry below

him, like his father, his sister, and a certain great-uncle William, it was perhaps because he never married at all.

The second brother, Thomas, who was employed in the General Post Office, followed in all material points the example of Stephen, married 'not very creditably,' and spent all the money he could lay his hands on. He died without issue; as did the fourth brother, John, who was of weak intellect and feeble health, and the fifth brother, William, whose brief career as one of Mrs. Buckner's satellites will fall to be considered later on. So soon, then, as the *Minotaur* had struck upon the Dogger Bank, Stowting and the line of the Jenkin family fell on the shoulders of the third brother, Charles.

Fleeming's grand- father.
Facility and self-indulgence are the family marks; facility (to judge by these imprudent marriages) being at once their quality and their defect; but in the case of Charles, a man of exceptional beauty and sweetness both of face and disposition, the family fault had quite grown to be a virtue, and we find him in consequence the drudge and milk-cow of his relatives. Born in 1766, Charles served at sea in his youth, and smelt both salt water and powder. The Jenkins had inclined hitherto, as far as I can make out, to the land service. Stephen's son had been a soldier; William (fourth of Stowting) had been an officer of the unhappy Braddock's in America, where, by the way, he owned and afterwards sold an estate on the James River, called after the parental seat; of which I should like well to hear if it still bears the name. It was probably by the influence of Captain Buckner, already connected with the family by his first marriage, that Charles Jenkin turned his mind in the direction of the navy; and it was in Buckner's own ship, the *Prothée*, 64, that the lad made his only campaign. It was in the days of Rodney's war, when the *Prothée*, we read, captured two large privateers to windward of Barbadoes, and was 'materially and distin- guishedly engaged' in both the actions with De Grasse. While at sea, Charles kept a journal, and made strange archaic pilot- book sketches, part plan, part elevation, some of which survive for the amusement of posterity. He did a good deal of surveying, so that here we may perhaps lay our finger on the beginning of

Fleeming's education as an engineer. What is still more strange, among the relics of the handsome midshipman and his stay in the gun-room of the *Prothée*, I find a code of signals graphically represented, for all the world as it would have been done by his grandson.

On the declaration of peace, Charles, because he had suffered from scurvy, received his mother's orders to retire; and he was not the man to refuse a request, far less to disobey a command. Thereupon he turned farmer, a trade he was to practise on a large scale; and we find him married to a Miss Schirr, a woman of some fortune, the daughter of a London merchant. Stephen, the not very reverend, was still alive, galloping about the country or skulking in his chancel. It does not appear whether he let or sold the paternal manor to Charles; one or other, it must have been; and the sailor-farmer settled at Stowting, with his wife, his mother, his unmarried sister, and his sick brother John. Out of the six people of whom his nearest family consisted, three were in his own house, and two others (the horse-leeches, Stephen and Thomas) he appears to have continued to assist with more amiability than wisdom. He hunted, belonged to the Yeomanry, owned famous horses, Maggie and Lucy, the latter coveted by royalty itself. 'Lord Rokeby, his neighbour, called him kinsman,' writes my artless chronicler, 'and altogether life was very cheery.' At Stowting his three sons, John, Charles, and Thomas Frewen, and his younger daughter, Anna, were all born to him; and the reader should here be told that it is through the report of this second Charles (born 1801) that he has been looking on at these confused passages of family history.

In the year 1805 the ruin of the Jenkins was begun. It was the work of a fallacious lady already mentioned, Aunt Anne Frewen, a sister of Mrs. John. Twice married, first to her cousin Charles Frewen, clerk to the Court of Chancery, Brunswick Herald, and Usher of the Black Rod, and secondly to Admiral Buckner, she was denied issue in both beds, and being very rich—she died worth about 60,000*l.*, mostly in land —she was in perpetual quest of an heir. The mirage of this fortune hung before successive members of the Jenkin family

Mrs. Buckner's fortune.

until her death in 1825, when it dissolved and left the latest
Alnaschar face to face with bankruptcy. The grandniece, Stephen's
daughter, the one who had not 'married imprudently,' appears
to have been the first; for she was taken abroad by the golden
aunt, and died in her care at Ghent in 1792. Next she adopted
William, the youngest of the five nephews; took him abroad
with her—it seems as if that were in the formula; was shut up
with him in Paris by the Revolution; brought him back to
Windsor, and got him a place in the King's Body Guard, where
he attracted the notice of George III. by his proficiency in
German. In 1797, being on guard at St. James's Palace,
William took a cold which carried him off; and Aunt Anne was
once more left heirless. Lastly, in 1805, perhaps moved by the
Admiral, who had a kindness for his old midshipman, perhaps
pleased by the good looks and the good nature of the man himself, Mrs. Buckner turned her eyes upon Charles Jenkin. He
was not only to be the heir, however, he was to be the chief
hand in a somewhat wild scheme of family farming. Mrs.
Jenkin, the mother, contributed 164 acres of land; Mrs. Buckner,
570, some at Northiam, some farther off; Charles let one-half
of Stowting to a tenant, and threw the other and various
scattered parcels into the common enterprise; so that the whole
farm amounted to near upon a thousand acres, and was scattered
over thirty miles of country. The ex-seaman of thirty-nine, on
whose wisdom and ubiquity the scheme depended, was to live
in the meanwhile without care or fear. He was to check himself
in nothing; his two extravagances, valuable horses and worthless
brothers, were to be indulged in comfort; and whether the year
quite paid itself or not, whether successive years left accumulated
savings or only a growing deficit, the fortune of the golden aunt
should in the end repair all.

On this understanding Charles Jenkin transported his
family to Church House, Northiam : Charles the second, then a
child of three, among the number. Through the eyes of the
boy we have glimpses of the life that followed : of Admiral and
Mrs. Buckner driving up from Windsor in a coach and six, two
post horses and their own four; of the house full of visitors, the

great roasts at the fire, the tables in the servants' hall laid for
thirty or forty for a month together; of the daily press of
neighbours, many of whom, Frewens, Lords, Bishops, Batchellors,
and Dynes, were also kinsfolk; and the parties 'under the
great spreading chestnuts of the old fore court,' where the
young people danced and made merry to the music of the
village band. Or perhaps, in the depth of winter, the father
would bid young Charles saddle his pony; they would ride
the thirty miles from Northiam to Stowting, with the snow to
the pony's saddle girths, and be received by the tenants like
princes.

This life of delights, with the continual visible comings
and goings of the golden aunt, was well qualified to relax the
fibre of the lads. John the heir, a yeoman and a fox-hunter,
'loud and notorious with his whip and spurs,' settled down into
a kind of Tony Lumpkin, waiting for the shoes of his father and
his aunt. Thomas Frewen, the youngest, is briefly dismissed as
'a handsome beau;' but he had the merit or the good fortune to
become a doctor of medicine, so that when the crash came he
was not empty-handed for the war of life. Charles, at the day-
school of Northiam, grew so well acquainted with the rod,
that his floggings became matter of pleasantry and reached the
ears of Admiral Buckner. Hereupon that tall, rough-voiced,
formidable uncle entered with the lad into a covenant: every time
that Charles was thrashed he was to pay the Admiral a penny;
every day that he escaped, the process was to be reversed. 'I
recollect,' writes Charles, 'going crying to my mother to be taken
to the Admiral to pay my debt.' It would seem by these terms
the speculation was a losing one; yet it is probable it paid
indirectly by bringing the boy under remark. The Admiral
was no enemy to dunces; he loved courage, and Charles, while
yet little more than a baby, would ride the great horse into the
pond. Presently it was decided that here was the stuff of a fine
sailor; and at an early period the name of Charles Jenkin was
entered on a ship's books.

From Northiam he was sent to another school at Boonshill,
near Rye, where the master took 'infinite delight' in strapping

*Fleem-
ing's
father.*

him. ' It keeps me warm and makes you grow,' he used to say.
And the stripes were not altogether wasted, for the dunce,
though still very 'raw,' made progress with his studies. It was
known, moreover, that he was going to sea, always a ground of
pre-eminence with schoolboys; and in his case the glory was
not altogether future, it wore a present form when he came
driving to Rye behind four horses in the same carriage with
an Admiral. ' I was not a little proud, you may believe,'
says he.

In 1814, when he was thirteen years of age, he was carried
by his father to Chichester to the Bishop's Palace. The Bishop
had heard from his brother the Admiral that Charles was likely
to do well, and had an order from Lord Melville for the lad's
admission to the Royal Naval College at Portsmouth. Both the
Bishop and the Admiral patted him on the head and said,
' Charles will restore the old family;' by which I gather with
some surprise that, even in these days of open house at Northiam
and golden hope of my aunt's fortune, the family was supposed
to stand in need of restoration. But the past is apt to look
brighter than nature, above all to those enamoured of their
genealogy; and the ravages of Stephen and Thomas must have
always given matter of alarm.

What with the flattery of bishops and admirals, the fine
company in which he found himself at Portsmouth, his visits
home, with their gaiety and greatness of life, his visits to Mrs.
Buckner (soon a widow) at Windsor, where he had a pony
kept for him and visited at Lord Melville's and Lord Harcourt's
and the Leveson-Gowers, he began to have ' bumptious notions,'
and his head was ' somewhat turned with fine people;' as to
some extent it remained throughout his innocent and honourable life

Goes to
sea.   In this frame of mind the boy was appointed to the *Conqueror*, Captain Davie, humorously known as Gentle Johnnie.
The captain had earned this name by his style of discipline,
which would have figured well in the pages of Marryat : ' Put
the prisoner's head in a bag and give him another dozen!'
survives as a specimen of his commands; and the men were

often punished twice or thrice in a week. On board the ship
of this disciplinarian, Charles and his father were carried in a
billy-boat from Sheerness in December 1816 : Charles with an
outfit suitable to his pretentions, a twenty-guinea sextant and
120 dollars in silver, which were ordered into the care of the
gunner. ' The old clerks and mates,' he writes, ' used to laugh
and jeer me for joining the ship in a billy-boat, and when they
found I was from Kent, vowed I was an old Kentish smuggler.
This to my pride, you will believe, was not a little offensive.'

The *Conqueror* carried the flag of Vice-Admiral Plampin, At St.
commanding at the Cape and St. Helena; and at that all- Helena.
important islet, in July 1817, she relieved the flagship of Sir
Pulteney Malcolm. Thus it befel that Charles Jenkin, coming
too late for the epic of the French wars, played a small part in
the dreary and disgraceful afterpiece of St. Helena. Life on the
guardship was onerous and irksome. The anchor was never
lifted, sail never made, the great guns were silent; none was
allowed on shore except on duty ; all day the movements of the
imperial captive were signalled to and fro ; all night the boats
rowed guard around the accessible portions of the coast. This
prolonged stagnation and petty watchfulness in what Napoleon
himself called that ' unchristian' climate, told cruelly on the
health of the ship's company. In eighteen months, according
to O'Meara, the *Conqueror* had lost one hundred and ten men
and invalided home one hundred and seven, ' being more than a
third of her complement.' It does not seem that our young
midshipman so much as once set eyes on Bonaparte ; and yet in
other ways Jenkin was more fortunate than some of his comrades. He drew in water-colour ; not so badly as his father, yet
ill enough ; and this art was so rare aboard the *Conqueror* that
even his humble proficiency marked him out and procured him
some alleviations. Admiral Plampin had succeeded Napoleon
at the Briars ; and here he had young Jenkin staying with him
to make sketches of the historic house. One of these is before
me as I write, and gives a strange notion of the arts in our old
English Navy. Yet it was again as an artist that the lad was
taken for a run to Rio, and apparently for a second outing in

a ten-gun brig. These, and a cruise of six weeks to windward
of the island undertaken by the *Conqueror* herself in quest of
health, were the only breaks in three years of murderous inaction; and at the end of that period Jenkin was invalided
home, having 'lost his health entirely.'

Meets
King Tom.

As he left the deck of the guard-ship the historic part of
his career came to an end. For forty-two years he continued to
serve his country obscurely on the seas, sometimes thanked for
inconspicuous and honourable services, but denied any opportunity of serious distinction. He was first two years in the *Larne*,
Captain Tait, hunting pirates and keeping a watch on the
Turkish and Greek squadrons in the Archipelago. Captain
Tait was a favourite with Sir Thomas Maitland, High Commis-
sioner of the Ionian Islands—King Tom as he was called—who
frequently took passage in the *Larne*. King Tom knew every
inch of the Mediterranean, and was a terror to the officers of the
watch. He would come on deck at night; and with his broad
Scotch accent, 'Well, sir,' he would say, 'what depth of water
have ye? Well now, sound; and ye'll just find so or so many
fathoms,' as the case might be; and the obnoxious passenger
was generally right. On one occasion, as the ship was going
into Corfu, Sir Thomas came up the hatchway and cast his eyes
towards the gallows. ' Bangham '—Charles Jenkin heard him
say to his aide-de-camp, Lord Bangham—' where the devil is
that other chap? I left four fellows hanging there ; now I can
only see three. Mind there is another there to-morrow.' And
sure enough there was another Greek dangling the next day.
' Captain Hamilton, of the *Cambrian*, kept the Greeks in order
afloat,' writes my author, ' and King Tom ashore.'

Services
in the
West
Indies.

From 1823 onward, the chief scene of Charles Jenkin's
activities was in the West Indies, where he was engaged off and
on till 1844, now as a subaltern, now in a vessel of his own,
hunting out pirates 'then very notorious' in the Leeward
Islands, cruising after slavers, or carrying dollars and provisions
for the Government. While yet a midshipman, he accompanied
Mr. Cockburn to Caraccas and had a sight of Bolivar. In
the brigantine *Griffon*, which he commanded in his last years

in the West Indies, he carried aid to Guadeloupe after the
earthquake, and twice earned the thanks of Government : once
for an expedition to Nicaragua to extort, under threat of a
blockade, proper apologies and a sum of money due to certain
British merchants ; and once during an insurrection in San
Domingo, for the rescue of certain others from a perilous im-
prisonment and the recovery of a ' chest of money ' of which
they had been robbed. Once, on the other hand, he earned his
share of public censure. This was in 1837, when he commanded
the *Romney* lying in the inner harbour of Havannah. The
Romney was in no proper sense a man-of-war ; she was a slave-
hulk, the bonded warehouse of the Mixed Slave Commission ;
where negroes, captured out of slavers under Spanish colours,
were detained provisionally, till the Commission should decide
upon their case and either set them free or bind them to apprenticeship. To this ship, already an eyesore to the authorities,
a Cuban slave made his escape. The position was invidious ;
on one side were the tradition of the British flag and the state
of public sentiment at home ; on the other, the certainty that
if the slave were kept, the *Romney* would be ordered at once
out of the harbour, and the object of the Mixed Commission
compromised. Without consultation with any other officer,
Captain Jenkin (then lieutenant) returned the man to shore and
took the Captain-General's receipt. Lord Palmerston approved
his course ; but the zealots of the anti-slave trade movement
(never to be named without respect) were much dissatisfied ;
and thirty-nine years later, the matter was again canvassed in
Parliament, and Lord Palmerston and Captain Jenkin defended
by Admiral Erskine in a letter to the *Times* (March 13, 1876).

In 1845, while still lieutenant, Charles Jenkin acted as End of his
Admiral Pigot's flag captain in the Cove of Cork, where there career.
were some thirty pennants ; and about the same time, closed his
career by an act of personal bravery. He had proceeded with
his boats to the help of a merchant vessel, whose cargo of com-
bustibles had taken fire and was smouldering under hatches ;
his sailors were in the hold, where the fumes were already heavy,
and Jenkin was on deck directing operations, when he found

his orders were no longer answered from below : he jumped down without hesitation and slung up several insensible men with his own hand. For this act, he received a letter from the Lords of the Admiralty expressing a sense of his gallantry ; and pretty soon after was promoted Commander, superseded, and could never again obtain employment.

The
Campbell-
Jacksons.
In 1828 or 1829, Charles Jenkin was in the same watch with another midshipman, Robert Colin Campbell Jackson, who introduced him to his family in Jamaica. The father, the Honourable Robert Jackson, Custos Rotulorum of Kingston, came of a Yorkshire family, said to be originally Scotch ; and on the mother's side, counted kinship with some of the Forbeses. The mother was Susan Campbell, one of the Campbells of Auchenbreck. Her father Colin, a merchant in Greenock, is said to have been the heir to both the estate and the baronetcy ; he claimed neither, which casts a doubt upon the fact ; but he had pride enough himself, and taught enough pride to his family, for any station or descent in Christendom. He had four daughters. One married an Edinburgh writer, as I have it on a first account—a minister, according to another—a man at least of reasonable station, but not good enough for the Campbells of Auchenbreck ; and the erring one was instantly discarded. Another married an actor of the name of Adcock, whom (as I receive the tale) she had seen acting in a barn ; but the phrase should perhaps be regarded rather as a measure of the family annoyance, than a mirror of the facts. The marriage was not in itself unhappy ; Adcock was a gentleman by birth and made a good husband ; the family reasonably prospered, and one of the daughters married no less a man than Clarkson Stanfield. But by the father, and the two remaining Miss Campbells, people of fierce passions and a truly Highland pride, the derogation was bitterly resented. For long the sisters lived estranged ; then, Mrs. Jackson and Mrs. Adcock were reconciled for a moment, only to quarrel the more fiercely ; the name of Mrs. Adcock was proscribed, nor did it again pass her sister's lips, until the morning when she announced : ' Mary Adcock is dead ; I saw her in her shroud last night.' Second

sight was hereditary in the house; and sure enough, as I have it reported, on that very night Mrs. Adcock had passed away. Thus, of the four daughters, two had, according to the idiotic notions of their friends, disgraced themselves in marriage; the others supported the honour of the family with a better grace, and married West Indian magnates of whom, I believe, the world has never heard and would not care to hear : So strange a thing is this hereditary pride. Of Mr. Jackson, beyond the fact that he was Fleeming's grandfather, I know naught. His wife, as I have said, was a woman of fierce passions; she would tie her house slaves to the bed and lash them with her own hand; and her conduct to her wild and down-going sons, was a mixture of almost insane self-sacrifice and wholly insane violence of temper. She had three sons and one daughter. Two of the sons went utterly to ruin, and reduced their mother to poverty. The third went to India, a slim, delicate lad, and passed so wholly from the knowledge of his relatives that he was thought to be long dead. Years later, when his sister was living in Genoa, a red-bearded man of great strength and stature, tanned by years in India, and his hands covered with barbaric gems, entered the room unannounced, as she was playing the piano, lifted her from her seat, and kissed her. It was her brother, suddenly returned out of a past that was never very clearly understood, with the rank of general, many strange gems, many cloudy stories of adventure, and next his heart, the daguerreotype of an Indian prince with whom he had mixed blood.

The last of this wild family, the daughter, Henrietta Camilla, *Fleeming's mother.* became the wife of the midshipman Charles, and the mother of the subject of this notice, Fleeming Jenkin. She was a woman of parts and courage. Not beautiful, she had a far higher gift, the art of seeming so; played the part of a belle in society, while far lovelier women were left unattended; and up to old age, had much of both the exigency and the charm that mark that character. She drew naturally, for she had no training, with unusual skill; and it was from her, and not from the two naval artists, that Fleeming inherited his eye and hand. She played on the harp and sang with something beyond the talent of an

amateur. At the age of seventeen, she heard Pasta in Paris;
flew up in a fire of youthful enthusiasm; and the next morning,
all alone and without introduction, found her way into the
presence of the *prima donna* and begged for lessons. Pasta
made her sing, kissed her when she had done, and though she
refused to be her mistress, placed her in the hands of a friend.
Nor was this all; for when Pasta returned to Paris, she sent for
the girl (once at least) to test her progress. But Mrs. Jenkin's
talents were not so remarkable as her fortitude and strength of
will; and it was in an art for which she had no natural taste
(the art of literature) that she appeared before the public. Her
novels, though they attained and merited a certain popularity
both in France and England, are a measure only of her courage.
They were a task, not a beloved task; they were written for
money in days of poverty, and they served their end. In the
least thing as well as in the greatest, in every province of life
as well as in her novels, she displayed the same capacity of
taking infinite pains, which descended to her son. When she
was about forty (as near as her age was known) she lost her
voice; set herself at once to learn the piano, working eight hours
a day; and attained to such proficiency that her collaboration
in chamber music was courted by professionals. And more
than twenty years later, the old lady might have been seen
dauntlessly beginning the study of Hebrew. This is the more
ethereal part of courage; nor was she wanting in the more
material. Once when a neighbouring groom, a married man,
had seduced her maid, Mrs. Jenkin mounted her horse, rode
over to the stable entrance and horsewhipped the man with her
own hand.

How a match came about between this talented and spirited
girl and the young midshipman, is not very easy to conceive.
Charles Jenkin was one of the finest creatures breathing;
loyalty, devotion, simple natural piety, boyish cheerfulness,
tender and manly sentiment in the old sailor fashion, were in
him inherent and inextinguishable either by age, suffering, or
injustice. He looked, as he was, every inch a gentleman; he
must have been everywhere notable, even among handsome men,

both for his face and his gallant bearing; not so much that of a sailor, you would have said, as like one of those gentle and graceful soldiers that, to this day, are the most pleasant of Englishmen to see. But though he was in these ways noble, the dunce scholar of Northiam was to the end no genius. Upon all points that a man must understand to be a gentleman, to be upright, gallant, affectionate and dead to self, Captain Jenkin was more knowing than one among a thousand; outside of that, his mind was very largely blank. He had indeed a simplicity that came near to vacancy; and in the first forty years of his married life, this want grew more accentuated. In both families imprudent marriages had been the rule; but neither Jenkin nor Campbell had ever entered into a more unequal union. It was the captain's good looks, we may suppose, that gained for him this elevation; and in some ways and for many years of his life, he had to pay the penalty. His wife, impatient of his incapacity and surrounded by brilliant friends, used him with a certain contempt. She was the managing partner; the life was hers, not his; after his retirement they lived much abroad, where the poor captain, who could never learn any language but his own, sat in the corner mumchance; and even his son, carried away by his bright mother, did not recognise for long the treasures of simple chivalry that lay buried in the heart of his father. Yet it would be an error to regard this marriage as unfortunate. It not only lasted long enough to justify itself in a beautiful and touching epilogue, but it gave to the world the scientific work and what (while time was) were of far greater value, the delightful qualities of Fleeming Jenkin. The Kentish-Welsh family, facile, extravagant, generous to a fault and far from brilliant, had given the father, an extreme example of its humble virtues. On the other side, the wild, cruel, proud and somewhat blackguard stock of the Scotch Campbell-Jacksons had put forth, in the person of the mother, all its force and courage.

The marriage fell in evil days. In 1823, the bubble of the Golden Aunt's inheritance had burst. She died holding the hand of the nephew she had so wantonly deceived; at the last she drew him down and seemed to bless him, surely with some remorseful feeling; for when the will was opened, there was not found so much as the mention of his name. He was deeply in debt; in debt even to the estate of his deceiver, so that he had to sell a piece of land to clear himself. ' My dear boy,' he said to Charles, ' there will be nothing left for you. I am a ruined man.' And here follows for me the strangest part of this story. From the death of the treacherous aunt, Charles Jenkin senior had still some nine years to live; it was perhaps too late for him to turn to saving, and perhaps his affairs were past restoration. But his family at least had all this while to prepare; they were still young men, and knew what they had to look for at their father's death; and yet when that happened in September 1831, the heir was still apathetically waiting. Poor

Fleeming's uncle John.

John, the days of his whips and spurs and Yeomanry dinners were quite over; and with that incredible softness of the Jenkin nature, he settled down, for the rest of a long life, into something not far removed above a peasant. The mill farm at Stowting had been saved out of the wreck; and here he built himself a house on the Mexican model, and made the two ends meet with rustic thrift, gathering dung with his own hands upon the road and not at all abashed at his employment. In dress, voice and manner, he fell into mere country plainness; lived without the least care for appearances, the least regret for the past or discontentment with the present; and when he came to die, died with Stoic cheerfulness, announcing that he had had a comfortable time and was yet well pleased to go. One would think there was little active virtue to be inherited from such a race; and yet in this same voluntary peasant, the special gift of Fleeming Jenkin was already half developed. The old man to the end was perpetually inventing; his strange, ill-spelled, unpunctuated correspondence is full (when he does not drop into cookery receipts) of pumps, road engines, steam-diggers, steam-ploughs, and steam-threshing machines; and I have it on Fleeming's word that what he did was full of ingenuity—only, as if by some cross destiny, useless. These disappointments he not only took with imperturbable good humour, but rejoiced with a particular relish over his nephew's success in the same field. ' I glory in

the professor,' he wrote to his brother; and to Fleeming himself, with a touch of simple drollery, 'I was much pleased with
your lecture but why did you hit me so hard with Conisure's'
(connoisseur's, *quasi* amateur's) 'engineering? Oh, what pre-
sumption!—either of you or *my*self!' A quaint, pathetic figure,
this of uncle John, with his dung cart and his inventions; and
the romantic fancy of his Mexican house ; and his craze about
the Lost Tribes, which seemed to the worthy man the key of all
perplexities; and his quiet conscience, looking back on a life not
altogether vain, for he was a good son to his father while his
father lived, and when evil days approached, he had proved
himself a cheerful Stoic.

It followed from John's inertia, that the duty of winding
up the estate fell into the hands of Charles. He managed it
with no more skill than might be expected of a sailor ashore,
saved a bare livelihood for John and nothing for the rest.
Eight months later, he married Miss Jackson; and with her
money, bought in some two-thirds of Stowting. In the beginning
of the little family history which I have been following to so
great an extent, the Captain mentions, with a delightful pride:
' A Court Baron and Court Leet are regularly held by the Lady
of the Manor, Mrs. Henrietta Camilla Jenkin;' and indeed the
pleasure of so describing his wife, was the most solid benefit of
the investment; for the purchase was heavily encumbered and
paid them nothing till some years before their death. In the
meanwhile, the Jackson family also, what with wild sons, an indulgent mother and the impending emancipation of the slaves, was
moving nearer and nearer to beggary ; and thus of two doomed
and declining houses, the subject of this memoir was born, heir
to an estate and to no money, yet with inherited qualities that
were to make him known and loved.

CHAPTER II.

1833-1851.

Birth and Childhood—Edinburgh—Frankfort-on-the-Main—Paris—The Revo-
lution of 1848—The Insurrection—Flight to Italy—Sympathy with Italy
—The Insurrection in Genoa—A Student in Genoa—The Lad and his
Mother.

Birth and childhood.

HENRY CHARLES FLEEMING JENKIN (Fleeming, pronounced Flem-
ming, to his friends and family) was born in a Government
building on the coast of Kent, near Dungeness, where his father
was serving at the time in the Coastguard, on March 25, 1833,
and named after Admiral Fleeming, one of his father's protectors
in the navy.

His childhood was vagrant like his life. Once he was left
in the care of his grandmother Jackson, while Mrs. Jenkin
sailed in her husband's ship and stayed a year at the Havannah.
The tragic woman was besides from time to time a member of
the family; she was in distress of mind and reduced in fortune
by the misconduct of her sons; her destitution and solitude made
it a recurring duty to receive her, her violence continually
enforced fresh separations. In her passion of a disappointed
mother, she was a fit object of pity; but her grandson, who
heard her load his own mother with cruel insults and reproaches,
conceived for her an indignant and impatient hatred, for which
he blamed himself in later life. It is strange from this point of
view to see his childish letters to Mrs. Jackson; and to think
that a man, distinguished above all by stubborn truthfulness,
should have been brought up to such dissimulation. But this
is of course unavoidable in life; it did no harm to Jenkin; and
whether he got harm or benefit from a so early acquaintance
with violent and hateful scenes, is more than I can guess. The

experience, at least, was formative; and in judging his character
it should not be forgotten. But Mrs. Jackson was not the only
stranger in their gates; the Captain's sister, Aunt Anna Jenkin,
lived with them until her death; she had all the Jenkin beauty
of countenance, though she was unhappily deformed in body
and of frail health; and she even excelled her gentle and
ineffectual family in all amiable qualities. So that each of the
two races from which Fleeming sprang, had an outpost by his
very cradle; the one he instinctively loved, the other hated;
and the lifelong war in his members had begun thus early by a
victory for what was best.

We can trace the family from one country place to another in
the south of Scotland; where the child learned his taste for sport
by riding home the pony from the moors. Before he was nine, he
could write such a passage as this about a Hallowe'en observance:
' I pulled a middling-sized cabbage-runt with a pretty sum of
gold about it. No witches would run after me when I was
sowing my hempseed this year: my nuts blazed away together
very comfortably to the end of their lives and when mamma put
hers in which were meant for herself and papa they blazed away
in the like manner.' Before he was ten he could write, with a
really irritating precocity, that he had been ' making some
pictures from a book called "Les Français peints par eux-
mêmes." . . . It is full of pictures of all classes, with a description of each in French. The pictures are a little caricatured,
but not much.' Doubtless this was only an echo from his
mother, but it shows the atmosphere in which he breathed. It
must have been a good change for this art critic to be the play-
mate of Mary Macdonald, their gardener's daughter at Barjarg,
and to sup with her family on potatoes and milk; and Fleeming
himself attached some value to this early and friendly experience
of another class.

His education, in the formal sense, began at Jedburgh. Edin-
Thence he went to the Edinburgh Academy, where Clerk Maxwell burgh and
was his senior and Tait his classmate; bore away many prizes; Frankfort.
and was once unjustly flogged by Rector Williams. He used
to insist that all his bad schoolfellows had died early, a belief

b 2

amusingly characteristic of the man's consistent optimism. In
1846 the mother and son proceeded to Frankfort-on-the-Main,
where they were soon joined by the father, now reduced to
inaction and to play something like third fiddle in his narrow
household. The emancipation of the slaves had deprived them
of their last resource beyond the half-pay of a captain; and life
abroad was not only desirable for the sake of Fleeming's education, it was almost enforced by reasons of economy. But it
was, no doubt, somewhat hard upon the captain. Certainly
that perennial boy found a companion in his son; they were
both active and eager, both willing to be amused, both young,
if not in years, then in character. They went out together on
excursions and sketched old castles, sitting side by side; they
had an angry rivalry in walking, doubtless equally sincere upon
both sides; and indeed we may say that Fleeming was excep-
tionally favoured, and that no boy had ever a companion more
innocent, engaging, gay and airy. But although in this case
it would be easy to exaggerate its import, yet, in the Jenkin
family also, the tragedy of the generations was proceeding, and
the child was growing out of his father's knowledge. His
artistic aptitude was of a different order. Already he had his
quick sight of many sides of life; he already overflowed with
distinctions and generalisations, contrasting the dramatic art
and national character of England, Germany, Italy, and France.
If he were dull, he would write stories and poems. ' I have
written,' he says at thirteen, ' a very long story in heroic
measure, 300 lines, and another Scotch story and innumerable
bits of poetry; ' and at the same age he had not only a keen
feeling for scenery, but could do something with his pen to call
it up. I feel I do always less than justice to the delightful
memory of Captain Jenkin; but with a lad of this character,
cutting the teeth of his intelligence, he was sure to fall into the
background.

Paris.
The Revo-
lution of
1848.
The family removed in 1847 to Paris, where Fleeming was
put to school under one Deluc. There he learned French, and
(if the captain is right) first began to show a taste for mathematics. But a far more important teacher than Deluc was at

hand ; the year 1848, so momentous for Europe, was momentous also for Fleeming's character. The family politics were Liberal ; Mrs. Jenkin, generous before all things, was sure to be upon the side of exiles ; and in the house of a Paris friend of hers, Mrs. Turner—already known to fame as Shelley's Cornelia de Boinville—Fleeming saw and heard such men as Manin, Gioberti, and the Ruffinis. He was thus prepared to sympathise with revolution ; and when the hour came, and he found himself in the midst of stirring and influential events, the lad's whole character was moved. He corresponded at that time with a young Edinburgh friend, one Frank Scott; and I am here going to draw somewhat largely on this boyish correspondence. It gives us at once a picture of the Revolution and a portrait of Jenkin at fifteen ; not so different (his friends will think) from the Jenkin of the end—boyish, simple, opinionated, delighting in action, delighting before all things in any generous sentiment.

'February 23, 1848.

' When at 7 o'clock to-day I went out, I met a large band going round the streets, calling on the inhabitants to illuminate their houses, and bearing torches. This was all very good fun, and everybody was delighted ; but as they stopped rather long and were rather turbulent in the Place de la Madeleine, near where we live ' [in the Rue Caumartin] ' a squadron of dragoons came up, formed, and charged at a hand gallop. This was a very pretty sight; the crowd was not too thick, so they easily got away ; and the dragoons only gave blows with the back of the sword, which hurt but did not wound. I was as close to them as I am now to the other side of the table; it was rather impressive, however. At the second charge they rode on the pavement and knocked the torches out of the fellows' hands; rather a shame, too—wouldn't be stood in England. . . . '

[At] ' ten minutes to ten . . . I went a long way along the Boulevards, passing by the office of Foreign Affairs, where Guizot lives, and where to-night there were about a thousand troops protecting him from the fury of the populace. After this was passed, the number of the people thickened, till about

half a mile further on, I met a troop of vagabonds, the wildest
vagabonds in the world—Paris vagabonds, well armed, having
probably broken into gunsmiths' shops and taken the guns and
swords. They were about a hundred. These were followed by
about a thousand (I am rather diminishing than exaggerating
numbers all through), indifferently armed with rusty sabres,
sticks, &c. An uncountable troop of gentlemen, workmen,
shopkeepers' wives (Paris women dare anything), ladies' maids,
common women—in fact, a crowd of all classes, though by far
the greater number were of the better dressed class—followed.
Indeed, it was a splendid sight: the mob in front chanting the
" *Marseillaise*," the national war hymn, grave and powerful,
sweetened by the night air—though night in these splendid
streets was turned into day, every window was filled with lamps,
dim torches were tossing in the crowd . . . for Guizot has late
this night given in his resignation, and this was an improvised
illumination.

'I and my father had turned with the crowd, and were close
behind the second troop of vagabonds. Joy was on every face.
I remarked to papa that "I would not have missed the scene
for anything, I might never see such a splendid one," when
plong went one shot—every face went pale—*r-r-r-r-r* went the
whole detachment, [and] the whole crowd of gentlemen and
ladies turned and cut. Such a scene!—ladies, gentlemen, and
vagabonds went sprawling in the mud, not shot but tripped up;
and those that went down could not rise, they were trampled
over. . . I ran a short time straight on and did not fall, then
turned down a side street, ran fifty yards and felt tolerably safe;
looked for papa, did not see him; so walked on quickly, giving
the news as I went.' [It appears, from another letter, the boy
was the first to carry word of the firing to the Rue St. Honoré;
and that his news wherever he brought it was received with
hurrahs. It was an odd entrance upon life for a little English
lad, thus to play the part of rumour in such a crisis of the history
of France.]

'But now a new fear came over me. I had little doubt but
my papa was safe, but my fear was that he should arrive at

home before me and tell the story; in that case I knew my mamma would go half mad with fright, so on I went as quick as possible. I heard no more discharges. When I got half way home, I found my way blocked up by troops. That way or the Boulevards I must pass. In the Boulevards they were fighting, and I was afraid all other passages might be blocked up . . . and I should have to sleep in a hotel in that case, and then my mamma—however, after a long *détour*, I found a passage and ran home, and in our street joined papa.

'. . . I'll tell you to-morrow the other facts gathered from newspapers and papa. . . . To-night I have given you what I have seen with my own eyes an hour ago, and began trembling with excitement and fear. If I have been too long on this one subject, it is because it is yet before my eyes.

'Monday, 24.

'It was that fire raised the people. There was fighting all through the night in the Rue Notre Dame de Lorette, on the Boulevards where they had been shot at, and at the Porte St. Denis. At ten o'clock, they resigned the house of the Minister of Foreign Affairs (where the disastrous volley was fired) to the people, who immediately took possession of it. I went to school but [was] hardly there when the row in that quarter commenced. Barricades began to be fixed. Everyone was very grave now; the *externes* went away, but no one came to fetch me, so I had to stay. No lessons could go on. A troop of armed men took possession of the barricades, so it was supposed I should have to sleep there. The revolters came and asked for arms, but Deluc (head-master) is a National Guard, and he said he had only his own and he wanted them; but he said he would not fire on them. Then they asked for wine, which he gave them. They took good care not to get drunk, knowing they would not be able to fight. They were very polite and behaved extremely well.

'About 12 o'clock a servant came for a boy who lived near me, [and] Deluc thought it best to send me with him. We heard a good deal of firing near, but did not come across any of

the parties. As we approached the railway, the barricades were
no longer formed of palings, planks, or stones; but they had
got all the omnibuses as they passed, sent the horses and pas-
sengers about their business, and turned them over. A double
row of overturned coaches made a capital barricade, with a few
paving stones.

'When I got home I found to my astonishment that in our
fighting quarter it was much quieter. Mamma had just been out
seeing the troops in the Place de la Concorde, when suddenly
the Municipal Guard, now fairly exasperated, prevented the
National Guard from proceeding, and fired at them; the National Guard had come with their musquets not loaded, but at
length returned the fire. Mamma saw the National Guard fire.
The Municipal Guard were round the corner. She was delighted
for she saw no person killed, though many of the Municipals
were. . . .

'I immediately went out with my papa (mamma had just
come back with him) and went to the Place de la Concorde. There
was an enormous quantity of troops in the Place. Suddenly the
gates of the gardens of the Tuileries opened : we rushed forward,
out gallopped an enormous number of cuirassiers, in the middle
of which were a couple of low carriages, said first to contain the
Count de Paris and the Duchess of Orleans, but afterwards they
said it was the King and Queen; and then I heard he had abdicated. I returned and gave the news.

'Went out again up the Boulevards. The house of the
Minister of Foreign Affairs was filled with people and " *Hôtel du
Peuple* " written on it; the Boulevards were barricaded with fine
old trees that were cut down and stretched all across the road.
We went through a great many little streets, all strongly barricaded, and sentinels of the people at the principal of them. The
streets are very unquiet, filled with armed men and women, for
the troops had followed the ex-King to Neuilly and left Paris in
the power of the people. We met the captain of the Third
Legion of the National Guard (who had principally protected the
people), badly wounded by a Municipal Guard, stretched on a
litter. He was in possession of his senses. He was surrounded

by a troop of men crying " Our brave captain—we have him yet
—he's not dead ! *Vive la Réforme !*" This cry was responded to
by all, and every one saluted him as he passed. I do not know
if he was mortally wounded. That Third Legion has behaved
splendidly.

' I then returned, and shortly afterwards went out again to
the garden of the Tuileries. They were given up to the people
and the palace was being sacked. The people were firing blank
cartridge to testify their joy and they had a cannon on the top
of the palace. It was a sight to see a palace sacked and armed
vagabonds firing out of the windows, and throwing shirts, papers,
and dresses of all kinds out of the windows. They are not
rogues, these French ; they are not stealing, burning, or doing
much harm. In the Tuileries they have dressed up some
of the statues, broken some, and stolen nothing but queer
dresses. I say, Frank, you must not hate the French ; hate
the Germans if you like. The French laugh at us a little and
call out *Goddam* in the streets ; but to-day, in civil war, when
they might have put a bullet through our heads, I never was
insulted once.

' At present we have a provisional Government, consisting
of Odion [*sic*] Barrot, Lamartine, Marast, and some others ;
among them a common workman, but very intelligent. This is
a triumph of liberty—rather !

' Now then, Frank, what do you think of it ? I in a revolution, and out all day. Just think, what fun ! So it was at first,
till I was fired at yesterday ; but to-day I was not frightened, but
it turned me sick at heart, I don't know why. There has been
no great bloodshed, [though] I certainly have seen men's blood
several times. But there's something shocking to see a whole
armed populace, though not furious, for not one single shop has
been broken open, except the gunsmiths' shops, and most of the
arms will probably be taken back again. For the French have
no cupidity in their nature ; they don't like to steal—it is not in
their nature. I shall send this letter in a day or two, when
I am sure the post will go again. I know I have been a long
time writing, but I hope you will find the matter of this letter

interesting, as coming from a person resident on the spot; though probably you don't take much interest in the French, but I can think, write, and speak on no other subject.

'Feb. 25.

' There is no more fighting, the people have conquered; but the barricades are still kept up, and the people are in arms, more than ever fearing some new act of treachery on the part of the ex-King. The fight where I was was the principal cause of the Revolution. I was in little danger from the shot, for there was an immense crowd in front of me, though quite within gunshot. [By another letter, a hundred yards from the troops.] I wished I had stopped there.

' The Paris streets are filled with the most extraordinary crowds of men, women and children, ladies and gentlemen. Every person joyful. The bands of armed men are perfectly polite. Mamma and aunt to-day walked through armed crowds alone, that were firing blank cartridge in all directions. Every person made way with the greatest politeness, and one common man with a blouse, coming by accident against her, immediately stopped to beg her pardon in the politest manner. There are few drunken men. The Tuileries is still being run over by the people; they only broke two things, a bust of Louis Philippe and one of Marshal Bugeaud, who fired on the people. . . .

' I have been out all day again to-day, and precious tired I am. The Republican party seem the strongest, and are going about with red ribbons in their button-holes. . . .

' The title of "Mister" is abandoned; they say nothing but "Citizen," and the people are shaking hands amazingly. They have got to the top of the public monuments, and, mingling with bronze or stone statues, five or six make a sort of *tableau vivant*, the top man holding up the red flag of the Republic; and right well they do it, and very picturesque they look. I think I shall put this letter in the post to-morrow as we got a letter to-night.

(On Envelope.)

' M. Lamartine has now by his eloquence conquered the whole armed crowd of citizens threatening to kill him if he did

not immediately proclaim the Republic and red flag. He said he
could not yield to the citizens of Paris alone, that the whole
country must be consulted, that he chose the tricolour, for it
had followed and accompanied the triumphs of France all over
the world, and that the red flag had only been dipped in the
blood of the citizens. For sixty hours he has been quieting
the people : he is at the head of everything. Don't be pre-
judiced, Frank, by what you see in the papers. The French
have acted nobly, splendidly ; there has been no brutality,
plundering, or stealing. . . . I did not like the French before ;
but in this respect they are the finest people in the world. I
am so glad to have been here.'

And there one could wish to stop with this apotheosis of
liberty and order read with the generous enthusiasm of a boy ;
but as the reader knows, it was but the first act of the piece.
The letters, vivid as they are, written as they were by a hand
trembling with fear and excitement, yet do injustice, in their
boyishness of tone, to the profound effect produced. At the
sound of these songs and shot of cannon, the boy's mind awoke.
He dated his own appreciation of the art of acting from the day
when he saw and heard Rachel recite the ' *Marseillaise* ' at the
Francais, the tricolor in her arms. What is still more strange,
he had been up to then invincibly indifferent to music, insomuch
that he could not distinguish ' God save the Queen ' from
' Bonnie Dundee ; ' and now, to the chanting of the mob, he
amazed his family by learning and singing ' *Mourir pour la
Patrie.*' But the letters, though they prepare the mind for no
such revolution in the boy's tastes and feelings, are yet full of
entertaining traits. Let the reader note Fleeming's eagerness
to influence his friend Frank, an incipient Tory (no less) as
further history displayed ; his unconscious indifference to his
father and devotion to his mother, betrayed in so many signi-
ficant expressions and omissions ; the sense of dignity of this
diminutive ' person resident on the spot,' who was so happy
as to escape insult ; and the strange picture of the household
—father, mother, son, and even poor Aunt Anna—all day in

the streets in the thick of this rough business, and the boy
packed off alone to school in a distant quarter on the very
morrow of the massacre.

They had all the gift of enjoying life's texture as it comes:
they were all born optimists. The name of liberty was honoured
in that family, its spirit also, but within stringent limits; and
some of the foreign friends of Mrs. Jenkin were, as I have said,
men distinguished on the Liberal side. Like Wordsworth, they
beheld

> France standing on the top of golden hours
> And human nature seeming born again.

At once, by temper and belief, they were formed to find their
element in such a decent and whiggish convulsion, spectacular
in its course, moderate in its purpose. For them,

> Bliss was it in that dawn to be alive,
> But to be young was very heaven.

And I cannot but smile when I think that (again like Wordsworth) they should have so specially disliked the consequence.

The Insur-
rection.

It came upon them by surprise. Liberal friends of the precise
right shade of colour had assured them, in Mrs. Turner's drawing-
room, that all was for the best; and they rose on February
28 without fear. About the middle of the day they heard
the sound of musketry, and the next morning they were
wakened by the cannonade. The French who had behaved so
'splendidly,' pausing, at the voice of Lamartine, just where
judicious Liberals could have desired—the French, who had 'no
cupidity in their nature,' were now about to play a variation on
the theme rebellion. The Jenkins took refuge in the house of
Mrs. Turner, the house of the false prophets, 'Anna going with
Mrs. Turner, that she might be prevented speaking English,
Fleeming, Miss H. and I ' (it is the mother who writes) ' walking
together. As we reached the Rue de Clichy, the report of the
cannon sounded close to our ears and made our hearts sick, I
assure you. The fighting was at the barrier Rochechouart, a
few streets off. All Saturday and Sunday we were a prey to

great alarm, there came so many reports that the insurgents were getting the upper hand. One could tell the state of affairs from the extreme quiet or the sudden hum in the street. When the news was bad, all the houses closed and the people disappeared ; when better, the doors half opened and you heard the sound of men again. From the upper windows we could see each discharge from the Bastille—I mean the smoke rising—and also the flames and smoke from the Boulevard la Chapelle. We were four ladies, and only Fleeming by way of a man, and difficulty enough we had to keep him from joining the National Guards—his pride and spirit were both fired. You cannot picture to yourself the multitudes of soldiers, guards, and armed men of all sorts we watched—not close to the window, however, for such havoc had been made among them by the firing from the windows, that as the battalions marched by, they cried, " Fermez vos fenêtres ! " and it was very painful to watch their looks of anxiety and suspicion as they marched by.'

'The Revolution,' writes Fleeming to Frank Scott, ' was quite delightful : getting popped at and run at by horses, and giving sous for the wounded into little boxes guarded by the raggedest, picturesquest, delightfullest, sentinels ; but the insurrection ! ugh, I shudder to think at [*sic*] it.' He found it ' not a bit of fun sitting boxed up in the house four days almost. . . . I was the only *gentleman* to four ladies, and didn't they keep me in order ! I did not dare to show my face at a window, for fear of catching a stray ball or being forced to enter the National Guard; [for] they would have it I was a man full-grown, French, and every way fit to fight. And my mamma was as bad as any of them ; she that told me I was a coward last time if I stayed in the house a quarter of an hour ! But I drew, examined the pistols, of which I found lots with caps, powder, and ball, while sometimes murderous intentions of killing a dozen insurgents and dying violently overpowered by numbers. . . .' We may drop this sentence here : under the conduct of its boyish writer, it was to reach no legitimate end.

Four days of such a discipline had cured the family of Paris ; Flight to Italy. the same year Fleeming was to write, in answer apparently to a

question of Frank Scott's, ' I could find no national game in France
but revolutions ; ' and the witticism was justified in their experience. On the first possible day, they applied for passports, and
were advised to take the road to Geneva. It appears it was scarce
safe to leave Paris for England. Charles Reade, with keen dramatic gusto, had just smuggled himself out of that city in the
bottom of a cab. English gold had been found on the insurgents, the name of England was in evil odour ; and it was thus
—for strategic reasons, so to speak—that Fleeming found himself on the way to that Italy where he was to complete his
education, and for which he cherished to the end a special
kindness.

Sympathy
with Italy. It was in Genoa they settled; partly for the sake of the
captain, who might there find naval comrades; partly because
of the Ruffinis, who had been friends of Mrs. Jenkin in their
time of exile and were now considerable men at home ;
partly, in fine, with hopes that Fleeming might attend the University ; in preparation for which he was put at once to school.
It was the year of Novara ; Mazzini was in Rome ; the dry bones
of Italy were moving; and for people of alert and liberal sympathies the time was inspiriting. What with exiles turned Ministers
of State, universities thrown open to Protestants, Fleeming himself the first Protestant student in Genoa, and thus, as his
mother writes, ' a living instance of the progress of liberal
ideas '—it was little wonder if the enthusiastic young woman
and the clever boy were heart and soul upon the side of Italy.
It should not be forgotten that they were both on their first
visit to that country ; the mother still ' child enough ' to be
delighted when she saw 'real monks;' and both mother and
son thrilling with the first sight of snowy Alps, the blue Mediterranean, and the crowded port and the palaces of Genoa. Nor
was their zeal without knowledge. Ruffini, deputy for Genoa
and soon to be head of the University, was at their side; and by
means of him the family appear to have had access to much
Italian society. To the end, Fleeming professed his admiration
of the Piedmontese and his unalterable confidence in the future
of Italy under their conduct ; for Victor Emanuel, Cavour, the

first La Marmora and Garibaldi, he had varying degrees of sympathy and praise : perhaps highest for the King, whose good sense and temper filled him with respect—perhaps least for Garibaldi, whom he loved but yet mistrusted.

But this is to look forward : these were the days not of Victor Emanuel but of Charles Albert ; and it was on Charles Albert that mother and son had now fixed their eyes as on the sword-bearer of Italy. On Fleeming's sixteenth birthday, they were, the mother writes, ' in great anxiety for news from the army. You can have no idea what it is to live in a country where such a struggle is going on. The interest is one that absorbs all others. We eat, drink, and sleep to the noise of drums and musketry. You would enjoy and almost admire Fleeming's enthusiasm and earnestness—and courage, I may say—for we are among the small minority of English who side with the Italians. The other day, at dinner at the Consul's, boy as he is, and in spite of my admonitions, Fleeming defended the Italian cause, and so well that he " tripped up the heels of his adversary " simply from being well-informed on the subject and honest. He is as true as steel, and for no one will he bend right or left. . . . Do not fancy him a Bobadil,' she adds, ' he is only a very true, candid boy. I am so glad he remains in all respects but information a great child.'

If this letter is correctly dated, the cause was already lost and the King had already abdicated when these lines were written. No sooner did the news reach Genoa, than there began ' tumultuous movements ; ' and the Jenkins received hints it would be wise to leave the city. But they had friends and interests ; even the captain had English officers to keep him company, for Lord Hardwicke's ship, the *Vengeance*, lay in port ; and supposing the danger to be real, I cannot but suspect the whole family of a divided purpose, prudence being possibly weaker than curiosity. Stay, at least, they did, and thus rounded their experience of the revolutionary year. On Sunday, April 1, Fleeming and the captain went for a ramble beyond the walls, leaving Aunt Anna and Mrs. Jenkin to walk on the bastions with some friends. On the way back, this party turned

The Insurrection of Genoa.

aside to rest in the Church of the Madonna delle Grazie. 'We
had remarked,' writes Mrs. Jenkin, 'the entire absence of
sentinels on the ramparts, and how the cannons were left in
solitary state ; and I had just remarked " How quiet everything
is ! " when suddenly we heard the drums begin to beat and
distant shouts. *Accustomed as we are* to revolutions, we never
thought of being frightened.' For all that, they resumed their
return home. On the way they saw men running and vociferat-
ing, but nothing to indicate a general disturbance, until, near
the Duke's palace, they came upon and passed a shouting mob
dragging along with it three cannon. It had scarcely passed
before they heard 'a rushing sound'; one of the gentlemen
thrust back the party of ladies under a shed, and the mob
passed again. A fine-looking young man was in their hands;
and Mrs. Jenkin saw him with his mouth open as if he sought
to speak, saw him tossed from one to another like a ball, and
then saw him no more. ' He was dead a few instants after, but
the crowd hid that terror from us. My knees shook under me
and my sight left me.' With this street tragedy, the curtain
rose upon their second revolution.

The attack on Spirito Santo, and the capitulation and
departure of the troops speedily followed. Genoa was in the
hands of the Republicans, and now came a time when the English
residents were in a position to pay some return for hospitality
received. Nor were they backward. Our Consul (the same
who had the benefit of correction from Fleeming) carried the
Intendente on board the *Vengeance*, escorting him through the
streets, getting along with him on board a shore boat, and when
the insurgents levelled their muskets, standing up and naming
himself, ' *Console Inglese.*' A friend of the Jenkins, Captain
Glynne, had a more painful, if a less dramatic part. One
Colonel Nosozzo had been killed (I read) while trying to prevent
his own artillery from firing on the mob; but in that hell's
cauldron of a distracted city, there were no distinctions made,
and the Colonel's widow was hunted for her life. In her grief
and peril, the Glynnes received and hid her; Captain Glynne
sought and found her husband's body among the slain, saved it

for two days, brought the widow a lock of the dead man's hair ; but at last, the mob still strictly searching, seems to have abandoned the body, and conveyed his guest on board the *Vengeance*. The Jenkins also had their refugees, the family of an *employé* threatened by a decree. ' You should have seen me making a Union Jack to nail over our door,' writes Mrs. Jenkin. ' I never worked so fast in my life. Monday and Tuesday,' she continues, ' were tolerably quiet, our hearts beating fast in the hope of La Marmora's approach, the streets barricaded, and none but foreigners and women allowed to leave the city.' On Wednesday, La Marmora came indeed, but in the ugly form of a bombardment ; and that evening the Jenkins sat without lights about their drawing-room window, ' watching the huge red flashes of the cannon ' from the Brigato and La Specula forts, and hearkening, not without some awful pleasure, to the thunder of the cannonade.

Lord Hardwicke intervened between the rebels and La Marmora ; and there followed a troubled armistice, filled with the voice of panic. Now the *Vengeance* was known to be cleared for action ; now it was rumoured that the galley slaves were to be let loose upon the town, and now that the troops would enter it by storm. Crowds, trusting in the Union Jack over the Jenkins' door, came to beg them to receive their linen and other valuables ; nor could their instances be refused ; and in the midst of all this bustle and alarm, piles of goods must be examined and long inventories made. At last the captain decided things had gone too far. He himself apparently remained to watch over the linen ; but at five o'clock on the Sunday morning, Aunt Anna, Fleeming and his mother were rowed in a pour of rain on board an English merchantman, to suffer ' nine mortal hours of agonising suspense.' With the end of that time, peace was restored. On Tuesday morning officers with white flags appeared on the bastions ; then, regiment by regiment, the troops marched in, two hundred men sleeping on the ground floor of the Jenkins' house, thirty thousand in all entering the city, but without disturbance, old La Marmora being a commander of a Roman sternness.

With the return of quiet, and the reopening of the universities, we behold a new character, Signor Flaminio: the
professors, it appears, made no attempt upon the Jenkin; and
thus readily italianised the Fleeming. He came well recommended; for their friend Ruffini was then, or soon after, raised to
be the head of the University; and the professors were very kind
and attentive, possibly to Ruffini's *protégé*, perhaps also to the
first Protestant student. It was no joke for Signor Flaminio
at first; certificates had to be got from Paris and from Rector
Williams; the classics must be furbished up at home that he
might follow Latin lectures; examinations bristled in the path,
the entrance examination with Latin and English essay, and
oral trials (much softened for the foreigner) in Horace, Tacitus
and Cicero, and the first University examination only three
months later, in Italian eloquence, no less, and other wider
subjects. On one point the first Protestant student was moved
to thank his stars: that there was no Greek required for the
degree. Little did he think, as he set down his gratitude, how
much, in later life and among cribs and dictionaries, he was to
lament this circumstance; nor how much of that later life he
was to spend acquiring, with infinite toil, a shadow of what he
might then have got with ease and fully. But if his Genoese
education was in this particular imperfect, he was fortunate in
the branches that more immediately touched on his career. The
physical laboratory was the best mounted in Italy. Bancalari,
the professor of natural philosophy, was famous in his day;
by what seems even an odd coincidence, he went deeply into
electro-magnetism; and it was principally in that subject that
Signor Flaminio, questioned in Latin and answering in Italian,
passed his Master of Arts degree with first-class honours.
That he had secured the notice of his teachers, one circumstance
sufficiently proves. A philosophical society was started under
ths presidency of Mamiani, ' one of the examiners and one of
the leaders of the Moderate party; ' and out of five promising
students brought forward by the professors to attend the sittings
and present essays, Signor Flaminio was one. I cannot find that
he ever read an essay; and indeed I think his hands were other-

wise too full. He found his fellow-students 'not such a bad set of chaps,' and preferred the Piedmontese before the Genoese; but I suspect he mixed not very freely with either. Not only were his days filled with university work, but his spare hours were fully dedicated to the arts under the eye of a beloved task-mistress. He worked hard and well in the art school, where he obtained a silver medal 'for a couple of legs the size of life drawn from one of Raphael's cartoons.' His holidays were spent in sketching; his evenings, when they were free, at the theatre. Here at the opera he discovered besides a taste for a new art, the art of music; and it was, he wrote, 'as if he had found out a heaven on earth.' 'I am so anxious that whatever he professes to know, he should really perfectly possess,' his mother wrote, 'that I spare no pains;' neither to him nor to myself, she might have added. And so when he begged to be allowed to learn the piano, she started him with characteristic barbarity on the scales; and heard in consequence 'heart-rending groans' and saw 'anguished claspings of hands' as he lost his way among their arid intricacies.

In this picture of the lad at the piano, there is something, for the period, girlish. He was indeed his mother's boy; and it was fortunate his mother was not altogether feminine. She gave her son a womanly delicacy in morals, to a man's taste— to his own taste in later life—too finely spun, and perhaps more elegant than healthful. She encouraged him besides in drawing-room interests. But in other points her influence was manlike. Filled with the spirit of thoroughness, she taught him to make of the least of these accomplishments a virile task; and the teaching lasted him through life. Immersed as she was in the day's movements and buzzed about by leading Liberals, she handed on to him her creed in politics: an enduring kindness for Italy, and a loyalty, like that of many clever women, to the Liberal party with but small regard to men or measures. This attitude of mind used often to disappoint me in a man so fond of logic; but I see now how it was learned from the bright eyes of his mother and to the sound of the cannonades of 1848. To some of her defects, besides, she made him heir. Kind as

The lad and his mother.

was the bond that united her to her son, kind and even pretty, she was scarce a woman to adorn a home; loving as she did to shine; careless as she was of domestic, studious of public graces. She probably rejoiced to see the boy grow up in somewhat of the image of herself, generous, excessive, enthusiastic, external; catching at ideas, brandishing them when caught; fiery for the right, but always fiery; ready at fifteen to correct a consul, ready at fifty to explain to any artist his own art.

The defects and advantages of such a training were obvious in Fleeming throughout life. His thoroughness was not that of the patient scholar, but of an untrained woman with fits of passionate study; he had learned too much from dogma, given indeed by cherished lips; and precocious as he was in the use of the tools of the mind, he was truly backward in knowledge of life and of himself. Such as it was at least, his home and school training was now complete; and you are to conceive the lad as being formed in a household of meagre revenue, among foreign surroundings, and under the influence of an imperious drawing-room queen; from whom he learned a great refinement of morals, a strong sense of duty, much forwardness of bearing, all manner of studious and artistic interests, and many ready-made opinions which he embraced with a son's and a disciple's loyalty.

MRS. JENKIN

FROM A SKETCH TAKEN BY HER SON AT GENOA

CHAPTER III.

1851–1858.

Return to England—Fleeming at Fairbairn's—Experience in a Strike—Dr.
Bell and Greek Architecture—The Gaskells—Fleeming at Greenwich—
The Austins—Fleeming and the Austins—His Engagement—Fleeming
and Sir W. Thomson.

IN 1851, the year of Aunt Anna's death, the family left Genoa
and came to Manchester, where Fleeming was entered in Fairbairn's works as an apprentice. From the palaces and Alps,
the Mole, the blue Mediterranean, the humming lanes and the
bright theatres of Genoa, he fell—and he was sharply conscious
of the fall—to the dim skies and the foul ways of Manchester.
England he found on his return 'a horrid place,' and there is
no doubt the family found it a dear one. The story of the
Jenkin finances is not easy to follow. The family, I am told,
did not practise frugality, only lamented that it should be
needful; and Mrs. Jenkin, who was always complaining of
'those dreadful bills,' was 'always a good deal dressed.' But
at this time of the return to England, things must have gone
further. A holiday tour of a fortnight, Fleeming feared would
be beyond what he could afford, and he only projected it 'to
have a castle in the air.' And there were actual pinches. Fresh
from a warmer sun, he was obliged to go without a greatcoat,
and learned on railway journeys to supply the place of one with
wrappings of old newspaper.

From half-past eight till six, he must 'file and chip vigorously
in a moleskin suit and infernally dirty.' The work was not
new to him, for he had already passed some time in a Genoese
shop; and to Fleeming no work was without interest. Whatever a man can do or know, he longed to know and do also.

Fleeming at Fair-bairn's.

'I never learned anything,' he wrote, 'not even standing on my head, but I found a use for it.' In the spare hours of his first telegraph voyage, to give an instance of his greed of knowledge, he meant 'to learn the whole art of navigation, every rope in the ship and how to handle her on any occasion; and once when he was shown a young lady's holiday collection of seaweeds, he must cry out, 'It showed me my eyes had been idle.' Nor was his the case of the mere literary smatterer, content if he but learn the names of things. In him, to do and to do well, was even a dearer ambition than to know. Anything done well, any craft, despatch, or finish, delighted and inspired him. I remember him with a twopenny Japanese box of three drawers, so exactly fitted that, when one was driven home, the others started from their places; the whole spirit of Japan, he told me, was pictured in that box; that plain piece of carpentry was as much inspired by the spirit of perfection as the happiest drawing or the finest bronze and he who could not enjoy it in the one was not fully able to enjoy it in the others. Thus, too, he found in Leonardo's engineering and anatomical drawings a perpetual feast; and of the former he spoke even with emotion. Nothing indeed annoyed Fleeming more than the attempt to separate the fine arts from the arts of handicraft; any definition or theory that failed to bring these two together, according to him, had missed the point; and the essence of the pleasure received lay in seeing things well done. Other qualities must be added; he was the last to deny that; but this, of perfect craft, was at the bottom of all. And on the other hand, a nail ill-driven, a joint ill-fitted, a tracing clumsily done, anything to which a man had set his hand and not set it aptly, moved him to shame and anger. With such a character, he would feel but little drudgery at Fairbairn's. There would be something daily to be done, slovenliness to be avoided, and a higher mark of skill to be attained; he would chip and file, as he had practised scales, impatient of his own imperfection but resolute to learn.

And there was another spring of delight. For he was now moving daily among those strange creations of man's brain, to

some so abhorrent, to him of an interest so inexhaustible : in
which iron, water and fire are made to serve as slaves, now
with a tread more powerful than an elephant's, and now with
a touch more precise and dainty than a pianist's. The taste
for machinery was one that I could never share with him, and
he had a certain bitter pity for my weakness. Once when I
had proved, for the hundredth time, the depth of this defect,
he looked at me askance : ' And the best of the joke,' said he,
' is that he thinks himself quite a poet.' For to him the struggle
of the engineer against brute forces and with inert allies, was
nobly poetic. Habit never dulled in him the sense of the
greatness of the aims and obstacles of his profession. Habit
only sharpened his inventor's gusto in contrivance, in triumphant
artifice, in the Odyssean subtleties, by which wires are taught to
speak, and iron hands to weave, and the slender ship to brave
and to outstrip the tempest. To the ignorant the great results
alone are admirable ; to the knowing, and to Fleeming in
particular, rather the infinite device and sleight of mind that
made them possible.

A notion was current at the time that, in such a shop as
Fairbairn's, a pupil would never be popular unless he drank
with the workmen and imitated them in speech and manner.
Fleeming, who would do none of these things, they accepted as
a friend and companion ; and this was the subject of remark in
Manchester, where some memory of it lingers till to-day. He
thought it one of the advantages of his profession to be brought in
a close relation with the working classes ; and for the skilled arti-
san he had a great esteem, liking his company, his virtues and
his taste in some of the arts. But he knew the classes too well
to regard them, like a platform speaker, in a lump. He drew,
on the other hand, broad distinctions ; and it was his profound
sense of the difference between one working man and another
that led him to devote so much time, in later days, to the
furtherance of technical education. In 1852 he had occasion to Experi-
see both men and masters at their worst, in the excitement of a ence of a
strike ; and very foolishly (after their custom) both would seem to strike.
have behaved. Beginning with a fair show of justice on either

side, the masters stultified their cause by obstinate impolicy, and
the men disgraced their order by acts of outrage. ' On Wednesday last,' writes Fleeming, ' about three thousand banded round
Fairbairn's door at 6 o'clock : men, women, and children, factory
boys and girls, the lowest of the low in a very low place. Orders
came that no one was to leave the works; but the men inside
(Knobsticks, as they are called) were precious hungry and
thought they would venture. Two of my companions and myself went out with the very first, and had the full benefit of every
possible groan and bad language.' But the police cleared a
lane through the crowd, the pupils were suffered to escape un-
hurt, and only the Knobsticks followed home and kicked with
clogs; so that Fleeming enjoyed, as we may say, for nothing,
that fine thrill of expectant valour with which he had sallied
forth into the mob. ' I never before felt myself so decidedly
somebody, instead of nobody,' he wrote.

Dr. Bell
and Greek
architec-
ture.

Outside as inside the works, he was ' pretty merry and well
to do,' zealous in study, welcome to many friends, unwearied in
loving-kindness to his mother. For some time he spent three
nights a week with Dr. Bell, ' working away at certain geometrical methods of getting the Greek architectural proportions : '
a business after Fleeming's heart, for he was never so pleased
as when he could marry his two devotions, art and science. This
was besides, in all likelihood, the beginning of that love and
intimate appreciation of things Greek, from the least to the
greatest, from the *Agamemnon* (perhaps his favourite tragedy)
down to the details of Grecian tailoring, which he used to express in his familiar phrase : ' The Greeks were the boys.' Dr.
Bell—the son of George Joseph, the nephew of Sir Charles, and
though he made less use of it than some, a sharer in the distinguished talents of his race—had hit upon the singular fact
that certain geometrical intersections gave the proportions of
the Doric order. Fleeming, under Dr. Bell's direction, applied
the same method to the other orders, and again found the proportions accurately given. Numbers of diagrams were prepared;
but the discovery was never given to the world, perhaps because
of the dissensions that arose between the authors. For Dr. Bell

believed that 'these intersections were in some way connected
with, or symbolical of, the antagonistic forces at work;' but his
pupil and helper, with characteristic trenchancy, brushed aside
this mysticism, and interpreted the discovery as 'a geometrical
method of dividing the spaces or (as might be said) of setting
out the work, purely empirical and, in no way connected with
any laws of either force or beauty.' 'Many a hard and pleasant fight we had over it,' wrote Jenkin, in later years; 'and
impertinent as it may seem, the pupil is still unconvinced by
the arguments of the master.' I do not know about the antagonistic forces in the Doric order; in Fleeming they were plain
enough; and the Bobadil of these affairs with Dr. Bell was still,
like the corrector of Italian consuls, 'a great child in everything
but information.' At the house of Colonel Cleather, he might be
seen with a family of children; and with these, there was no word
of the Greek orders; with these Fleeming was only an uproarious
boy and an entertaining draughtsman; so that his coming was
the signal for the young people to troop into the playroom
where sometimes the roof rang with romping, and sometimes
they gathered quietly about him as he amused them with his
pencil.

In another Manchester family, whose name will be familiar The Gaskells.
to my readers—that of the Gaskells, Fleeming was a frequent
visitor. To Mrs. Gaskell, he would often bring his new ideas, a
process that many of his later friends will understand and, in
their own cases, remember. With the girls, he had 'constant
fierce wrangles,' forcing them to reason out their thoughts and
to explain their prepossessions; and I hear from Miss Gaskell
that they used to wonder how he could throw all the ardour of
his character into the smallest matters, and to admire his un-
selfish devotion to his parents. Of one of these wrangles, I
have found a record most characteristic of the man. Fleeming
had been laying down his doctrine that the end justifies the
means, and that it is quite right 'to boast of your six men-
servants to a burglar or to steal a knife to prevent a murder;'
and the Miss Gaskells, with girlish loyalty to what is current,
had rejected the heresy with indignation. From such passages-

at-arms, many retire mortified and ruffled; but Fleeming had
no sooner left the house than he fell into delighted admiration
of the spirit of his adversaries. From that it was but a step to
ask himself ' what truth was sticking in their heads; ' for even
the falsest form of words (in Fleeming's life-long opinion) re-
posed upon some truth, just as he could ' not even allow that
people admire ugly things, they admire what is pretty in the
ugly thing.' And before he sat down to write his letter, he
thought he had hit upon the explanation. ' I fancy the true
idea,' he wrote, ' is that you must never do yourself or anyone
else a moral injury—make any man a thief or a liar—for any
end; ' quite a different thing, as he would have loved to point
out, from never stealing or lying. But this perfervid disputant
was not always out of key with his audience. One whom he
met in the same house announced that she would never again
be happy. ' What does that signify ? ' cried Fleeming. ' We
are not here to be happy but to be good.' And the words (as
his hearer writes to me) became to her a sort of motto during
life.

Fleeming
at Green-
wich.

From Fairbairn's and Manchester, Fleeming passed to a
railway survey in Switzerland, and thence again to Mr. Penn's
at Greenwich, where he was engaged as draughtsman. There
in 1856, we find him in ' a terribly busy state, finishing up
engines for innumerable gun-boats and steam frigates for the
ensuing campaign.' From half-past eight in the morning till
nine or ten at night, he worked in a crowded office among un-
congenial comrades, ' saluted by chaff, generally low personal
and not witty,' pelted with oranges and apples, regaled with
dirty stories, and seeking to suit himself with his surroundings
or (as he writes it) trying to be as little like himself as possible.
His lodgings were hard by ' across a dirty green and through
some half-built streets of two storied houses; ' he had Carlyle
and the poets, engineering and mathematics, to study by himself in such spare time as remained to him; and there were
several ladies, young and not so young, with whom he liked to
correspond. But not all of these could compensate for the
absence of that mother, who had made herself so large a figure

in his life, for sorry surroundings, unsuitable society, and work
that leaned to the mechanical. ' Sunday,' says he, ' I generally
visit some friends in town and seem to swim in clearer water,
but the dirty green seems all the dirtier when I get back.
Luckily I am fond of my profession, or I could not stand this
life.' It is a question in my mind, if he could have long continued to stand it without loss. ' We are not here to be happy,
but to be good,' quoth the young philosopher; but no man had
a keener appetite for happiness than Fleeming Jenkin. There
is a time of life besides when, apart from circumstances, few
men are agreeable to their neighbours and still fewer to themselves; and it was at this stage that Fleeming had arrived, later
than common and even worse provided. The letter from which
I have quoted is the last of his correspondence with Frank Scott,
and his last confidential letter to one of his own sex. ' If you
consider it rightly,' he wrote long after, ' you will find the want
of correspondence no such strange want in men's friendships.
There is, believe me, something noble in the metal which
does not rust though not burnished by daily use.' It is well
said; but the last letter to Frank Scott is scarcely of a noble
metal. It is plain the writer has outgrown his old self, yet
not made acquaintance with the new. This letter from a busy
youth of three and twenty, breathes of seventeen : the sickening
alternations of conceit and shame, the expense of hope *in
vacuo*, the lack of friends, the longing after love; the whole
world of egoism under which youth stands groaning, a voluntary
Atlas.

With Fleeming this disease was never seemingly severe.
The very day before this (to me) distasteful letter, he had
written to Miss Bell of Manchester in a sweeter strain; I do
not quote the one, I quote the other ; fair things are the best.
' I keep my own little lodgings,' he writes, ' but come up every
night to see mamma' (who was then on a visit to London) ' if
not kept too late at the works; and have singing lessons
once more, and sing " *Donne l'amore è scaltro pargoletto* " ; and
think and talk about you ; and listen to mamma's projects *de*
Stowting. Everything turns to gold at her touch, she's a fairy

and no mistake. We go on talking till I have a picture in my head, and can hardly believe at the end that the original is Stowting. Even you don't know half how good mamma is; in other things too, which I must not mention. She teaches me how it is not necessary to be very rich to do much good. I begin to understand that mamma would find useful occupation and create beauty at the bottom of a volcano. She has little weaknesses, but is a real generous-hearted woman, which I suppose is the finest thing in the world.' Though neither mother nor son could be called beautiful, they make a pretty picture; the ugly, generous, ardent woman weaving rainbow illusions; the ugly, clear-sighted, loving son sitting at her side in one of his rare hours of pleasure, half-beguiled, half-amused, wholly admiring, as he listens. But as he goes home, and the fancy pictures fade, and Stowting is once more burthened with debt, and the noisy companions and the long hours of drudgery once more approach, no wonder if the dirty green seems all the dirtier or if Atlas must resume his load.

But in healthy natures, this time of moral teething passes quickly of itself, and is easily alleviated by fresh interests; and already, in the letter to Frank Scott, there are two words of hope: his friends in London, his love for his profession. The last might have saved him; for he was erelong to pass into a new sphere, where all his faculties were to be tried and exercised, and his life to be filled with interest and effort. But it was not left to engineering: another and more influential aim was to be set before him. He must, in any case, have fallen in love; in any case, his love would have ruled his life; and the question of choice was, for the descendant of two such families, a thing of paramount importance. Innocent of the world, fiery, generous, devoted as he was, the son of the wild Jacksons and the facile Jenkins might have been led far astray. By one of those partialities that fill men at once with gratitude and wonder, his choosing was directed well. Or are we to say that by a man's choice in marriage, as by a crucial merit, he deserves his fortune? One thing at least reason may discern: that a man but partly chooses, he also partly forms, his helpmate; and he must in part deserve her, or the treasure is but won for a moment to be lost. Fleeming chanced if you will (and indeed all these opportunities are as ' random as blind man's buff') upon a wife who was worthy of him; but he had the wit to know it, the courage to wait and labour for his prize, and the tenderness and chivalry that are required to keep such prizes precious. Upon this point he has himself written well, as usual with fervent optimism, but as usual (in his own phrase) with a truth sticking in his head.

' Love,' he wrote, ' is not an intuition of the person most suitable to us, most required by us ; of the person with whom life flowers and bears fruit. If this were so, the chances of our meeting that person would be small indeed ; our intuition would often fail; the blindness of love would then be fatal as it is proverbial. No, love works differently, and in its blindness lies its strength. Man and woman, each strongly desires to be loved, each opens to the other that heart of ideal aspirations which they have often hid till then ; each, thus knowing the ideal of the other, tries to fulfil that ideal, each partially succeeds. The greater the love, the greater the success ; the nobler the idea of each, the more durable, the more beautiful the effect. Meanwhile the blindness of each to the other's defects enables the transformation to proceed [unobserved,] so that when the veil is withdrawn (if it ever is, and this I do not know) neither knows that any change has occurred in the person whom they loved. Do not fear, therefore. I do not tell you that your friend will not change, but as I am sure that her choice cannot be that of a man with a base ideal, so I am sure the change will be a safe and a good one. Do not fear that anything you love will vanish, he must love it too.'

Among other introductions in London, Fleeming had presented a letter from Mrs. Gaskell to the Alfred Austins. This was a family certain to interest a thoughtful young man. Alfred, the youngest and least known of the Austins, had been a beautiful golden-haired child, petted and kept out of the way of both sport and study by a partial mother. Bred an attorney, he had (like both his brothers) changed his way of life, and

was called to the bar when past thirty. A Commission of
Enquiry into the state of the poor in Dorsetshire gave him an
opportunity of proving his true talents ; and he was appointed
a Poor Law Inspector, first at Worcester, next at Manchester,
where he had to deal with the potato famine and the Irish im-
migration of the 'forties, and finally in London, where he again
distinguished himself during an epidemic of cholera. He was
then advanced to the Permanent Secretaryship of Her Majesty's
Office of Works and Public Buildings ; a position which he filled
with perfect competence, but with an extreme of modesty ; and on
his retirement, in 1868, he was made a Companion of the Bath.
While apprentice to a Norwich attorney, Alfred Austin was a
frequent visitor in the house of Mr. Barron, a rallying place in
those days of intellectual society. Edward Barron, the son of a
rich saddler or leather merchant in the Borough, was a man
typical of the time. When he was a child, he had once been
patted on the head in his father's shop by no less a man than
Samuel Johnson, as the Doctor went round the Borough can-
vassing for Mr. Thrale ; and the child was true to this early
consecration. 'A life of lettered ease spent in provincial retirement,' it is thus that the biographer of that remarkable man,
William Taylor announces his subject ; and the phrase is equally
descriptive of the life of Edward Barron. The pair were close
friends : ' W. T. and a pipe render everything agreeable,' writes
Barron in his diary in 1828; and in 1833, after Barron had
moved to London and Taylor had tasted the first public failure
of his powers, the latter wrote : ' To my ever dearest Mr. Barron
say, if you please, that I miss him more than I regret him—
that I acquiesce in his retirement from Norwich, because I
could ill brook his observation of my increasing debility of mind.'
This chosen companion of William Taylor must himself have
been no ordinary man ; and he was the friend besides of Borrow,
whom I find him helping in his Latin. But he had no desire
for popular distinction, lived privately, married a daughter of
Dr. Enfield of Enfield's *Speaker*, and devoted his time to the
education of his family, in a deliberate and scholarly fashion,
and with certain traits of stoicism, that would surprise a mo-

dern. From these children we must single out his youngest
daughter, Eliza, who learned under his care to be a sound Latin,
an elegant Grecian, and to suppress emotion without outward
sign after the manner of the Godwin school. This was the
more notable, as the girl really derived from the Enfields;
whose high-flown romantic temper, I wish I could find space to
illustrate. She was but seven years old, when Alfred Austin remarked and fell in love with her; and the union thus early
prepared was singularly full. Where the husband and wife
differed, and they did so on momentous subjects, they differed
with perfect temper and content; and in the conduct of life,
and in depth and durability of love, they were at one. Each full
of high spirits, each practised something of the same repression:
no sharp word was uttered in their house. The same point of
honour ruled them: a guest was sacred and stood within the
pale from criticism. It was a house, besides, of unusual intellectual tension. Mrs. Austin remembered, in the early days
of the marriage, the three brothers, John, Charles and Alfred,
marching to and fro, each with his hands behind his back, and
'reasoning high' till morning; and how, like Dr. Johnson, they
would cheer their speculations with as many as fifteen cups of
tea. And though, before the date of Fleeming's visit, the
brothers were separated, Charles long ago retired from the
world at Brandeston, and John already near his end in the
'rambling old house' at Weybridge, Alfred Austin and his wife
were still a centre of much intellectual society, and still, as
indeed they remained until the last, youthfully alert in mind.
There was but one child of the marriage, Anne, and she was herself something new for the eyes of the young visitor; brought
up, as she had been, like her mother before her, to the standard
of a man's acquirements. Only one art had she been denied,
she must not learn the violin—the thought was too monstrous
even for the Austins; and indeed it would seem as if that tide
of reform which we may date from the days of Mary Wollstone-
craft had in some degree even receded; for though Miss Austin
was suffered to learn Greek, the accomplishment was kept secret
like a piece of guilt. But whether this stealth was caused by

a backward movement in public thought since the time of Edward Barron, or by the change from enlightened Norwich to barbarian London, I have no means of judging.

When Fleeming presented his letter, he fell in love at first sight with Mrs. Austin and the life and atmosphere of the house. There was in the society of the Austins, outward, stoical conformers to the world, something gravely suggestive of essential eccentricity, something unpretentiously breathing of intellectual effort, that could not fail to hit the fancy of this hot-brained boy. The unbroken enamel of courtesy, the self-restraint, the dignified kindness of these married folk, had besides a particular attraction for their visitor. He could not but compare what he saw, with what he knew of his mother and himself. Whatever virtues Fleeming possessed, he could never count on being civil; whatever brave, true-hearted qualities he was able to admire in Mrs. Jenkin, mildness of demeanour was not one of them. And here he found persons who were the equals of his mother and himself in intellect and width of interest, and the equals of his father in mild urbanity of disposition. Show Fleeming an active virtue, and he always loved it. He went away from that house struck through with admiration, and vowing to himself that his own married life should be upon that pattern, his wife (whoever she might be) like Eliza Barron, himself such another husband as Alfred Austin. What is more strange, he not only brought away, but left behind him, golden opinions. He must have been—he was, I am told—a trying lad; but there shone out of him such a light of innocent candour, enthusiasm, intelligence and appreciation, that to persons already some way forward in years, and thus able to enjoy indulgently the perennial comedy of youth, the sight of him was delightful. By a pleasant coincidence, there was one person in the house whom he did not appreciate and who did not appreciate him : Anne Austin, his future wife. His boyish vanity ruffled her; his appearance, never impressive, was then, by reason of obtrusive boyishness, still less so; she found occasion to put him in the wrong by correcting a false quantity; and when Mr. Austin, after doing his visitor the almost unheard-of honour of accompanying him to the door, announced 'That was what young men were like in my time'—she could only reply, looking on her handsome father, 'I thought they had been better looking.'

This first visit to the Austins took place in 1855; and it seems it was some time before Fleeming began to know his mind; and yet longer ere he ventured to show it. The corrected quantity, to those who knew him well, will seem to have played its part; he was the man always to reflect over a correction and to admire the castigator. And fall in love he did; not hurriedly but step by step, not blindly but with critical discrimination; not in the fashion of Romeo, but before he was done, with all Romeo's ardour and more than Romeo's faith. The high favour to which he presently rose in the esteem of Alfred Austin and his wife, might well give him ambitious notions; but the poverty of the present and the obscurity of the future were there to give him pause; and when his aspirations began to settle round Miss Austin, he tasted, perhaps for the only time in his life, the pangs of diffidence. There was indeed opening before him a wide door of hope. He had changed into the service of Messrs. Liddell & Gordon; these gentlemen had begun to dabble in the new field of marine telegraphy; and Fleeming was already face to face with his life's work. That impotent sense of his own value, as of a ship aground, which makes one of the agonies of youth, began to fall from him. New problems which he was endowed to solve, vistas of new enquiry which he was fitted to explore, opened before him continually. His gifts had found their avenue and goal. And with this pleasure of effective exercise, there must have sprung up at once the hope of what is called by the world success. But from these low beginnings, it was a far look upward to Miss Austin: the favour of the loved one seems always more than problematical to any lover; the consent of parents must be always more than doubtful to a young man with a small salary and no capital except capacity and hope. But Fleeming was not the lad to lose any good thing for the lack of trial; and at length, in the autumn of 1857, this boyish-sized, boyish-mannered and superlatively ill-dressed young engineer, entered the house of the Austins, with such sinkings as we may fancy, and asked leave to pay his addresses to the daughter. Mrs. Austin already loved him like a son, she was but too glad to give him her consent; Mr. Austin reserved the right to inquire into his character; from neither was there a word about his prospects, by neither was his income mentioned. ' Are these people,' he wrote, struck with wonder at this dignified disinterestedness, ' are these people the same as other people?' It was not till he was armed with this permission, that Miss Austin even suspected the nature of his hopes: so strong, in this unmannerly boy, was the principle of true courtesy; so powerful, in this impetuous nature, the springs of self-repression. And yet a boy he was; a boy in heart and mind; and it was with a boy's chivalry and frankness that he won his wife. His conduct was a model of honour, hardly of tact; to conceal love from the loved one, to court her parents, to be silent and discreet till these are won, and then without preparation to approach the lady—these are not arts that I would recommend for imitation. They lead to final refusal. Nothing saved Fleeming from that fate, but one circumstance that cannot be counted upon—the hearty favour of the mother, and one gift that is inimitable and that never failed him throughout life, the gift of a nature essentially noble and outspoken. A happy and high-minded anger flashed through his despair: it won for him his wife.

Nearly two years passed before it was possible to marry: two years of activity, now in London; now at Birkenhead, fitting out ships, inventing new machinery for new purposes, and dipping into electrical experiment; now in the *Elba* on his first telegraph cruise between Sardinia and Algiers: a busy and delightful period of bounding ardour, incessant toil, growing hope and fresh interests, with behind and through all, the image of his beloved. A few extracts, from his correspondence with his betrothed will give the note of these truly joyous years. ' My profession gives me all the excitement and interest I ever hope for, but the sorry jade is obviously jealous of you.'—

' " Poor Fleeming," in spite of wet, cold and wind, clambering
over moist, tarry slips, wandering among pools of slush in
waste places inhabited by wandering locomotives, grows visibly
stronger, has dismissed his office cough and cured his toothache.'—'The whole of the paying out and lifting machinery
must be designed and ordered in two or three days, and I am
half crazy with work. I like it though : it's like a good ball,
the excitement carries you through.'—'I was running to and
from the ships and warehouse through fierce gusts of rain and
wind till near eleven, and you cannot think what a pleasure it
was to be blown about and think of you in your pretty dress.'—
' I am at the works till ten and sometimes till eleven. But 1
have a nice office to sit in, with a fire to myself, and bright
brass scientific instruments all round me, and books to read,
and experiments to make, and enjoy myself amazingly. I find
the study of electricity so entertaining that I am apt to neglect
my other work.' And for a last taste, ' Yesterday I had some
charming electrical experiments. What shall I compare them
to—a new song ? a Greek play ? '

It was at this time besides that he made the acquaintance Fleeming
of Professor, now Sir William, Thomson. To describe the part and Sir W.
Thomson.
played by these two in each other's lives would lie out of my way.
They worked together on the Committee on Electrical Standards ;
they served together at the laying down or the repair of many
deep-sea cables ; and Sir William was regarded by Fleeming,
not only with the ' worship ' (the word is his own) due to great
scientific gifts, but with an ardour of personal friendship not
frequently excelled. To their association, Fleeming brought
the valuable element of a practical understanding ; but he never
thought or spoke of himself where Sir William was in question ;
and I recall quite in his last days, a singular instance of this
modest loyalty to one whom he admired and loved. He drew
up a paper, in a quite personal interest, of his own services ;
yet even here he must step out of his way, he must add, where
it had no claim to be added, his opinion that, in their joint work,
the contributions of Sir William had been always greatly the
most valuable. Again, I shall not readily forget with what

emotion he once told me an incident of their associated travels.
On one of the mountain ledges of Madeira, Fleeming's pony bolted
between Sir William and the precipice above ; by strange good
fortune and thanks to the steadiness of Sir William's horse,
no harm was done ; but for the moment, Fleeming saw his
friend hurled into the sea, and almost by his own act : it was a
memory that haunted him.

CHAPTER IV.

1859–1868.

Fleeming's Marriage—His Married Life—Professional Difficulties—Life at Claygate—Illness of Mrs. F. Jenkin—and of Fleeming—Appointment to the Chair at Edinburgh.

ON Saturday, Feb. 26, 1859, profiting by a holiday of four days, Fleeming was married to Miss Austin at Northiam : a place connected not only with his own family but with that of his bride as well. By Tuesday morning, he was at work again, fitting out cableships at Birkenhead. Of the walk from his lodgings to the works, I find a graphic sketch in one of his letters : ' Out over the railway bridge, along a wide road raised to the level of a ground floor above the land, which, not being built upon, harbours puddles, ponds, pigs, and Irish hovels ;—so to the dock warehouses, four huge piles of building with no windows, surrounded by a wall about twelve feet high ;—in through the large gates, round which hang twenty or thirty rusty Irish, playing pitch and toss and waiting for employment ; —on along the railway, which came in at the same gates and which branches down between each vast block—past a pilot-engine butting refractory trucks into their places—on to the last block, [and] down the branch, sniffing the guano-scented air and detecting the old bones. The hartshorn flavour of the guano becomes very strong, as I near the docks where, across the *Elba's* decks, a huge vessel is discharging her cargo of the brown dust, and where huge vessels have been discharging that same cargo for the last five months.' This was the walk he took his young wife on the morrow of his return. She had been used to the society of lawyers and civil servants, moving in that circle which seems to itself the pivot of the nation and

Fleeming's marriage.

is in truth only a clique like another ; and Fleeming was to her
the nameless assistant of a nameless firm of engineers, doing
his inglorious business, as she now saw for herself, among
unsavoury surroundings. But when their walk brought them
within view of the river, she beheld a sight to her of the most
novel beauty : four great, sea-going ships dressed out with flags.
' How lovely ! ' she cried. ' What is it for ? '—' For you,' said
Fleeming. Her surprise was only equalled by her pleasure.
But perhaps, for what we may call private fame, there is no
life like that of the engineer ; who is a great man in out of the
way places, by the dockside or on the desert island or in populous
ships, and remains quite unheard of in the coteries of London
And Fleeming had already made his mark among the few who
had an opportunity of knowing him.

His
married
life.

His marriage was the one decisive incident of his career ; from
that moment until the day of his death, he had one thought to
which all the rest were tributary, the thought of his wife.
No one could know him even slightly, and not remark the absorbing greatness of that sentiment ; nor can any picture of the
man be drawn that does not in proportion dwell upon it. This
is a delicate task ; but if we are to leave behind us (as we wish)
some presentment of the friend we have lost, it is a task that
must be undertaken.

For all his play of mind and fancy, for all his indulgence —and,
as time went on, he grew indulgent—Fleeming had views of duty
that were even stern. He was too shrewd a student of his
fellow men to remain long content with rigid formulæ of conduct. Iron-bound, impersonal ethics, the procrustean bed of
rules, he soon saw at their true value as the deification of aver-
ages. ' As to Miss (I declare I forget her name) being bad,' I
find him writing, ' people only mean that she has broken the
Decalogue—which is not at all the same thing. People who
have kept in the high road of Life really have less opportunity
for taking a comprehensive view of it than those who have leaped
over the hedges and strayed up the hills ; not but what the hedges
are very necessary, and our stray travellers often have a weary
time of it. So, you may say, have those in the dusty roads.'

Yet he was himself a very stern respecter of the hedgerows; sought safety and found dignity in the obvious path of conduct; and would palter with no simple and recognised duty of his epoch. Of marriage in particular, of the bond so formed, of the obligations incurred, of the debt men owe to their children, he conceived in a truly antique spirit: not to blame others, but to constrain himself. It was not to blame, I repeat, that he held these views; for others, he could make a large allowance; and yet he tacitly expected of his friends and his wife a high standard of behaviour. Nor was it always easy to wear the armour of that ideal.

Acting upon these beliefs; conceiving that he had indeed 'given himself' (in the full meaning of these words) for better, for worse; painfully alive to his defects of temper and deficiency in charm; resolute to make up for these; thinking last of himself: Fleeming was in some ways the very man to have made a noble, uphill fight of an unfortunate marriage. In other ways, it is true he was one of the most unfit for such a trial. And it was his beautiful destiny to remain to the last hour the same absolute and romantic lover, who had shown to his new bride the flag-draped vessels in the Mersey. No fate is altogether easy; but trials are our touchstone, trials overcome our reward; and it was given to Fleeming to conquer. It was given to him to live for another, not as a task, but till the end as an enchanting pleasure. 'People may write novels,' he wrote in 1869, 'and other people may write poems, but not a man or woman among them can write to say how happy a man may be, who is desperately in love with his wife after ten years of marriage.' And again in 1885, after more than twenty-six years of marriage, and within but five weeks of his death: 'Your first letter from Bournemouth,' he wrote, 'gives me heavenly pleasure—for which I thank Heaven and you too—who are my heaven on earth.' The mind hesitates whether to say that such a man has been more good or more fortunate.

Any woman (it is the defect of her sex) comes sooner to the stable mind of maturity than any man; and Jenkin was to the end of a most deliberate growth. In the next chapter, when I come to deal with his telegraphic voyages and give some taste of his correspondence, the reader will still find him at twenty-five an arrant schoolboy. His wife besides was more thoroughly educated than he. In many ways she was able to teach him, and he proud to be taught; in many ways she outshone him, and he delighted to be outshone. All these superiorities, and others that, after the manner of lovers, he no doubt forged for himself, added as time went on to the humility of his original love. Only once, in all I know of his career, did he show a touch of smallness. He could not learn to sing correctly; his wife told him so and desisted from her lessons; and the mortification was so sharply felt that for years he could not be induced to go to a concert, instanced himself as a typical man without an ear, and never sang again. I tell it; for the fact that this stood singular in his behaviour, and really amazed all who knew him, is the happiest way I can imagine to commend the tenor of his simplicity; and because it illustrates his feeling for his wife. Others were always welcome to laugh at him; if it amused them, or if it amused him, he would proceed undisturbed with his occupation, his vanity invulnerable. With his wife it was different: his wife had laughed at his singing; and for twenty years the fibre ached. Nothing, again, was more notable than the formal chivalry of this unmannered man to the person on earth with whom he was the most familiar. He was conscious of his own innate and often rasping vivacity and roughness; and he was never forgetful of his first visit to the Austins and the vow he had registered on his return. There was thus an artificial element in his punctilio that at times might almost raise a smile. But it stood on noble grounds; for this was how he sought to shelter from his own petulance the woman who was to him the symbol of the household and to the end the beloved of his youth.

Pro-
fessional
diffi-
culties.
    I wish in this chapter to chronicle small beer; taking a hasty glance at some ten years of married life and of professional struggle; and reserving till the next all the more interesting matter of his cruises. Of his achievements and their worth, it is not for me to speak: his friend and partner, Sir William

Thomson, has contributed a note on the subject, which will be found in the Appendix and to which I must refer the reader. He is to conceive in the meanwhile for himself Fleeming's manifold engagements : his service on the Committee on Electrical Standards, his lectures on electricity at Chatham, his chair at the London University, his partnership with Sir William Thomson and Mr. Varley in many ingenious patents, his growing credit with engineers and men of science ; and he is to bear in mind that of all this activity and acquist of reputation, the immediate profit was scanty. Soon after his marriage, Fleeming had left the service of Messrs Liddell & Gordon, and entered into a general engineering partnership with Mr. Forde, a gentleman in a good way of business. It was a fortunate partnership in this, that the parties retained their mutual respect unlessened and separated with regret ; but men's affairs, like men, have their times of sickness, and by one of these unaccountable variations, for hard upon ten years the business was disappointing and the profits meagre. 'Inditing drafts of German railways which will never get made : ' it is thus I find Fleeming, not without a touch of bitterness, describe his occupation. Even the patents hung fire at first. There was no salary to rely on ; children were coming and growing up ; the prospect was often anxious. In the days of his courtship, Fleeming had written to Miss Austin a dissuasive picture of the trials of poverty, assuring her these were no figments but truly bitter to support ; he told her this, he wrote, beforehand, so that when the pinch came and she suffered, she should not be disappointed in herself nor tempted to doubt her own magnamimity : a letter of admirable wisdom and solicitude. But now that the trouble came, he bore it very lightly. It was his principle, as he once prettily expressed it, ' to enjoy each day's happiness, as it arises, like birds or children.' His optimism, if driven out at the door, would come in again by the window ; if it found nothing but blackness in the present, would hit upon some ground of consolation in the future or the past. And his courage and energy were indefatigable. In the year 1863, soon after the birth of their first son, they moved into a cottage at Claygate near

Esher; and about this time, under manifold troubles both of money and health, I find him writing from abroad: 'The country will give us, please God, health and strength. I will love and cherish you more than ever, you shall go where you wish, you shall receive whom you wish—and as for money you shall have that too. I cannot be mistaken. I have now measured myself with many men. I do not feel weak, I do not feel that I shall fail. In many things I have succeeded, and I will in this. And meanwhile the time of waiting, which, please Heaven, shall not be long, shall also not be so bitter. Well, well, I promise much, and do not know at this moment how you and the dear child are. If he is but better, courage, my girl, for I see light.'

Life at Claygate.

This cottage at Claygate, stood just without the village, well surrounded with trees and commanding a pleasant view. A piece of the garden was turfed over to form a croquet green, and Fleeming became (I need scarce say) a very ardent player. He grew ardent, too, in gardening. This he took up at first to please his wife, having no natural inclination; but he had no sooner set his hand to it, than like everything else he touched it became with him a passion. He budded roses, he potted cuttings in the coach-house; if there came a change of weather at night, he would rise out of bed to protect his favourites; when he was thrown with a dull companion, it was enough for him to discover in the man a fellow gardener; on his travels, he would go out of his way to visit nurseries and gather hints; and to the end of his life, after other occupations prevented him putting his own hand to the spade, he drew up a yearly programme for his gardener, in which all details were regulated. He had begun by this time to write. His paper on Darwin, which had the merit of convincing on one point the philosopher himself, had indeed been written before this in London lodgings; but his pen was not idle at Claygate; and it was here he wrote (among other things) that review of '*Fecundity, Fertility, Sterility, and Allied Topics,*' which Dr. Matthews Duncan prefixed by way of introduction to the second edition of the work. The mere act of writing seems to cheer the vanity of the most

FLEEMING JENKIN

AGED 26

FROM A SKETCH BY HIMSELF

incompetent; but a correction accepted by Darwin, and a whole review borrowed and reprinted by Matthews Duncan, are compliments of a rare strain, and to a man still unsuccessful must have been precious indeed. There was yet a third of the same kind in store for him; and when Munro himself owned that he had found instruction in the paper on Lucretius, we may say that Fleeming had been crowned in the capitol of reviewing.

Croquet, charades, Christmas magic lanterns for the village children, an amateur concert or a review article in the evening; plenty of hard work by day; regular visits to meetings of the British Association, from one of which I find him characteristically writing : 'I cannot say that I have had any amusement yet, but I am enjoying the dulness and dry bustle of the whole thing'; occasional visits abroad on business, when he would find the time to glean (as I have said) gardening hints for himself, and old folksongs or new fashions of dress for his wife; and the continual study and care of his children : these were the chief elements of his life. Nor were friends wanting. Captain and Mrs. Jenkin, Mr. and Mrs. Austin, Clerk Maxwell, Miss Bell of Manchester, and others came to them on visits. Mr. Hertslet of the Foreign Office, his wife and his daughter, were neighbours and proved kind friends; in 1867 the Howitts came to Claygate and sought the society of 'the two bright, clever young people';[1] and in a house close by Mr. Frederick Ricketts came to live with his family. Mr. Ricketts was a valued friend during his short life; and when he was lost with every circumstance of heroism in the *La Plata*, Fleeming mourned him sincerely.

I think I shall give the best idea of Fleeming in this time of his early married life, by a few sustained extracts from his letters to his wife, while she was absent on a visit in 1864.

'*Nov.* 11.—Sunday was too wet to walk to Isleworth, for which I was sorry, so I staid and went to Church and thought of you at Ardwick all through the Commandments, and heard Dr. —— expound in a remarkable way a prophecy of St. Paul's about Roman Catholics, which *mutatis mutandis* would

Letters from Claygate.

[1] *Reminiscences of My Later Life*, by Mary Howitt, *Good Words*, May 1886.

do very well for Protestants in some parts. Then I made a little nursery of Borecole and Enfield market cabbage, grubbing in wet earth with leggings and gray coat on. Then I tidied up the coach-house to my own and Christine's admiration. Then encouraged by *bouts-rimés* I wrote you a copy of verses; high time I think; I shall just save my tenth year of knowing my lady love without inditing poetry or rhymes to her.

'Then I rummaged over the box with my father's letters and found interesting notes from myself. One I should say my first letter, which little Austin I should say would rejoice to see and shall see—with a drawing of a cottage and a spirited "cob." What was more to the purpose, I found with it a paste-cutter which Mary begged humbly for Christine and I generously gave this morning.

'Then I read some of Congreve. There are admirable scenes in the manner of Sheridan; all wit and no character, or rather one character in a great variety of situations and scenes. I could show you some scenes, but others are too coarse even for my stomach hardened by a course of French novels.

'All things look so happy for the rain.

'*Nov.* 16.—Verbenas looking well. . . . I am but a poor creature without you; I have naturally no spirit or fun or enterprise in me. Only a kind of mechanical capacity for ascertaining whether two really is half four, etc.; but when you are near me I can fancy that I too shine, and vainly suppose it to be my proper light; whereas by my extreme darkness when you are not by, it clearly can only be by a reflected brilliance that I seem aught but dull. Then for the moral part of me: if it were not for you and little Odden, I should feel by no means sure that I had any affection power in me. . . . Even the muscular me suffers a sad deterioration in your absence. I don't get up when I ought to, I have snoozed in my chair after dinner; I do not go in at the garden with my wonted vigour, and feel ten times as tired as usual with a walk in your absence; so you see, when you are not by, I am a person without ability, affections or vigour, but droop dull, selfish and spiritless; can you wonder that I love you?

'*Nov.* 17.—. . . I am very glad we married young. I would not have missed these five years, no, not for any hopes; they are my own.

'*Nov.* 30.—I got through my Chatham lecture very fairly though almost all my apparatus went astray. I dined at the mess, and got home to Isleworth the same evening; your father very kindly sitting up for me.

'*Dec.* 1.—Back at dear Claygate. Many cuttings flourish, especially those which do honour to your hand. Your Californian annuals are up and about. Badger is fat, the grass green. . . .

'*Dec.* 3.—Odden will not talk of you, while you are away, having inherited, as I suspect, his father's way of declining to consider a subject which is painful, as your absence is. . . . I certainly should like to learn Greek and I think it would be a capital pastime for the long winter evenings. . . . How things are misrated! I declare croquet is a noble occupation compared to the pursuits of business men. As for so-called idleness—that is, one form of it—I vow it is the noblest aim of man. When idle, one can love, one can be good, feel kindly to all, devote oneself to others, be thankful for existence, educate one's mind, one's heart, one's body. When busy, as I am busy now or have been busy to-day, one feels just as you sometimes felt when you were too busy, owing to want of servants.

'*Dec.* 5.—On Sunday I was at Isleworth, chiefly engaged in playing with Odden. We had the most enchanting walk together through the brickfields. It was very muddy, and, as he remarked, not fit for Nanna, but fit for us *men*. The dreary waste of bared earth, thatched sheds and standing water, was a paradise to him; and when we walked up planks to deserted mixing and crushing mills, and actually saw where the clay was stirred with long iron prongs, and chalk or lime ground with 'a tind of a mill,' his expression of contentment and triumphant heroism knew no limit to its beauty. Of course on returning I found Mrs. Austin looking out at the door in an anxious manner, and thinking we had been out quite long enough. . . . I am reading Don Quixote chiefly and am his fervent admirer, but I am so sorry he did not place his affections on a Dulcinea of somewhat worthier stamp. In fact I think there must be a mistake about it. Don Quixote might and would serve his lady in most preposterous fashion, but I am sure he would have chosen a lady of merit. He imagined her to be such no doubt, and drew a charming picture of her occupations by the banks of the river; but in his other imaginations, there was some kind of peg on which to hang the false costumes he created; windmills are big, and wave their arms like giants; sheep in the distance are somewhat like an army; a little boat on the river-side must look much the same whether enchanted or belonging to millers; but except that Dulcinea is a woman, she bears no resemblance at all to the damsel of his imagination.'

At the time of these letters, the oldest son only was born to them. In September of the next year, with the birth of the second, Charles Frewen, there befell Fleeming a terrible alarm and what proved to be a lifelong misfortune. Mrs. Jenkin was taken suddenly and alarmingly ill; Fleeming ran a matter of two miles to fetch the doctor, and drenched with sweat as he was, returned with him at once in an open gig. On their arrival at the house, Mrs. Jenkin half unconsciously took and kept hold of her husband's hand. By the doctor's orders, windows and doors were set open to create a thorough draught, and the patient was on no account to be disturbed. Thus, then, did Fleeming pass the whole of that night, crouching on the floor in the draught, and not daring to move lest he should wake the sleeper. He had never been strong; energy had stood him instead of vigour; and the result of that night's exposure was

flying rheumatism varied by settled sciatica. Sometimes it quite disabled him, sometimes it was less acute; but he was rarely free from it until his death. I knew him for many years; for more than ten we were closely intimate; I have lived with him for weeks; and during all this time, he only once referred to his infirmity and then perforce, as an excuse for some trouble he put me to, and so slightly worded that I paid no heed. This is a good measure of his courage under sufferings of which none

THE CHILDREN

FROM A LETTER WRITTEN AT CLAYGATE

but the untried will think lightly. And I think it worth noting how this optimist was acquainted with pain. It will seem strange only to the superficial. The disease of pessimism springs never from real troubles, which it braces men to bear, which it delights men to bear well. Nor does it readily spring at all, in minds that have conceived of life as a field of ordered duties, not as a chase in which to hunt for gratifications. ' We are not here to be happy but to be good '; I wish he had mended the phrase : ' We are not here to be happy, but to try to be good,' comes nearer the modesty of truth. With such old-fashioned morality, it is possible to get through life, and see the worst of it, and feel some of the worst of it, and still acquiesce piously and even gladly in man's fate. Feel some of the worst of it, I say ; for some of the rest of the worst is, by this simple faith, excluded.

It was in the year 1868, that the clouds finally rose. The business in partnership with Mr. Forde began suddenly to pay well; about the same time the patents showed themselves a valuable property ; and but a little after, Fleeming was appointed to the new chair of engineering in the University of Edinburgh. Thus, almost at once, pecuniary embarrassments passed for ever out of his life. Here is his own epilogue to the time at Claygate, and his anticipations of the future in Edinburgh.

His appointment to the chair at Edinburgh.

' . . . The dear old house at Claygate is not let and the pretty garden a mass of weeds. I feel rather as if we had behaved unkindly to them. We were very happy there, but now that it is over I am conscious of the weight of anxiety as to money which I bore all the time. With you in the garden, with Austin in the coach-house, with pretty songs in the little low white room, with the moonlight in the dear room upstairs, ah, it was perfect; but the long walk, wondering, pondering, fearing, scheming, and the dusty jolting railway, and the horrid fusty office with its endless disappointments, they are well gone. It is well enough to fight and scheme and bustle about in the eager crowd here [in London] for a while now and then, but not for a lifetime. What I have now is just perfect. Study for winter, action for summer, lovely country for recreation, a pleasant town for talk. . . .'

CHAPTER V.

NOTES OF TELEGRAPH VOYAGES, 1858 TO 1873.

But it is now time to see Jenkin at his life's work. I have before me certain imperfect series of letters written, as he says, 'at hazard, for one does not know at the time what is important and what is not': the earlier addressed to Miss Austin, after the betrothal; the later to Mrs. Jenkin the young wife. I should premise that I have allowed myself certain editorial freedoms, leaving out and splicing together, much as he himself did with the Bona cable: thus edited the letters speak for themselves, and will fail to interest none who love adventure or activity. Addressed as they were to her whom he called his 'dear engineering pupil,' they give a picture of his work so clear that a child may understand, and so attractive that I am half afraid their publication may prove harmful, and still further crowd the ranks of a profession already overcrowded. But their most engaging quality is the picture of the writer; with his indomitable self-confidence and courage, his readiness in every pinch of circumstance or change of plan, and his ever fresh enjoyment of the whole web of human experience, nature, adventure, science, toil and rest, society and solitude. It should be borne in mind that the writer of these buoyant pages was, even while he wrote, harassed by responsibility, stinted in sleep and often struggling with the prostration of sea-sickness. To this last enemy, which he never overcame, I have omitted, in my search after condensation, a good many references; if they were all left, such was the man's temper, they would not represent one hundredth part of what he suffered, for he was never given to complaint. But indeed he had met this

ugly trifle, as he met every thwart circumstance of life, with a
certain pleasure of pugnacity ; and suffered it not to check him,
whether in the exercise of his profession or the pursuit of
amusement.

I.

'Birkenhead : April 18, 1858.

' Well, you should know, Mr. —— having a contract to lay
down a submarine telegraph from Sardinia to Africa failed three
times in the attempt. The distance from land to land is about
140 miles. On the first occasion, after proceeding some 70
miles, he had to cut the cable—the cause I forget ; he tried
again, same result ; then picked up about 20 miles of the lost
cable, spliced on a new piece, and very nearly got across that
time, but ran short of cable, and when but a few miles off Galita
in very deep water, had to telegraph to London for more cable
to be manufactured and sent out whilst he tried to stick to the
end : for five days, I think, he lay there sending and receiving
messages, but heavy weather coming on, the cable parted and
Mr. —— went home in despair—at least I should think so.
' He then applied to those eminent engineers, R. S. Newall
& Co., who made and laid down a cable for him last autumn—
Fleeming Jenkin (at the time in considerable mental agitation)
having the honour of fitting out the *Elba* for that purpose.'
[On this occasion, the *Elba* has no cable to lay ; but] ' is going
out in the beginning of May to endeavour to fish up the cables
Mr. —— lost. There are two ends at or near the shore : the
third will probably not be found within 20 miles from land.
One of these ends will be passed over a very big pulley or sheave
at the bows, passed six times round a big barrel or drum ; which
will be turned round by a steam engine on deck, and thus wind
up the cable, while the *Elba* slowly steams ahead. The cable is
not wound round and round the drum as your silk is wound on
its reel, but on the contrary never goes round more than six
times, going off at one side as it comes on at the other, and
going down into the hold of the *Elba* to be coiled along in a big
coil or skein.

'I went down to Gateshead to discuss with Mr. Newall the form which this tolerably simple idea should take, and have been busy since I came here drawing, ordering and putting up the machinery—uninterfered with, thank goodness, by anyone. I own I like responsibility; it flatters one and then, your father might say, I have more to gain than to lose. Moreover I do like this bloodless, painless combat with wood and iron, forcing the stubborn rascals to do my will, licking the clumsy cubs into an active shape, seeing the child of to-day's thought working to-morrow in full vigour at his appointed task.

'May 12.

'By dint of bribing, bullying, cajoling, and going day by day to see the state of things ordered, all my work is very nearly ready now; but those who have neglected these precautions are of course disappointed. Five hundred fathoms of chain [were] ordered by —— some three weeks since, to be ready by the 10th without fail; he sends for it to-day—150 fathoms all they can let us have by the 15th—and how the rest is to be got, who knows? He ordered a boat a month since and yesterday we could see nothing of her but the keel and about two planks. I could multiply instances without end. At first one goes nearly mad with vexation at these things; but one finds so soon that they are the rule, that then it becomes necessary to feign a rage one does not feel. I look upon it as the natural order of things, that if I order a thing, it will not be done—if by accident it gets done, it will certainly be done wrong: the only remedy being to watch the performance at every stage.

'To-day was a grand field day. I had steam up and tried the engine against pressure or resistance. One part of the machinery is driven by a belt or strap of leather. I always had my doubts this might slip; and so it did, wildly. I had made provision for doubling it, putting on two belts instead of one. No use—off they went, slipping round and off the pulleys instead of driving the machinery. Tighten them—no use. More strength there—down with the lever—smash something, tear the belts, but get them tight—now then, stand clear, on with the steam ;—and the belts slip away as if nothing held them. Men begin to look queer; the circle of quidnuncs make sage remarks. Once more—no use. I begin to know I ought to feel sheepish and beat, but somehow I feel cocky instead. I laugh and say " Well, I am bound to break something down "— and suddenly see. " Oho, there's the place; get weight on there, and the belt won't slip." With much labour, on go the belts again. " Now then, a spar thro' there and six men's weight on; mind you're not carried away."—" Ay, ay, sir." But evidently no one believes in the plan. " Hurrah, round she goes—stick to your spar. All right, shut off steam." And the difficulty is vanquished.

' This or such as this (not always quite so bad) occurs hour after hour, while five hundred tons of coal are rattling down into the holds and bunkers, riveters are making their infernal row all round, and riggers bend the sails and fit the rigging :— a sort of Pandemonium, it appeared to young Mrs. Newall, who was here on Monday and half-choked with guano ; but it suits the likes o' me.

' S.S. *Elba*, River Mersey: May 17.

' We are delayed in the river by some of the ship's papers not being ready. Such a scene at the dock gates. Not a sailor will join till the last moment ; and then, just as the ship forges ahead through the narrow pass, beds and baggage fly on board, the men half tipsy clutch at the rigging, the captain swears, the women scream and sob, the crowd cheer and laugh, while one or two pretty little girls stand still and cry outright, regard-less of all eyes.

' These two days of comparative peace have quite set me on my legs again. I was getting worn and weary with anxiety and work. As usual I have been delighted with my shipwrights. I gave them some beer on Saturday, making a short oration. To-day when they went ashore and I came on board, they gave three cheers. whether for me or the ship I hardly know, but I had just bid them good-bye, and the ship was out of hail; but I was startled and hardly liked to claim the compliment by acknowledging it.

'S.S. *Elba*: May 25.

'My first intentions of a long journal have been fairly frustrated by sea-sickness. On Tuesday last about noon we started from the Mersey in very dirty weather, and were hardly out of the river when we met a gale from the south-west and a heavy sea, both right in our teeth; and the poor *Elba* had a sad shaking. Had I not been very sea-sick, the sight would have been exciting enough, as I sat wrapped in my oilskins on the bridge; [but] in spite of all my efforts to talk, to eat and to grin, I soon collapsed into imbecility; and I was heartily thankful towards evening to find myself in bed.

'Next morning, I fancied it grew quieter and, as I listened, heard, "Let go the anchor," whereon I concluded we had run into Holyhead Harbour, as was indeed the case. All that day we lay in Holyhead, but I could neither read nor write nor draw. The captain of another steamer which had put in came on board, and we all went for a walk on the hill; and in the evening there was an exchange of presents. We gave some tobacco I think, and received a cat, two pounds of fresh butter, a Cumberland ham, *Westward Ho!* and Thackeray's *English Humourists*. I was astonished at receiving two such fair books from the captain of a little coasting screw. Our captain said he [the captain of the screw] had plenty of money, five or six hundred a year at least.—"What in the world makes him go rolling about in such a craft, then?"—"Why, I fancy he's reckless; he's desperate in love with that girl I mentioned, and she won't look at him." Our honest, fat, old captain says this very grimly in his thick, broad voice.

'My head won't stand much writing yet, so I will run up and take a look at the blue night sky off the coast of Portugal.

'May 26.

'A nice lad of some two and twenty, A—— by name, goes out in a nondescript capacity as part purser, part telegraph clerk, part generally useful person. A—— was a great comfort during the miseries [of the gale]; for when with a dead head wind and a heavy sea, plates, books, papers, stomachs were being rolled about in sad confusion, we generally managed to lie on our backs, and grin, and try discordant staves of the *Flowers of the Forest* and the *Low-backed Car.* We could sing and laugh, when we could do nothing else ; though A—— was ready to swear after each fit was past, that that was the first time he had felt anything, and at this moment would declare in broad Scotch that he'd never been sick at all, qualifying the oath with " except for a minute now and then." He brought a cornet-a-piston to practise on, having had three weeks' instructions on that melodious instrument ; and if you could hear the horrid sounds that come ! especially at heavy rolls. When I hint he is not improving, there comes a confession : "I don't feel quite right yet, you see !" But he blows away manfully, and in self-defence I try to roar the tune louder.

'11.30 P.M.

' Long past Cape St. Vincent now. We went within about 400 yards of the cliffs and lighthouse in a calm moonlight, with porpoises springing from the sea, the men crooning long ballads as they lay idle on the forecastle and the sails flapping uncertain on the yards. As we passed, there came a sudden breeze from land, hot and heavy scented ; and now as I write its warm rich flavour contrasts strongly with the salt air we have been breathing.

' I paced the deck with H——, the second mate, and in the quiet night drew a confession that he was engaged to be married, and gave him a world of good advice. He is a very nice, active, little fellow, with a broad Scotch tongue and ": dirty, little rascal " appearance. He had a sad disappointment at starting. Having been second mate on the last voyage, when the first mate was discharged, he took charge of the *Elba* all the time she was in port, and of course looked forward to being chief mate this trip. Liddell promised him the post. He had not authority to do this ; and when Newall heard of it, he appointed another man. Fancy poor H—— having told all the men and most of all, his sweetheart ! But more remains behind ; for when it

came to signing articles, it turned out that O——, the new first
mate, had not a certificate which allowed him to have a second
mate. Then came rather an affecting scene. For H—— proposed to sign as chief (he having the necessary higher certificate)
but to act as second for the lower wages. At first O—— would
not give in but offered to go as second. But our brave little
H—— said, no : " The owners wished Mr. O—— to be chief
mate, and chief mate he should be." So he carried the day,
signed as chief and acts as second. Shakespeare and Byron are
his favourite books. I walked into Byron a little, but can well
understand his stirring up a rough, young sailor's romance. I
lent him *Westward Ho* from the cabin; but to my astonishment he did not care much for it ; he said it smelt of the shilling
railway library ; perhaps I had praised it too highly. Scott is
his standard for novels. I am very happy to find good taste by
no means confined to gentlemen, H—— having no pretensions
to that title. He is a man after my own heart.

'Then I came down to the cabin and heard young A——'s
schemes for the future. His highest picture is a commission in
the Prince of Vizianagram's irregular horse. His eldest brother is tutor to his Highness's children, and grand vizier, and
magistrate, and on his Highness's household staff, and seems
to be one of those Scotch adventurers one meets with and hears
of in queer berths—raising cavalry, building palaces, and using
some petty Eastern king's long purse with their long Scotch
heads.

'Off Bona : June 4.

'I read your letter carefully, leaning back in a Maltese boat
to present the smallest surface of my body to a grilling sun, and
sailing from the *Elba* to Cape Hamrah about three miles distant.
How we fried and sighed! At last, we reached land under
Fort Genova, and I was carried ashore pick-a-back, and plucked
the first flower I saw for Annie. It was a strange scene, far
more novel than I had imagined : the high, steep banks covered
with rich, spicy vegetation of which I hardly knew one plant.
The dwarf palm with fan-like leaves, growing about two feet
high, formed the staple of the verdure. As we brushed through

them, the gummy leaves of a cistus stuck to the clothes ; and with its small white flower and yellow heart, stood for our English dog-rose. In place of heather, we had myrtle and lentisque with leaves somewhat similar. That large bulb with long flat leaves? Do not touch it if your hands are cut; the Arabs use it as blisters for their horses. Is that the same sort? No, take that one up ; it is the bulb of a dwarf palm, each layer of the onion peels off, brown and netted, like the outside of a cocoanut. It is a clever plant that; from the leaves we get a vegetable horsehair;—and eat the bottom of the centre spike. All the leaves you pull have the same aromatic scent. But here a little patch of cleared ground shows old friends, who seem to cling by abused civilisation :—fine, hardy thistles, one of them bright yellow, though ;—honest, Scotch-looking, large daisies or gowans ;—potatoes here and there, looking but sickly ; and dark sturdy fig-trees looking cool and at their ease in the burning sun.

'Here we are at Fort Genova, crowning the little point, a small old building, due to my old Genoese acquaintance who fought and traded bravely once upon a time. A broken cannon of theirs forms the threshold ; and through a dark, low arch, we enter upon broad terraces sloping to the centre, from which rain water may collect and run into that well. Large-breeched French troopers lounge about and are most civil ; and the whole party sit down to breakfast in a little white-washed room, from the door of which the long, mountain coastline and the sparkling sea show of an impossible blue through the openings of a white-washed rampart. I try a sea-egg, one of those prickly fellows—sea-urchins, they are called sometimes ; the shell is of a lovely purple, and when opened, there are rays of yellow adhering to the inside ; these I eat, but they are very fishy.

'We are silent and shy of one another, and soon go out to watch while turbaned, blue-breeched, bare-legged Arabs dig holes for the land telegraph posts on the following principle : one man takes a pick and bangs lazily at the hard earth ; when a little is loosened, his mate with a small spade lifts it on one side ;

and *da capo.* They have regular features and look quite in place among the palms. Our English workmen screw the earthenware insulators on the posts, strain the wire, and order Arabs about by the generic term of Johnny. I find W—— has nothing for me to do; and that in fact no one has anything to do. Some instruments for testing have stuck at Lyons, some at Cagliari; and nothing can be done—or at any rate, is done. I wander about, thinking of you and staring at big, green grass-hoppers—locusts, some people call them—and smelling the rich brushwood. There was nothing for a pencil to sketch, and I soon got tired of this work, though I have paid willingly much money for far less strange and lovely sights.

'Off Cape Spartivento: June 8.

'At two this morning, we left Cagliari; at five cast anchor here. I got up and began preparing for the final trial; and shortly afterwards everyone else of note on board went ashore to make experiments on the state of the cable, leaving me with the prospect of beginning to lift at 12 o'clock. I was not ready by that time; but the experiments were not concluded and moreover the cable was found to be imbedded some four or five feet in sand, so that the boat could not bring off the end. At three, Messrs. Liddell, &c., came on board in good spirits, having found two wires good or in such a state as permitted messages to be transmitted freely. The boat now went to grapple for the cable some way from shore while the *Elba* towed a small lateen craft which was to take back the consul to Cagliari some distance on its way. On our return we found the boat had been unsuccessful; she was allowed to drop astern, while we grappled for the cable in the *Elba* [without more success]. The coast is a low mountain range covered with brushwood or heather—pools of water and a sandy beach at their feet. I have not yet been ashore, my hands having been very full all day.

'June 9.

'Grappling for the cable outside the bank had been voted too uncertain; [and the day was spent in] efforts to pull the cable off through the sand which has accumulated over it. By getting the cable tight on to the boat, and letting the swell pitch her about till it got slack, and then tightening again with blocks and pulleys, we managed to get out from the beach towards the ship at the rate of about twenty yards an hour. When they had got about 100 yards from shore, we ran round in the *Elba* to try and help them, letting go the anchor in the shallowest possible water, this was about sunset. Suddenly someone calls out he sees the cable at the bottom : there it was sure enough, apparently wriggling about as the waves rippled. Great excitement ; still greater when we find our own anchor is foul of it and has been the means of bringing it to light. We let go a grapnel, get the cable clear of the anchor on to the grapnel —the captain in an agony lest we should drift ashore meanwhile —hand the grappling line into the big boat, steam out far enough, and anchor again. A little more work and one end of the cable is up over the bows round my drum. I go to my engine and we start hauling in. All goes pretty well, but it is quite dark. Lamps are got at last, and men arranged. We go on for a quarter of a mile or so from shore and then stop at about half-past nine with orders to be up at three. Grand work at last ! A number of the *Saturday Review* here ; it reads so hot and feverish, so tomblike and unhealthy, in the midst of dear Nature's hills and sea, with good wholesome work to do. Pray that all go well to-morrow.

'June 10.

'Thank heaven for a most fortunate day. At three o'clock this morning in a damp, chill mist all hands were roused to work. With a small delay, for one or two improvements I had seen to be necessary last night, the engine started and since that time I do not think there has been half an hour's stoppage. A rope to splice, a block to change, a wheel to oil, an old rusted anchor to disengage from the cable which brought it up, these have been our only obstructions. Sixty, seventy, eighty, a hundred, a hundred and twenty revolutions at last, my little engine tears away. The even black rope comes straight out of the blue heaving water; passes slowly round an open-hearted, good-tempered looking pulley, five feet diameter; aft past a vicious nipper, to bring all up should anything go wrong; through a gentle guide; on to a huge bluff drum, who wraps him round his body and says " Come you must," as plain as drum can speak : the chattering pauls say " I've got him, I've got him, he can't get back ": whilst black cable, much slacker and easier in mind and body, is taken by a slim V-pulley and passed down into the huge hold, where half a dozen men put him comfortably to bed after his exertion in rising from his long bath. In good sooth, it is one of the strangest sights I know to see that black fellow rising up so steadily in the midst of the blue sea. We are more than half way to the place where we expect the fault; and already the one wire, supposed previously to be quite bad near the African coast, can be spoken through. I am very glad I am here, for my machines are my own children and I look on their little failings with a parent's eye and lead them into the path of duty with gentleness and firmness. I am naturally in good spirits, but keep very quiet for misfortunes may arise at any instant; moreover to-morrow my paying-out apparatus will be wanted should all go well, and that will be another nervous operation. Fifteen miles are safely in; but no one knows better than I do that nothing is done till all is done.

' June 11.

' 9 A.M.—We have reached the splice supposed to be faulty, and no fault has been found. The two men learned in electri-city, L—— and W——, squabble where the fault is.

' *Evening.*—A weary day in a hot broiling sun ; no air. After the experiments, L—— said the fault might be ten miles ahead ; by that time, we should be according to a chart in about a thousand fathoms of water—rather more than a mile. It was most difficult to decide whether to go on or not. I made pre-parations for a heavy pull, set small things to rights and went to sleep. About four in the afternoon, Mr. Liddell decided to proceed, and we are now (at seven) grinding it in at the rate of a mile and three-quarters per hour, which appears a grand speed to us. If the paying-out only works well! I have just thought of a great improvement in it; I can't apply it this time however.—The sea is of an oily calm, and a perfect fleet of brigs and ships surrounds us, their sails hardly filling in the lazy breeze. The sun sets behind the dim coast of the Isola San Pietro, the coast of Sardinia high and rugged becomes softer and softer in the distance, while to the westward still the isolated rock of Toro springs from the horizon.—It would amuse you to see how cool (in head) and jolly everybody is. A testy word now and then shows the wires are strained a little, but everyone laughs and makes his little jokes as if it were all in fun : yet we are all as much in earnest as the most earnest of the earnest bastard German school or demonstrative of Frenchmen. I enjoy it very much.

'June 12.

' 5.30 A.M.—Out of sight of land : about thirty nautical miles in the hold ; the wind rising a little ; experiments being made for a fault, while the engine slowly revolves to keep us hanging at the same spot : depth supposed about a mile. The machinery has behaved admirably. Oh! that the paying-out were over! The new machinery there is but rough, meant for an experiment in shallow water, and here we are in a mile of water.

' 6.30.—I have made my calculations and find the new paying-out gear cannot possibly answer at this depth, some portion would give way. Luckily, I have brought the old things with me and am getting them rigged up as fast as may be. Bad news from the cable. Number four has given in some portion of the last ten miles : the fault in number three is still at the bottom of the sea : number two is now the only good wire ; and the hold is getting in such a mess, through keeping bad bits out and cutting for splicing and testing, that there will be great risk in paying out. The cable is somewhat strained in its ascent from one mile below us ; what it will be when we get to two miles is a problem we may have to determine.

' 9 P.M.—A most provoking unsatisfactory day. We have

MEMOIR

done nothing. The wind and sea have both risen. Too little
notice has been given to the telegraphists who accompany this
expedition ; they had to leave all their instruments at Lyons in
order to arrive at Bona in time ; our tests are therefore of the
roughest, and no one really knows where the faults are. Mr.
L—— in the morning lost much time; then he told us, after
we had been inactive for about eight hours, that the fault in
number three was within six miles ; and at six o'clock in the
evening, when all was ready for a start to pick up these six
miles, he comes and says there must be a fault about thirty
miles from Bona ! By this time it was too late to begin paying
out to-day, and we must lie here moored in a thousand fathoms
till light to-morrow morning. The ship pitches a good deal,
but the wind is going down.

‘ June 13, Sunday.

‘ The wind has not gone down however. It now (at 10.30)
blows a pretty stiff gale, the sea has also risen ; and the
Elba's bows rise and fall about 9 feet. We make twelve
pitches to the minute, and the poor cable must feel very sea-
sick by this time. We are quite unable to do anything, and
continue riding at anchor in one thousand fathoms, the
engines going constantly so as to keep the ship's bows up to
the cable, which by this means hangs nearly vertical and
sustains no strain but that caused by its own weight and the
pitching of the vessel. We were all up at four but the weather
entirely forbade work for to-day, so some went to bed and most
lay down, making up our leeway as we nautically term our loss
of sleep. I must say Liddell is a fine fellow and keeps his
patience and temper wonderfully ; and yet how he does fret and
fume about trifles at home ! This wind has blown now for
36 hours, and yet we have telegrams from Bona to say the
sea there is as calm as a mirror. It makes one laugh to
remember one is still tied to the shore. Click, click, click, the
pecker is at work : I wonder what Herr P—— says to Herr
L—— —tests, tests, tests, nothing more. This will be a very
anxious day.

‘ June 14.

‘ Another day of fatal inaction.

'June 15.

' 9.30.—The wind has gone down a deal ; but even now there are doubts whether we shall start to-day. When shall I get back to you ?

' 9 P.M.—Four miles from land. Our run has been successful and eventless. Now the work is nearly over I feel a little out of spirits—why, I should be puzzled to say—mere wantonness, or reaction perhaps after suspense.

'June 16.

' Up this morning at three, coupled my self-acting gear to the break and had the satisfaction of seeing it pay out the last four miles in very good style. With one or two little improvements, I hope to make it a capital thing. The end has just gone ashore in two boats, three out of four wires good. Thus ends our first expedition. By some odd chance a *Times* of June the 7th has found its way on board through the agency of a wretched old peasant who watches the end of the line here. A long account of breakages in the Atlantic trial trip. To-night we grapple for the heavy cable, eight tons to the mile. I long to have a tug at him ; he may puzzle me, and though misfortunes or rather difficulties are a bore at the time, life when working with cables is tame without them.

' 2 P.M.—Hurrah, he is hooked, the big fellow, almost at the first cast. He hangs under our bows looking so huge and imposing that I could find it in my heart to be afraid of him.

'June 17.

' We went to a little bay called Chia, where a fresh-water stream falls into the sea and took in water. This is rather a long operation so I went a walk up the valley with Mr. Liddell. The coast here consists of rocky mountains 800 to 1,000 feet high covered with shrubs of a brilliant green. On landing our first amusement was watching the hundreds of large fish who lazily swam in shoals about the river ; the big canes on the further side hold numberless tortoises, we are told, but see none, for just now they prefer taking a siesta. A little further on, and what is this with large pink flowers in such abundance?—the oleander in full flower. At first I fear to pluck them, thinking they must be cultivated and valuable; but soon the banks show a long line of thick tall shrubs, one mass of glorious pink and green. Set these in a little valley, framed by mountains whose rocks gleam out blue and purple colours such as prae-Raphaelites only dare attempt, shining out hard and weirdlike amongst the clumps of castor-oil plants, cistus, arbor vitae and many other evergreens, whose names, alas! I know not; the cistus is brown now, the rest all deep or brilliant green. Large herds of cattle browse on the baked deposit at the foot of these large crags. One or two half-savage herdsmen in sheepskin kilts &c. ask for cigars; partridges whirr up on either side of us; pigeons coo and nightingales sing amongst the blooming oleander. We get six sheep and many fowls, too, from the priest of the small village; and then run back to Spartivento and make preparations for the morning.

'June 18.

'The big cable is stubborn and will not behave like his smaller brother. The gear employed to take him off the drum is not strong enough; he gets slack on the drum and plays the mischief. Luckily for my own conscience, the gear I had wanted was negatived by Mr. Newall. Mr. Liddell does not exactly blame me, but he says we might have had a silver pulley cheaper than the cost of this delay. He has telegraphed for more men to Cagliari, to try to pull the cable off the drum into the hold, by hand. I look as comfortable as I can, but feel as if people were blaming me. I am trying my best to get something rigged which may help us; I wanted a little difficulty, and feel much better.—The short length we have picked up was covered at places with beautiful sprays of coral, twisted and twined with shells of those small, fairy animals we saw in the aquarium at home; poor little things, they died at once, with their little bells and delicate bright tints.

'12 *o'Clock.*—Hurrah, victory! for the present anyhow. Whilst in our first dejection, I thought I saw a place where a

flat roller would remedy the whole misfortune; but a flat roller
at Cape Spartivento, hard, easily unshipped, running freely!
There was a grooved pulley used for the paying-out machinery
with a spindle wheel, which might suit me. I filled him up
with tarry spunyarn, nailed sheet copper round him, bent some
parts in the fire; and we are paying-in without more trouble
now. You would think some one would praise me; no, no
more praise than blame before; perhaps now they think better
of me, though.

'10 P.M.—We have gone on very comfortably for nearly six
miles. An hour and a half was spent washing down; for along
with many coloured polypi, from corals, shells and insects, the
big cable brings up much mud and rust, and makes a fishy
smell by no means pleasant: the bottom seems to teem with
life.—But now we are startled by a most unpleasant, grinding
noise; which appeared at first to come from the large low pulley,
but when the engines stopped, the noise continued; and we
now imagine it is something slipping down the cable, and the
pulley but acts as sounding-board to the big fiddle. Whether
it is only an anchor or one of the two other cables, we know
not. We hope it is not the cable just laid down.

'June 19.

'10 A.M.—All our alarm groundless, it would appear: the
odd noise ceased after a time, and there was no mark sufficiently strong on the large cable to warrant the suspicion that we
had cut another line through. I stopped up on the look-out
till three in the morning, which made 23 hours between sleep
and sleep. One goes dozing about, though, most of the day,
for it is only when something goes wrong that one has to look
alive. Hour after hour, I stand on the forecastle-head, picking
off little specimens of polypi and coral, or lie on the saloon deck
reading back numbers of the *Times*—till something hitches, and
then all is hurly-burly once more. There are awnings all along
the ship, and a most ancient, fish-like smell beneath.

'1 *o'Clock*.—Suddenly a great strain in only 95 fathoms
of water—belts surging and general dismay; grapnels being
thrown out in the hope of finding what holds the cable.—

Should it prove the young cable! We are apparently crossing
its path—not the working one, but the lost child; Mr. Liddell
would start the big one first though it was laid first: he wanted
to see the job done, and meant to leave us to the small one
unaided by his presence.

' 3.30.—Grapnel caught something, lost it again; it left
its marks on the prongs. Started lifting gear again ; and after
hauling in some 50 fathoms—grunt, grunt, grunt—we hear the
other cable slipping down our big one, playing the selfsame tune
we heard last night—louder however.

' 10 P.M.—The pull on the deck engines became harder and
harder I got steam up in a boiler on deck, and another little
engine starts hauling at the grapnel. I wonder if there ever
was such a scene of confusion : Mr. Liddell and W—— and the
captain all giving orders contradictory &c. on the forecastle ;
D——, the foreman of our men, the mates, &c. following the
example of our superiors; the ship's engine and boilers below,
a 50-horse engine on deck, a boiler 14 feet long on deck beside
it, a little steam winch tearing round ; a dozen Italians (20
have come to relieve our hands, the men we telegraphed for to
Cagliari) hauling at the rope ; wiremen, sailors, in the crevices
left by ropes and machinery; everything that could swear
swearing—I found myself swearing like a trooper at last. We
got the unknown difficulty within ten fathoms of the surface ;
but then the forecastle got frightened that, if it was the small
cable which we had got hold of, we should certainly break it by
continuing the tremendous and increasing strain. So at last
Mr. Liddell, decided to stop ; cut the big cable, buoying its end ;
go back to our pleasant watering-place at Chia, take more
water and start lifting the small cable. The end of the large
one has even now regained its sandy bed ; and three buoys—
one to grapnel foul of the supposed small cable, two to the big
cable—are dipping about on the surface. One more—a flag-
buoy—will soon follow, and then straight for shore.

' June 20.

' It is an ill-wind &c. I have an unexpected opportunity of
forwarding this engineering letter ; for the craft which brought

out our Italian sailors must return to Cagliari to-night, as the little cable will take us nearly to Galita, and the Italian skipper could hardly find his way from thence. To-day—Sunday—not much rest. Mr. Liddell is at Spartivento telegraphing. We are at Chia, and shall shortly go to help our boat's crew in getting the small cable on board. We dropped them some time since in order that they might dig it out of the sand as far as possible.

'June 21.

' Yesterday—Sunday as it was—all hands were kept at work all day, coaling, watering and making a futile attempt to pull the cable from the shore on board through the sand. This attempt was rather silly after the experience we had gained at Cape Spartivento. This morning we grappled, hooked the cable at once, and have made an excellent start. Though I have called this the small cable, it is much larger than the Bona one.—Here comes a break down and a bad one.

'June 22.

' We got over it however; but it is a warning to me that my future difficulties will arise from parts wearing out. Yesterday the cable was often a lovely sight, coming out of the water one large incrustation of delicate, net-like corals and long, white curling shells. No portion of the dirty black wires was visible; instead we had a garland of soft pink with little scarlet sprays and white enamel intermixed. All was fragile however, and could hardly be secured in safety ; and inexorable iron crushed the tender leaves to atoms.—This morning at the end of my watch, about 4 o'clock, we came to the buoys, proving our anticipations right concerning the crossing of the cables. I went to bed for four hours, and on getting up, found a sad mess. A tangle of the six-wire cable hung to the grapnel which had been left buoyed, and the small cable had parted and is lost for the present. Our hauling of the other day must have done the mischief.

'June 23.

' We contrived to get the two ends of the large cable and to pick the short end up. The long end, leading us seaward, was next put round the drum and a mile of it picked up; but then, fearing another tangle, the end was cut and buoyed, and we returned to grapple for the three-wire cable. All this is very tiresome for me. The buoying and dredging are managed entirely by W———, who has had much experience in this sort of thing; so I have not enough to do and get very homesick. At noon the wind freshened and the sea rose so high that we had to run for land and are once more this evening anchored at Chia.

'June 24.

'The whole day spent in dredging without success. This operation consists in allowing the ship to drift slowly across the line where you expect the cable to be, while at the end of a long rope, fast either to the bow or stern, a grapnel drags along the ground. This grapnel is a small anchor, made like four pothooks tied back to back. When the rope gets taut, the ship is stopped and the grapnel hauled up to the surface in the hopes of finding the cable on its prongs.—I am much discontented with myself for idly lounging about and reading *Westward Ho* for the second time, instead of taking to electricity or picking up nautical information. I am uncommonly idle. The sea is not quite so rough but the weather is squally and the rain comes in frequent gusts.

'June 25.

'To-day about 1 o'clock we hooked the three-wire cable, buoyed the long sea end, and picked up the short [or shore] end. Now it is dark and we must wait for morning before lifting the buoy we lowered to-day and proceeding seawards. —The depth of water here is about 600 feet, the height of a respectable English hill; our fishing line was about a quarter of a mile long. It blows pretty fresh and there is a great deal of sea.

'26th.

'This morning it came on to blow so heavily that it was impossible to take up our buoy. The *Elba* recommenced rolling in true Baltic style and towards noon we ran for land.

'27th, Sunday.

'This morning was a beautiful calm. We reached the buoys at about 4.30 and commenced picking up at 6.30. Shortly a new cause of anxiety arose. Kinks came up in great quantities, about thirty in the hour. To have a true conception of a kink, you must see one: it is a loop drawn tight, all the wires get twisted and the gutta-percha inside pushed out. These much diminish the value of the cable, as they must all be cut out, the gutta-percha made good, and the cable spliced. They arise from the cable having been badly laid down so that it forms folds and tails at the bottom of the sea. These kinks have another disadvantage: they weaken the cable very much.—At about six o'clock [P.M.] we had some twelve miles lifted, when I went to the bows; the kinks were exceedingly tight and were giving way in a most alarming manner. I got a cage rigged up to prevent the end (if it broke) from hurting anyone, and sat down on the bowsprit, thinking I should describe kinks to Annie:—suddenly I saw a great many coils and kinks altogether at the surface. I jumped to the gutta-percha pipe, by blowing through which the signal is given to stop the engine. I blow, but the engine does not stop; again—no answer: the coils and kinks jam in the bows and I rush aft shouting stop! Too late: the cable had parted and must lie in peace at the bottom. Someone had pulled the gutta-percha tube across a bare part of the steam pipe and melted it. It had been used hundreds of times in the last few days and gave no symptoms of failing. I believe the cable must have gone at any rate; however, since it went in my watch and since I might have secured the tubing more strongly, I feel rather sad. . . .

'June 28.

'Since I could not go to Annie I took down Shakespeare, and by the time I had finished *Antony and Cleopatra*, read the second half of *Troilus* and got some way in *Coriolanus*, I felt it was childish to regret the accident had happened in my watch, and moreover I felt myself not much to blame in the tubing matter—it had been torn down, it had not fallen down; so I went to bed, and slept without fretting, and woke this morning in the same good mood—for which thank you and our friend Shakespeare. I am happy to say Mr. Liddell said the loss of the cable did not much matter; though this would have been no consolation had I felt myself to blame.—This morning we have grappled for and found another length of small cable which Mr. —— dropped in 100 fathoms of water. If this also gets full of kinks, we shall probably have to cut it after 10 miles or so, or more probably still it will part of its own free will or weight.

'10 P.M.—This second length of three-wire cable soon got into the same condition as its fellow—i.e. came up twenty kinks an hour—and after seven miles were in, parted on the pulley over the bows at one of the said kinks; during my watch again, but this time no earthly power could have saved it. I had taken all manner of precautions to prevent the end doing any damage when the smash came, for come I knew it must. We now return to the six-wire cable. As I sat watching the cable to-night, large phosphorescent globes kept rolling from it and fading in the black water.

'29th.

'To-day we returned to the buoy we had left at the end of the six-wire cable, and after much trouble from a series of tangles, got a fair start at noon. You will easily believe a tangle of iron rope inch and a half diameter is not easy to un-ravel, especially with a ton or so hanging to the ends. It is now eight o'clock and we have about six and a half miles safe: it becomes very exciting however, for the kinks are coming fast and furious.

'July 2.

'Twenty-eight miles safe in the hold. The ship is now so deep, that the men are to be turned out of their aft hold, and the remainder coiled there; so the good *Elba's* nose need not burrow too far into the waves. There can only be about 10 or 12 miles more, but these weigh 80 or 100 tons.

'July 5.

'Our first mate was much hurt in securing a buoy on the evening of the 2nd. As interpreter [with the Italians] I am useful in all these cases ; but for no fortune would I be a doctor to witness these scenes continually. Pain is a terrible thing.— Our work is done: the whole of the six-wire cable has been recovered ; only a small part of the three-wire, but that wire was bad and, owing to its twisted state, the value small. We may therefore be said to have been very successful.'

II.

I have given this cruise nearly in full. From the notes, unhappily imperfect, of two others, I will take only specimens ; for in all there are features of similarity and it is possible to have too much even of submarine telegraphy and the romance of engineering. And first from the cruise of 1859 in the Greek Islands and to Alexandria, take a few traits, incidents and pictures.

' May 10, 1859.

' We had a fair wind and we did very well, seeing a little bit of Cerig or Cythera, and lots of turtle-doves wandering about over the sea and perching, tired and timid, in the rigging of our little craft. Then Falconera, Antimilo and Milo, topped with huge white clouds, barren, deserted, rising bold and mysterious from the blue chafing sea ;—Argentiera, Siphano, Scapho, Paros, Antiparos and late at night Syra itself. *Adam Bede* in one hand, a sketch-book in the other, lying on rugs under an awning, I enjoyed a very pleasant day.

' May 14.

' Syra is semi-eastern. The pavement, huge shapeless blocks sloping to a central gutter ; from this bare two-storeyed houses, sometimes plaster many coloured, sometimes rough-hewn marble, rise, dirty and ill-finished to straight, plain, flat roofs ; shops guiltless of windows, with signs in Greek letters ; dogs, Greeks in blue, baggy, Zouave breeches and a fez, a few narghi-lehs and a sprinkling of the ordinary continental shopboys.—In the evening I tried one more walk in Syra with A——, but in vain endeavoured to amuse myself or to spend money ; the first effort resulting in singing *Doodah* to a passing Greek or two,

the second in spending, no, in making A—— spend, threepence
on coffee for three.

'May 16.

'On coming on deck, I found we were at anchor in Canea
bay, and saw one of the most lovely sights man could witness.
Far on either hand stretch bold mountain capes, Spada and
Maleka, tender in colour, bold in outline; rich sunny levels
lie beneath them, framed by the azure sea. Right in front, a
dark brown fortress girdles white mosques and minarets. Rich
and green, our mountain capes here join to form a setting for
the town, in whose dark walls—still darker—open a dozen high-
arched caves in which the huge Venetian galleys used to lie in
wait. High above all, higher and higher yet, up into the firma-
ment, range after range of blue and snow-capped mountains.
I was bewildered and amazed, having heard nothing of this
great beauty. The town when entered is quite eastern. The
streets are formed of open stalls under the first story, in which
squat tailors, cooks, sherbet vendors and the like, busy at their
work or smoking narghilehs. Cloths stretched from house to
house keep out the sun. Mules rattle through the crowd; curs
yelp between your legs; negroes are as hideous and bright
clothed as usual; grave Turks with long chibouques continue to
march solemnly without breaking them; a little Arab in one
dirty rag pokes fun at two splendid little Turks with brilliant
fezzes; wiry mountaineers in dirty, full, white kilts, shouldering
long guns and one hand on their pistols, stalk untamed past a
dozen Turkish soldiers, who look sheepish and brutal in worn
cloth jacket and cotton trousers. A headless, wingless lion of
St. Mark still stands upon a gate, and has left the mark of his
strong clutch. Of ancient times when Crete was Crete, not a
trace remains; save perhaps in the full, well-cut nostril and firm
tread of that mountaineer, and I suspect that even his sires were
Albanians, mere outer barbarians.

'May 17.

'I spent the day at the little station where the cable was
landed, which has apparently been first a Venetian monastery
and then a Turkish mosque. At any rate the big dome

is very cool, and the little ones hold [our electric] batteries capitally. A handsome young Bashibazouk guards it, and a still handsomer mountaineer is the servant; so I draw them and the monastery and the hill, till I'm black in the face with heat and come on board to hear the Canea cable is still bad.

'May 23.

' We arrived in the morning at the east end of Candia, and had a glorious scramble over the mountains which seem built of adamant. Time has worn away the softer portions of the rock, only leaving sharp jagged edges of steel. Sea eagles soaring above our heads; old tanks, ruins and desolation at our feet. The ancient Arsinoe stood here; a few blocks of marble with the cross attest the presence of Venetian Christians; but now— the desolation of desolations. Mr. Liddell and I separated from the rest, and when we had found a sure bay for the cable, had a tremendous lively scramble back to the boat. These are the bits of our life which I enjoy, which have some poetry, some grandeur in them.

'May 29 (?).

' Yesterday we ran round to the new harbour [of Alexandria], landed the shore end of the cable close to Cleopatra's bath, and made a very satisfactory start about one in the afternoon. We had scarcely gone 200 yards when I noticed that the cable ceased to run out, and I wondered why the ship had stopped. People ran aft to tell me not to put such a strain on the cable; I answered indignantly that there was no strain; and suddenly it broke on every one in the ship at once that we were aground. Here was a nice mess. A violent scirocco blew from the land; making one's skin feel as if it belonged to some one else and didn't fit, making the horizon dim and yellow with fine sand, oppressing every sense and raising the thermometer 20 degrees in an hour, but making calm water round us which enabled the ship to lie for the time in safety. The wind might change at any moment, since the scirocco was only accidental; and at the first wave from seaward bump would go the poor ship, and there would [might] be an end of our voyage. The captain, without waiting to sound, began to make an effort to put the ship over what was supposed to be a sandbank ; but by the time soundings were made, this was found to be impossible, and he had only been jamming the poor *Elba* faster on a rock. Now every effort was made to get her astern, an anchor taken out, a rope brought to a winch I had for the cable, and the engines backed ; but all in vain. A small Turkish Government steamer, which is to be our consort, came to our assistance but of course very slowly, and much time was occupied before we could get a hawser to her. I could do no good after having made a chart of the soundings round the ship, and went at last on to the bridge to sketch the scene. But at that moment the strain from the winch and a jerk from the Turkish steamer got off the boat, after we had been some hours aground. The carpenter reported that she had made only two inches of water in one compartment ; the cable was still uninjured astern, and our spirits rose ; when, will you believe it ? after going a short distance astern, the pilot ran us once more fast aground on what seemed to me nearly the same spot. The very same scene was gone through as on the first occasion, and dark came on whilst the wind shifted, and we were still aground. Dinner was served up but poor Mr. Liddell could eat very little ; and bump, bump, grind, grind, went the ship fifteen or sixteen times as we sat at dinner. The slight sea however did enable us to bump off. This morning we appear not to have suffered in any way ; but a sea is rolling in, which a few hours ago would have settled the poor old *Elba*.

'June —.

'The Alexandria cable has again failed ; after paying out two thirds of the distance successfully, an unlucky touch in deep water snapped the line. Luckily the accident occurred in Mr. Liddell's watch. Though personally it may not really concern me, the accident weighs like a personal misfortune. Still I am glad I was present : a failure is probably more instructive than a success ; and this experience may enable us to avoid misfortune in still greater undertakings.

'We left Syra the morning after our arrival on Saturday the 4th. This we did (first) because we were in a hurry to do something and (second) because, coming from Alexandria, we had four days' quarantine to perform. We were all mustered along the side while the doctor counted us; the letters were popped into a little tin box and taken away to be smoked; the guardians put on board to see that we held no communication with the shore —without them we should still have had four more days' quarantine; and with twelve Greek sailors besides, we started merrily enough picking up the Canea cable. . . . To our utter dismay, the yarn covering began to come up quite decayed, and the cable, which when laid should have borne half a ton, was now in danger of snapping with a tenth part of that strain. We went as slow as possible in fear of a break at every instant. My watch was from eight to twelve in the morning, and during that time we had barely secured three miles of cable. Once it broke inside the ship, but I seized hold of it in time —the weight being hardly anything—and the line for the nonce was saved. Regular nooses were then planted inboard with men to draw them taut, should the cable break inboard. A——, who should have relieved me, was unwell, so I had to continue my look-out; and about one o'clock the line again parted but was again caught in the last noose, with about four inches to spare. Five minutes afterwards it again parted and was yet once more caught. Mr. Liddell (whom I had called) could stand this no longer; so we buoyed the line and ran into a bay in Siphano, waiting for calm weather, though I was by no means of opinion that the slight sea and wind had been the cause of our failures.—All next day (Monday) we lay off Siphano, amusing ourselves on shore with fowling pieces and navy revolvers. I need not say we killed nothing; and luckily we did not wound any of ourselves. A guardiano accompanied us, his functions being limited to preventing actual contact with the natives, for they might come as near and talk as much as they pleased. These isles of Greece are sad, interesting places.

They are not really barren all over, but they are quite destitute
of verdure; and tufts of thyme, wild mastic or mint, though
they sound well, are not nearly so pretty as grass. Many little
churches, glittering white, dot the islands; most of them, I
believe, abandoned during the whole year with the exception of
one day sacred to their patron saint. The villages are mean,
but the inhabitants do not look wretched and the men are good
sailors. There is something in this Greek race yet; they will
become a powerful Levantine nation in the course of time.—What
a lovely moonlight evening that was! the barren island cutting
the clear sky with fantastic outline, marble cliffs on either hand
fairly gleaming over the calm sea. Next day, the wind still
continuing, I proposed a boating excursion and decoyed A——,
L—— and S—— into accompanying me. We took the
little gig, and sailed away merrily enough round a point to a
beautiful white bay, flanked with two glistening little churches,
fronted by beautiful distant islands; when suddenly, to my
horror, I discovered the *Elba* steaming full speed out from the
island. Of course we steered after her; but the wind that instant
ceased, and we were left in a dead calm. There was nothing for it
but to unship the mast, get out the oars and pull. The ship
was nearly certain to stop at the buoy; and I wanted to learn
how to take an oar, so here was a chance with a vengeance!
L—— steered, and we three pulled—a broiling pull it was about
half way across to Palikandro—still we did come in, pulling an
uncommon good stroke, and I had learned to hang on my oar.
L—— had pressed me to let him take my place; but though I
was very tired at the end of the first quarter of an hour, and
then every successive half hour, I would not give in. I nearly
paid dear for my obstinacy however; for in the evening I had
alternate fits of shivering and burning.'

III.

The next extracts, and I am sorry to say the last, are from
Fleeming's letters of 1860, when he was back at Bona and
Spartivento and for the first time at the head of an expedition.

Unhappily these letters are not only the last, but the series is quite imperfect; and this is the more to be lamented as he had now begun to use a pen more skilfully, and in the following notes there is at times a touch of real distinction in the manner.

'Cagliari : October 5, 1860.

' All Tuesday I spent examining what was on board the *Elba*, and trying to start the repairs of the Spartivento land line, which has been entirely neglected, and no wonder, for no one has been paid for three months, no, not even the poor guards who have to keep themselves, their horses and their families, on their pay. Wednesday morning, I started for Spartivento and got there in time to try a good many experiments. Spartivento looks more wild and savage than ever, but is not without a strange deadly beauty : the hills covered with bushes of a metallic green with coppery patches of soil in between ; the valleys filled with dry salt mud and a little stagnant water ; where that very morning the deer had drunk, where herons, curlews and other fowl abound, and where, alas! malaria is breeding with this rain. (No fear for those who do not sleep on shore.) A little iron hut had been placed there since 1858 ; but the windows had been carried off, the door broken down, the roof pierced all over. In it, we sat to make experiments ; and how it recalled Birkenhead ! There was Thomson, there was my testing board, the strings of gutta percha ; Harry P—— even, battering with the batteries ; but where was my darling Annie ? Whilst I sat feet in sand, with Harry alone inside the hut— mats, coats and wood to darken the window—the others visited the murderous old friar, who is of the order of Scaloppi and for whom I brought a letter from his superior, ordering him to pay us attention ; but he was away from home, gone to Cagliari in a boat with the produce of the farm belonging to his convent. Then they visited the tower of Chia, but could not get in because the door is thirty feet off the ground ; so they came back and pitched a magnificent tent which I brought from the *Bahiana* a long time ago—and where they will live (if I mistake not) in preference to the friar's, or the owl- and bat-haunted tower. MM. T—— and S—— will be left there : T—— an intelligent, hard-working Frenchman with whom I am well pleased; he can speak English and Italian well, and has been two years at Genoa. S—— is a French German with a face like an ancient Gaul, who has been sergeant-major in the French line and who is, I see, a great, big, muscular *fainéant*. We left the tent pitched and some stores in charge of a guide, and ran back to Cagliari.

'Certainly, being at the head of things is pleasanter than being subordinate. We all agree very well; and I have made the testing office into a kind of private room where I can come and write to you undisturbed, surrounded by my dear, bright brass things which all of them remind me of our nights at Birkenhead. Then I can work here, too, and try lots of experiments; you know how I like that! and now and then I read— Shakespeare principally. Thank you so much for making me bring him : I think I must get a pocket edition of *Hamlet* and *Henry the Fifth*, so as never to be without them.

'Cagliari : October 7.

'[The town was full ?] . . . of red-shirted English Garibaldini. A very fine looking set of fellows they are, too : the officers rather raffish, but with medals Crimean and Indian; the men a very sturdy set, with many lads of good birth I should say. They still wait their consort the Emperor and will, I fear, be too late to do anything. I meant to have called on them, but they are all gone into barracks some way from the town, and I have been much too busy to go far.

'The view from the ramparts was very strange and beautiful. Cagliari rises on a very steep rock, at the mouth of a wide plain circled by large hills and three-quarters filled with lagoons; it looks, therefore, like an old island citadel. Large heaps of salt mark the border between the sea and the lagoons; thousands of flamingoes whiten the centre of the huge shallow marsh; hawks hover and scream among the trees under the high mouldering battlements.—A little lower down, the band played. Men and ladies bowed and pranced, the costumes posed, church bells

tinkled, processions processed, the sun set behind thick clouds
capping the hills ; I pondered on you and enjoyed it all.

'Decidedly I prefer being master to being man : boats at all
hours, stewards flying for marmalade, captain enquiring when
ship is to sail, clerks to copy my writing, the boat to steer when
we go out—I have run her nose on several times; decidedly, I
begin to feel quite a little king. Confound the cable, though!
I shall never be able to repair it.

'Bona: October 14.

'We left Cagliari at 4.30 on the 9th and soon got to Spartivento. I repeated some of my experiments, but found Thomson,
who was to have been my grand stand-by, would not work on
that day in the wretched little hut. Even if the windows and
door had been put in, the wind which was very high made the
lamp flicker about and blew it out; so I sent on board and got old
sails, and fairly wrapped the hut up in them ; and then we were
as snug as could be, and I left the hut in glorious condition
with a nice little stove in it. The tent which should have been
forthcoming from the curé's for the guards, had gone to Cagliari;
but I found another, [a] green, Turkish tent, in the *Elba* and
soon had him up. The square tent left on the last occasion was
standing all right and tight in spite of wind and rain. We
landed provisions, two beds, plates, knives, forks, candles, cook-
ing utensils, and were ready for a start at 6 P.M.; but the wind
meanwhile had come on to blow at such a rate that I thought
better of it, and we stopped. T—— and S—— slept ashore
however, to see how they liked it ; at least they tried to sleep,
for S—— the ancient sergeant-major had a toothache, and
T—— thought the tent was coming down every minute. Next
morning they could only complain of sand and a leaky coffee-
pot, so I leave them with a good conscience. The little en-
campment looked quite picturesque : the green round tent, the
square white tent and the hut all wrapped up in sail , on a
sand hill. looking on the sea and masking those confounded
marshes at the back. One would have thought the Cagliaritans
were in a conspiracy to frighten the two poor fellows, who (I
believe) will be safe enough if they do not go into the marshes

after nightfall. S—— brought a little dog to amuse them, such a jolly, ugly little cur without a tail, but full of fun; he will be better than quinine.

'The wind drove a barque which had anchored near us for shelter, out to sea. We started, however, at 2 P.M., and had a quick passage but a very rough one, getting to Bona by daylight [on the 11th]. Such a place as this is for getting anything done! The health boat went away from us at 7.30 with W—— on board; and we heard nothing of them till 9.30, when W—— came back with two fat Frenchmen who are to look on on the part of the Government. They are exactly alike : only one has four bands and the other three round his cap, and so I know them. Then I sent a boat round to Fort Gênois [Fort Genova of 1858] where the cable is landed, with all sorts of things and directions, whilst I went ashore to see about coals and a room at the fort. We hunted people in the little square in their shops and offices, but only found them in cafés. One amiable gentleman wasn't up at 9.30, was out at 10, and as soon as he came back the servant said he would go to bed and not get up till 3 : he came however to find us at a café, and said that, on the contrary, two days in the week he did not do so! Then my two fat friends must have their breakfast after their " something " at a café; and all the shops shut from 10 to 2 ; and the post does not open till 12 ; and there was a road to Fort Gênois, only a bridge had been carried away, &c. At last I got off, and we rowed round to Fort Gênois, where my men had put up a capital gipsy tent with sails, and there was my big board and Thomson's number 5 in great glory. I soon came to the conclusion there was a break. Two of my faithful Cagliaritans slept all night in the little tent, to guard it and my precious instruments ; and the sea, which was rather rough, silenced my Frenchmen.

'Next day I went on with my experiments, whilst a boat grappled for the cable a little way from shore and buoyed it where the *Elba* could get hold. I brought all back to the *Elba*, tried my machinery and was all ready for a start next morning. But the wretched coal had not come yet ; Government permission from Algiers to be got ; lighters, men, baskets, and I know not what forms to be got or got through—and everybody asleep ! Coals or no coals, I was determined to start next morning ; and start we did at four in the morning, picked up the buoy with our deck engine, popped the cable across a boat, tested the wires to make sure the fault was not behind us, and started picking up at 11. Everything worked admirably, and about 2 P.M., in came the fault. There is no doubt the cable was broken by coral fishers ; twice they have had it up to their own knowledge.

'Many men have been ashore to-day and have come back tipsy, and the whole ship is in a state of quarrel from top to bottom, and they will gossip just within my hearing. And we have had moreover three French gentlemen and a French lady to dinner, and I had to act host and try to manage the mixtures to their taste. The good-natured little Frenchwoman was most amusing ; when I asked her if she would have some apple tart— " *Mon Dieu*," with heroic resignation, " *je veux bien* ; " or a little *plombodding*—" *Mais ce que vous voudrez, Monsieur !* "

'S.S. *Elba*, somewhere not far from Bona : Oct. 19.

'Yesterday [after three previous days of useless grappling] was destined to be very eventful. We began dredging at day-break and hooked at once every time in rocks ; but by capital luck, just as we were deciding it was no use to continue in that place, we hooked the cable : up it came, was tested, and lo ! another complete break, a quarter of a mile off. I was amazed at my own tranquillity under these disappointments, but I was not really half so fussy as about getting a cab. Well, there was nothing for it but grappling again and, as you may imagine, we were getting about six miles from shore. But the water did not deepen rapidly ; we seemed to be on the crest of a kind of submarine mountain in prolongation of Cape de Gonde, and pretty havoc we must have made with the crags. What rocks we did hook ! No sooner was the grapnel down than the ship was anchored ; and then came such a business : ship's engines going, deck engine thundering, belt slipping, fear of

breaking ropes: actually breaking grapnels. It was always an
hour or more before we could get the grapnel down again. At
last we had to give up the place, though we knew we were close
to the cable, and go further to sea in much deeper water; to my
great fear, as I knew the cable was much eaten away and would
stand but little strain. Well, we hooked the cable first dredge
this time, and pulled it slowly and gently to the top, with much
trepidation. Was it the cable? was there any weight on? it
was evidently too small. Imagine my dismay when the cable
did come up, but hanging loosely, thus

instead of taut, thus

showing certain signs of a break close by. For a moment I felt
provoked, as I thought "Here we are in deep water, and the
cable will not stand lifting!" I tested at once, and by the very
first wire found it had broken towards shore and was good
towards sea. This was of course very pleasant; but from that
time to this, though the wires test very well, not a signal has
come from Spartivento. I got the cable into a boat, and a
gutta-percha line from the ship to the boat, and we signalled
away at a great rate—but no signs of life. The tests however
make me pretty sure one wire at least is good; so I determined
to lay down cable from where we were to the shore, and go to
Spartivento to see what had happened there. I fear my men
are ill. The night was lovely, perfectly calm; so we lay close
to the boat and signals were continually sent, but with no
result. This morning I laid the cable down to Fort Gênois in
style; and now we are picking up odds and ends of cable between the different breaks, and getting our buoys on board, &c.
To-morrow I expect to leave for Spartivento.'

IV.

And now I am quite at an end of journal keeping; diaries and diary letters being things of youth which Fleeming had at length outgrown. But one or two more fragments from his correspondence may be taken, and first this brief sketch of the laying of the Norderney cable; mainly interesting as showing under what defects of strength and in what extremities of pain, this cheerful man must at times continue to go about his work.

'I slept on board 29th September having arranged everything to start by daybreak from where we lay in the roads : but at daybreak a heavy mist hung over us so that nothing of land or water could be seen. At midday it lifted suddenly and away we went with perfect weather, but could not find the buoys Forde left, that evening. I saw the captain was not strong in navigation, and took matters next day much more into my own hands and before nine o'clock found the buoys ; (the weather had been so fine we had anchored in the open sea near Texel). It took us till the evening to reach the buoys, get the cable on board, test the first half, speak to Lowestoft, make the splice, and start. H—— had not finished his work at Norderney, so I was alone on board for Reuter. Moreover the buoys to guide us in our course were not placed, and the captain had very vague ideas about keeping his course ; so I had to do a good deal, and only lay down as I was for two hours in the night. I managed to run the course perfectly. Everything went well, and we found Norderney just where we wanted it next afternoon, and if the shore end had been laid, could have finished there and then, October 1st. But when we got to Norderney, we found the *Caroline* with shore end lying apparently aground, and could not understand her signals ; so we had to anchor suddenly and I went off in a small boat with the captain to the *Caroline*. It was cold by this time, and my arm was rather stiff and I was tired ; I hauled myself up on board the *Caroline* by a rope and found H—— and two men on board. All the rest were trying to get the shore end on shore, but had failed and apparently had stuck on shore, and the waves were getting up. We had anchored in the right place and next morning we hoped the shore end would be laid, so we had only to go back. It was of course still colder and quite night. I went to bed and hoped to sleep, but, alas, the rheumatism got into the joints and caused me terrible pain so that I could not sleep. I bore it as long as I could in order to disturb no one, for all were tired; but at last I could bear it no longer and managed to wake the steward and got a mustard poultice which took the pain from the shoulder; but then the elbow got very bad, and I had to call the second steward and get a second poultice, and then it was daylight, and I felt very ill and feverish. The sea was now rather rough—too rough rather for small boats, but luckily a sort of thing called a scoot came out, and we got on board her with some trouble, and got on shore after a good tossing about which made us all sea-sick. The cable sent from the *Caroline* was just 60 yards too short and did not reach the shore, so although the *Caroline* did make the splice late that night, we could neither test nor speak. Reuter was at Norderney, and I had to do the best I could which was not much, and went to bed early; I thought I should never sleep again, but in sheer desperation got up in the middle of the night and gulped a lot of raw whiskey and slept at last. But not long. A Mr. F—— washed my face and hands and dressed me; and we hauled the cable out of the sea, and got it joined to the telegraph station, and on October 3rd telegraphed to Lowestoft first and then to London. Miss Clara Volkman a niece of Mr. Reuter's sent the first message to Mrs. Reuter who was waiting (Varley used Miss Clara's hand as a kind of key) and I sent one of the first messages to Odden. I thought a message addressed to him would not frighten you, and that he would enjoy a message through Papa's cable. I hope he did. They were all very merry, but I had been so lowered by pain that I could not enjoy myself in spite of the success.'

V.

Of the 1869 cruise in the *Great Eastern*, I give what I am able; only sorry it is no more, for the sake of the ship itself, already almost a legend even to the generation that saw it launched.

'*June* 17, 1869.—Here are the names of our staff in whom I expect you to be interested, as future *Great Eastern* stories may be full of them : Theophilus Smith, a man of Latimer Clark's; Leslie C. Hill, my prizeman at University College ; Lord Sackville Cecil ; King, one of the Thomsonian Kings ; Laws, goes for Willoughby Smith, who will also be on board ; Varley, Clark, and Sir James Anderson make up the sum of all you know anything of. A Captain Halpin commands the big ship. There are four smaller vessels. The *Wm. Cory* which laid the Norderney cable has already gone to St. Pierre to lay the shore ends. The *Hawk* and *Chiltern* have gone to Brest to lay shore ends. The *Hawk* and *Scanderia* go with us across the Atlantic and we shall at St. Pierre be transhipped into one or the other.

'*June* 18. *Somewhere in London.*—The shore end is laid, as you may have seen and we are all under pressing orders to march, so we start from London to-night at 5.10.

'*June* 20. *Off Ushant.*—I am getting quite fond of the big ship. Yesterday morning in the quiet sunlight, she turned so slowly and lazily in the great harbour at Portland, and bye and bye slipped out past the long pier with so little stir, that I could hardly believe we were really off. No men drunk, no women crying, no singing or swearing, no confusion or bustle on deck—nobody apparently aware that they had anything to do. The look of the thing was that the ship had been spoken to civilly and had kindly undertaken to do everything that was necessary without any further interference. I have a nice cabin with plenty of room for my legs in my berth and have slept two nights like a top. Then we have the ladies' cabin set apart as an engineer's office, and I think this decidedly the nicest place in the ship : 35 ft. × 20 ft. broad—four tables, three great mirrors, plenty of air and no heat from the funnels which spoil the great dining-room. I saw a whole library of books on the walls when here last, and this made me less anxious to provide light literature; but alas, to-day, I find that they are every one bibles or prayer-books. Now one cannot read many hundred bibles. . . . As for the motion of the ship it is not very much, but 'twill suffice. Thomson shook hands and wished me well. I *do* like Thomson. . . . Tell Austin that the *Great Eastern* has six masts and four funnels. When I get back I will make a little model of her for all the chicks and pay out cotton reels. . . . Here we are at 4.20 at Brest. We leave probably to-morrow morning.

'*July* 12. *Great Eastern.*—Here as I write we run our last course for the buoy at the St. Pierre shore end. It blows and lightens, and our good ship rolls, and buoys are hard to find; but we must soon now finish our work, and then this letter will start for home. . . . Yesterday we were mournfully groping our way through the wet grey fog, not at all sure where we were, with one consort lost and the other faintly answering the roar of our great whistle through the mist. As to the ship which was to meet us, and pioneer us up the deep channel, we did not know if we should come within twenty miles of her; when suddenly up went the fog, out came the sun and there, straight ahead was the *Wm. Cory* our pioneer, and a little dancing boat, the *Gulnare* sending signals of welcome with many-coloured flags. Since then we have been steaming in a grand procession; but now at 2 A.M. the fog has fallen, and the great roaring whistle calls up the distant answering notes all around us. Shall we, or shall we not find the buoy?

'*July* 13.—All yesterday we lay in the damp dripping fog, with whistles all round and guns firing so that we might not bump up against one another. This little delay has let us get our reports into tolerable order. We are now at 7 o'clock getting the cable end again, with the main cable buoy close to us.'

A telegram of July 20: 'I have received your four welcome letters. The Americans are charming people.'

SHIPMATES

FROM SKETCHES IN NOTE-BOOKS

VI.

And here to make an end are a few random bits about the cruise to Pernambuco:—

'*Plymouth, June* 21, 1873.—I have been down to the seashore and smelt the salt sea and like it; and I have seen the *Hooper* pointing her great bow sea-ward, while light smoke rises from her funnels telling that the fires are being lighted; and sorry as I am to be without you, something inside me answers to the call to be off and doing.

'*Lalla Rookh. Plymouth, June* 22.—We have been a little cruise in the yacht over to the Eddystone lighthouse, and my sea-legs seem very well on. Strange how alike all these starts are—first on shore, steaming hot days with a smell of bone-dust and tar and salt water; then the little puffing panting steam-launch that bustles out across a port with green woody sides, little yachts sliding about, men-of-war training-ships, and then a great big black hulk of a thing with a mass of smaller vessels sticking to it like parasites; and that is one's home being coaled. Then comes the Champagne lunch where every one says all that is polite to everyone else, and then the uncertainty when to start. So far as we know *now*, we are to start to-morrow morning at daybreak; letters that come later are to be sent to Pernambuco by first mail. . . . My father has sent me the heartiest sort of Jack Tar's cheer.

'*S.S. Hooper. Off Funchal, June* 29.—Here we are off Madeira at seven o'clock in the morning. Thomson has been sounding with his special toy ever since half-past three (1087 fathoms of water). I have been watching the day break, and long jagged islands start into being out of the dull night. We are still some miles from land; but the sea is calmer than Loch Eil often was, and the big *Hooper* rests very contentedly after a pleasant voyage and favourable breezes. I have not been able to do any real work except the testing [of the cable] for though not sea-sick, I get a little giddy when I try to think on board. . . . The ducks have just had their daily souse and are quacking and

gabbling in a mighty way outside the door of the captain's deck
cabin where I write. The cocks are crowing, and new-laid eggs
are said to be found in the coops. Four mild oxen have been
untethered and allowed to walk along the broad iron decks—a
whole drove of sheep seem quite content while licking big lumps
of bay salt. Two exceedingly impertinent goats lead the cook
a perfect life of misery. They steal round the galley and *will*
nibble the carrots or turnips if his back is turned for one
minute; and then he throws something at them and misses
them; and they scuttle off laughing impudently, and flick one
ear at him from a safe distance. This is the most impudent
gesture I ever saw. Winking is nothing to it. The ear
normally hangs down behind; the goat turns sideways to her
enemy—by a little knowing cock of the head flicks one ear
over one eye, and squints from behind it for half a minute—
tosses her head back, skips a pace or two further off, and
repeats the manœuvre. The cook is very fat and cannot run
after that goat much.

'*Pernambuco, Aug.* 1.—We landed here yesterday, all well
and cable sound, after a good passage. . . . I am on familiar
terms with cocoa-nuts, mangoes and bread-fruit trees, but I
think I like the negresses best of anything I have seen. In
turbans and loose sea-green robes, with beautiful black-brown
complexions and a stately carriage, they really are a satisfaction
to my eye. The weather has been windy and rainy; the
Hooper has to lie about a mile from the town, in an open road-
stead, with the whole swell of the Atlantic driving straight on
shore. The little steam launch gives all who go in her a good
ducking, as she bobs about on the big rollers; and my old gym-
nastic practice stands me in good stead on boarding and leaving
her. We clamber down a rope ladder hanging from the high
stern, and then taking a rope in one hand, swing into the launch
at the moment when she can contrive to steam up under us—
bobbing about like an apple thrown into a tub all the while.
The President of the province and his suite tried to come off to
a State luncheon on board on Sunday; but the launch being
rather heavily laden, behaved worse than usual, and some

green seas stove in the President's hat and made him wetter than he had probably ever been in his life; so after one or two rollers, he turned back; and indeed he was wise to do so, for I don't see how he could have got on board. . . . Being fully convinced that the world will not continue to go round unless I pay it personal attention, I must run away to my work.'

CHAPTER VI.

1869-1885.

Edinburgh—Colleagues—*Farrago Vitæ*—I. The Family Circle—Fleeming and his Sons—Highland Life—The Cruise of the Steam Launch—Summer in Styria—Rustic Manners—II. The Drama—Private Theatricals—III. Sanitary Associations—The Phonograph—IV. Fleeming's Acquaintance with a Student—His late Maturity of Mind--Religion and Morality—His Love of Heroism—Taste in Literature—V. His Talk—His late Popularity—Letter from M. Trélat.

THE remaining external incidents of Fleeming's life, pleasures, honours, fresh interests, new friends, are not such as will bear to be told at any length or in the temporal order. And it is now time to lay narration by, and to look at the man he was and the life he lived, more largely.

Edinburgh, which was thenceforth to be his home, is a metropolitan small town; where college professors and the lawyers of the Parliament House give the tone, and persons of leisure, attracted by educational advantages, make up much of the bulk of society. Not, therefore, an unlettered place, yet not pedantic, Edinburgh will compare favourably with much larger cities. A hard and disputatious element has been commented on by strangers: it would not touch Fleeming, who was himself regarded, even in this metropolis of disputation, as a thorny tablemate. To golf unhappily he did not take, and golf is a cardinal virtue in the city of the winds. Nor did he become an archer of the Queen's Body Guard, which is the Chiltern Hundreds of the distasted golfer. He did not even frequent the Evening Club, where his colleague Tait (in my day) was so punctual and so genial. So that in some ways he stood outside of the lighter and kindlier life of his new home. I should not like to say that he was generally popular; but there as elsewhere, those who knew him well enough to love him, loved him well. And he, upon his side, liked a place where a dinner party was not of necessity unintellectual, and where men stood up to him in argument.

The presence of his old classmate, Tait, was one of his early *Colleagues.* attractions to the chair; and now that Fleeming is gone again, Tait still remains, ruling and really teaching his great classes. Sir Robert Christison was an old friend of his mother's; Sir Alexander Grant, Kelland and Sellar, were new acquaintances and highly valued; and these too, all but the last, have been taken from their friends and labours. Death has been busy in the Senatus. I will speak elsewhere of Fleeming's demeanour to his students; and it will be enough to add here that his relations with his colleagues in general were pleasant to himself.

Edinburgh, then, with its society, its university work, its *Farrago* delightful scenery and its skating in the winter, was thenceforth *Vitæ.* his base of operations. But he shot meanwhile erratic in many directions: twice to America, as we have seen, on telegraph voyages; continually to London on business; often to Paris; year after year to the Highlands to shoot, to fish, to learn reels and Gaelic, to make the acquaintance and fall in love with the character of Highlanders; and once to Styria, to hunt chamois and dance with peasant maidens. All the while, he was pursuing the course of his electrical studies, making fresh inventions, taking up the phonograph, filled with theories of graphic representation; reading, writing, publishing, founding sanitary associations, interested in technical education, investigating the laws of metre, drawing, acting, directing private theatricals, going a long way to see an actor—a long way to see a picture; in the very bubble of the tideway of contemporary interests. And all the while he was busied about his father and mother, his wife, and in particular his sons; anxiously watching, anxiously guiding these, and plunging with his whole fund of youthfulness into their sports and interests. And all the while he was himself maturing—not in character or body, for these remained young—but in the stocked mind, in the tolerant knowledge of life and man, in pious acceptance of the universe. Here is a farrago for a chapter : here is a world of interests and activities, human, artistic, social, scientific, at each of which he sprang with impetuous pleasure, on each of which he squandered energy, the arrow drawn to the head, the whole intensity of his spirit bent, for the moment, on the momentary purpose. It was this that lent such unusual interest to his society, so that no friend of his can forget that figure of Fleeming coming charged with some new discovery : it is this that makes his character so difficult to represent. Our fathers, upon some difficult theme, would invoke the Muse ; I can but appeal to the imagination of the reader. When I dwell upon some one thing, he must bear in mind it was only one of a score ; that the unweariable brain was teeming at the very time with other thoughts; that the good heart had left no kind duty forgotten.

I.

The family circle. In Edinburgh, for a considerable time, Fleeming's family, to three generations, was united : Mr. and Mrs. Austin at Hailes, Captain and Mrs. Jenkin in the suburb of Merchiston, Fleeming himself in the city. It is not every family that could risk with safety such close interdomestic dealings ; but in this also Fleeming was particularly favoured. Even the two extremes, Mr. Austin and the Captain, drew together. It is pleasant to find that each of the old gentlemen set a high value on the good looks of the other, doubtless also on his own ; and a fine picture they made as they walked the green terrace at Hailes, conversing by the hour. What they talked of is still a mystery to those who knew them ; but Mr. Austin always declared that on these occasions he learned much. To both of these families of elders, due service was paid of attention ; to both, Fleeming's easy circumstances had brought joy ; and the eyes of all were on the grandchildren. In Fleeming's scheme of duties, those of the family stood first; a man was first of all a child, nor did he cease to be so, but only took on added obligations, when he became in turn a father. The care of his parents was always a first thought with him, and their gratification his delight.

And the care of his sons, as it was always a grave subject of study with him, and an affair never neglected, so it brought him a thousand satisfactions. 'Hard work they are,' as he once wrote, 'but what fit work!' And again: 'O, it's a cold house where a dog is the only representative of a child!' Not that dogs were despised; we shall drop across the name of Jack, the harum-scarum Irish terrier, ere we have done; his own dog Plato went up with him daily to his lectures, and still (like other friends) feels the loss and looks visibly for the reappearance of his master; and Martin the cat Fleeming has himself immortalised, to the delight of Mr. Swinburne, in the columns of the *Spectator*. Indeed there was nothing in which men take interest, in which he took not some; and yet always most in the strong human bonds, ancient as the race and woven of delights and duties.

He was even an anxious father; perhaps that is the part where optimism is hardest tested. He was eager for his sons; eager for their health, whether of mind or body; eager for their education; in that, I should have thought, too eager. But he kept a pleasant face upon all things, believed in play, loved it himself, shared boyishly in theirs, and knew how to put a face of entertainment upon business and a spirit of education into entertainment. If he was to test the progress of the three boys, this advertisement would appear in their little manuscript paper:—'Notice: The Professor of Engineering in the University of Edinburgh intends at the close of the scholastic year to hold examinations in the following subjects: (1) For boys in the fourth class of the Academy—Geometry and Algebra; (2) For boys at Mr. Henderson's school—Dictation and Recitation; (3) For boys taught exclusively by their mothers—Arithmetic and Reading.' Prizes were given; but what prize would be so conciliatory as this boyish little joke? It may read thin here; it would smack racily in the playroom. Whenever his sons 'started a new fad' (as one of them writes to me) they 'had only to tell him about it, and he was at once interested and keen to help.' He would discourage them in nothing unless it was hopelessly too hard for them; only, if there was

Fleeming and his sons.

any principle of science involved, they must understand the principle; and whatever was attempted, that was to be done thoroughly. If it was but play, if it was but a puppetshow they were to build, he set them the example of being no sluggard in play. When Frewen the second son embarked on the ambitious design to make an engine for a toy steamboat, Fleeming made him begin with a proper drawing—doubtless to the disgust of the young engineer; but once that foundation laid, helped in the work with unflagging gusto, 'tinkering away,' for hours, and assisted at the final trial 'in the big bath' with no less excitement than the boy. 'He would take any amount of trouble to help us,' writes my correspondent. 'We never felt an affair was complete till we had called him to see, and he would come at any time, in the middle of any work.' There was indeed one recognised playhour, immediately after the despatch of the day's letters; and the boys were to be seen waiting on the stairs until the mail should be ready and the fun could begin. But at no other time did this busy man suffer his work to interfere with that first duty to his children; and there is a pleasant tale of the inventive Master Frewen, engaged at the time upon a toy crane, bringing to the study where his father sat at work a half-wound reel that formed some part of his design, and observing, 'Papa, you might finiss windin' this for me; I am so very busy to-day.'

I put together here a few brief extracts from Fleeming's letters, none very important in itself, but all together building up a pleasant picture of the father with his sons.

'*Jan. 15th*, 1875.—Frewen contemplates suspending soap bubbles by silk threads for experimental purposes. I don't think he will manage that. Bernard' [the youngest] 'volunteered to blow the bubbles with enthusiasm.'

'*Jan. 17th.*—I am learning a great deal of electrostatics in consequence of the perpetual cross-examination to which I am subjected. I long for you on many grounds but one is that I may not be obliged to deliver a running lecture on abstract points of science, subject to cross-examination by two acute students. Bernie does not cross-examine much; but if anyone gets discomfited, he laughs a sort of little silver-whistle giggle, which is trying to the unhappy blunderer.'

' *May 9th.*—Frewen is deep in parachutes. I beg him not to drop from the top landing in one of his own making.'

' *June 6th*, 1876.—Frewen's crank axle is a failure just at present—but he bears up.'

' *June 14th.*—The boys enjoy their riding. It gets them whole funds of adventures. One of their caps falling off is matter for delightful reminiscences ; and when a horse breaks his step, the occurrence becomes a rear, a shy or a plunge as they talk it over. Austin, with quiet confidence, speaks of the greater pleasure in riding a spirited horse, even if he does give a little trouble. It is the stolid brute that he dislikes. (N.B. You can still see six inches between him and the saddle when his pony trots.) I listen and sympathise and throw out no hint that their achievements are not really great.'

' *June 18th.*—Bernard is much impressed by the fact that I can be useful to Frewen about the steamboat ' [which the latter irrepressible inventor was making]. ' He says quite with awe, " He would not have got on nearly so well if you had not helped him." '

' *June 27th.*—I do not see what I could do without Austin. He talks so pleasantly and is so truly good all through.'

' *July 7th.*—My chief difficulty with Austin is to get him measured for a pair of trousers. Hitherto I have failed, but I keep a stout heart and mean to succeed. Frewen the observer, in describing the paces of two horses, says " Polly takes twenty-seven steps to get round the school. I couldn't count Sophy, but she takes more than a hundred." '

' *Feb. 18th*, 1877.—We all feel very lonely without you. Frewen had to come up and sit in my room for company last night and I actually kissed him, a thing that has not occurred for years. Jack, poor fellow, bears it as well as he can, and has taken the opportunity of having a fester on his foot, so he is lame and has it bathed, and this occupies his thoughts a good deal.'

' *Feb. 19th.*—As to Mill, Austin has not got the list yet. I

think it will prejudice him very much against Mill—but that
is not my affair. Education of that kind! . . . I would as soon
cram my boys with food and boast of the pounds they had eaten,
as cram them with literature.'

But if Fleeming was an anxious father, he did not suffer his
anxiety to prevent the boys from any manly or even dangerous
pursuit. Whatever it might occur to them to try, he would
carefully show them how to do it, explain the risks, and then
either share the danger himself or, if that were not possible,
stand aside and wait the event with that unhappy courage of
the looker-on. He was a good swimmer, and taught them to
swim. He thoroughly loved all manly exercises; and during
their holidays, and principally in the Highlands, helped and
encouraged them to excel in as many as possible : to shoot, to
fish, to walk, to pull an oar, to hand, reef and steer, and to run
a steam launch. In all of these, and in all parts of Highland
life, he shared delightedly. He was well on to forty when he
took once more to shooting, he was forty-three when he killed
his first salmon, but no boy could have more single-mindedly
rejoiced in these pursuits. His growing love for the Highland
character, perhaps also a sense of the difficulty of the task, led
him to take up at forty-one the study of Gaelic ; in which he
made some shadow of progress, but not much : the fastnesses of
that elusive speech retaining to the last their independence.
At the house of his friend Mrs. Blackburn, who plays the part
of a Highland lady as to the manner born, he learned the
delightful custom of kitchen dances, which became the rule at
his own house and brought him into yet nearer contact with his
neighbours. And thus at forty-two, he began to learn the reel ;
a study, to which he brought his usual smiling earnestness ; and
the steps, diagramatically represented by his own hand, are
before me as I write.

The cruise of the steam launch.

It was in 1879 that a new feature was added to the Highland life : a steam launch, called the *Purgle*, the Styrian corrup-
tion of Walpurga, after a friend to be hereafter mentioned. 'The
steam launch goes,' Fleeming wrote. 'I wish you had been
present to describe two scenes of which she has been the occasion

FRAU MOSER

FROM THE STYRIAN SKETCH BOOK

already: one during which the population of Ullapool, to a baby, was harnessed to her hurrahing—and the other in which the same population sat with its legs over a little pier, watching Frewen and Bernie getting up steam for the first time.' The *Purgle* was got with educational intent; and it served its purpose so well, and the boys knew their business so practically, that when the summer was at an end, Fleeming, Mrs. Jenkin, Frewen the engineer, Bernard the stoker, and Kenneth Robertson a Highland seaman, set forth in her to make the passage south. The first morning they got from Loch Broom into Gruinard bay, where they lunched upon an island; but the wind blowing up in the afternoon, with sheets of rain, it was found impossible to beat to sea; and very much in the situation of castaways upon an unknown coast, the party landed at the mouth of Gruinard river. A shooting lodge was spied among the trees; there Fleeming went; and though the master Mr. Murray was from home, though the two Jenkin boys were of course as black as colliers, and all the castaways so wetted through that, as they stood in the passage, pools formed about their feet and ran before them into the house, yet Mrs. Murray kindly entertained them for the night. On the morrow, however, visitors were to arrive; there would be no room and, in so out of the way a spot, most probably no food for the crew of the *Purgle*; and on the morrow about noon, with the bay white with spindrift and the wind so strong that one could scarcely stand against it, they got up steam and skulked under the land as far as Sanda Bay. Here they crept into a seaside cave, and cooked some food; but the weather now freshening to a gale, it was plain they must moor the launch where she was, and find their way overland to some place of shelter. Even to get their baggage from on board was no light business; for the dingy was blown so far to leeward every trip, that they must carry her back by hand along the beach. But this once managed, and a cart procured in the neighbourhood, they were able to spend the night in a pot-house at Ault Bea. Next day, the sea was unapproachable; but the next they had a pleasant passage to Poolewe, hugging the cliffs, the falling swell bursting close by them in the gullies, and the black scarts that sat like ornaments on the top of every stack and pinnacle, looking down into the *Purgle* as she passed. The climate of Scotland had not done with them yet : for three days they lay storm-stayed in Poolewe, and when they put to sea on the morning of the fourth, the sailors prayed them for God's sake not to attempt the passage. Their setting out was indeed merely tentative ; but presently they had gone too far to return, and found themselves committed to double Rhu Reay with a foul wind and a cross sea. From half-past eleven in the morning until half-past five at night, they were in immediate and unceasing danger. Upon the least mishap, the *Purgle* must either have been swamped by the seas or bulged upon the cliffs of that rude headland. Fleeming and Robertson took turns baling and steering ; Mrs. Jenkin, so violent was the commotion of the boat, held on with both hands ; Frewen, by Robertson's direction, ran the engine, slacking and pressing her to meet the seas ; and Bernard, only twelve years old, deadly sea-sick, and continually thrown against the boiler, so that he was found next day to be covered with burns, yet kept an even fire. It was a very thankful party that sat down that evening to meat in the Hotel at Gairloch. And perhaps, although the thing was new in the family, no one was much surprised when Fleeming said grace over that meal. Thenceforward he continued to observe the form, so that there was kept alive in his house a grateful memory of peril and deliverance. But there was nothing of the muff in Fleeming ; he thought it a good thing to escape death, but a becoming and a healthful thing to run the risk of it ; and what is rarer, that which he thought for himself, he thought for his family also. In spite of the terrors of Rhu Reay, the cruise was persevered in and brought to an end under happier conditions.

Summer in Styria. One year, instead of the Highlands, Alt Aussee, in the Steiermark, was chosen for the holidays ; and the place, the people, and the life delighted Fleeming. He worked hard at German, which he had much forgotten since he was a boy ; and what is highly characteristic, equally hard at the patois, in which he learned to excel. He won a prize at a Schützen-fest ; and though he hunted chamois without much success, brought down

WALPURGA

FROM THE STYRIAN SKETCH BOOK

more interesting game in the shape of the Styrian peasants, and in particular of his gillie, Joseph. This Joseph was much of a character; and his appreciations of Fleeming have a fine note of their own. The bringing up of the boys he deigned to approve of: '*fast so gut wie ein Bauer*,' was his trenchant criticism. The attention and courtly respect with which Fleeming surrounded his wife, was something of a puzzle to the philosophic gillie ; he announced in the village that Mrs. Jenkin—*die silberne Frau*, as the folk had prettily named her from some silver ornaments—was a '*geborene Gräfin*' who had married beneath her ; and when Fleeming explained what he called the English theory (though indeed it was quite his own) of married relations, Joseph, admiring but unconvinced, avowed it was '*gar schön*.' Joseph's cousin, Walpurga Moser, to an orchestra of clarionet and zither, taught the family the country dances, the Steierisch and the Ländler, and gained their hearts during the lessons. Her sister Loys, too, who was up at the Alp with the cattle, came down to church on Sundays, made acquaintance with the Jenkins, and must have them up to see the sunrise from her house upon the Loser, where they had supper and all slept in the loft among the hay. The Mosers were not lost sight of; Walpurga still corresponds with Mrs. Jenkin, and it was a late pleasure of Fleeming's to choose and despatch a wedding present for his little mountain friend. This visit was brought to an end by a ball in the big inn parlour ; the refreshments chosen, the list of guests drawn up, by Joseph ; the best music of the place in attendance ; and hosts and guests in their best clothes. The ball was opened by Mrs. Jenkin dancing Steierisch with a lordly Bauer, in gray and silver and with a plumed hat; and Fleeming followed with Walpurga Moser.

There ran a principle through all these holiday pleasures. *Rustic manners.* In Styria as in the Highlands, the same course was followed : Fleeming threw himself as fully as he could into the life and occupations of the native people, studying everywhere their dances and their language, and conforming, always with pleasure, to their rustic etiquette. Just as the ball at Alt Aussee was designed for the taste of Joseph, the parting feast at Atta-dale was ordered in every particular to the taste of Murdoch the Keeper. Fleeming was not one of the common, so-called gentlemen, who take the tricks of their own coterie to be eternal principles of taste. He was aware, on the other hand, that rustic people dwelling in their own places, follow ancient rules with fastidious precision, and are easily shocked and embarrassed by what (if they used the word) they would have to call the vulgarity of visitors from town. And he, who was so cavalier with men of his own class, was sedulous to shield the more tender feelings of the peasant; he, who could be so· trying in a drawing-room, was even punctilious in the cottage. It was in all respects a happy virtue. It renewed his life, during these holidays, in all particulars. It often enter-tained him with the discovery of strange survivals; as when, by the orders of Murdoch, Mrs. Jenkin must publicly taste of every dish before it was set before her guests. And thus to throw himself into a fresh life and a new school of manners was a grateful exercise of Fleeming's mimetic instinct; and to the pleasures of the open air, of hardships supported, of dexterities improved and displayed, and of plain and elegant society, added a spice of drama.

II.

The drama.

Fleeming was all his life a lover of the play and all that belonged to it. Dramatic literature he knew fully. He was one of the not very numerous people who can read a play: a knack, the fruit of much knowledge and some imagination, comparable to that of reading score. Few men better under-stood the artificial principles on which a play is good or bad; few more unaffectedly enjoyed a piece of any merit of construc-tion. His own play (which the reader will find reprinted farther on) was conceived with a double design; for he had long been filled with his theory of the true story of Griselda; used to gird at Father Chaucer for his misconception; and was, perhaps first of all, moved by the desire to do justice to the Marquis of

Saluces, and perhaps only in the second place, by the wish to treat a story (as he phrased it) like a sum in arithmetic. I do not think he quite succeeded; but I must own myself no fit judge. Fleeming and I were teacher and taught as to the principles, disputatious rivals in the practice, of dramatic writing.

Acting had always, ever since Rachel and the *Marseillaise*, a particular power on him. 'If I do not cry at the play,' he used to say, 'I want to have my money back.' Even from a poor play with poor actors he could draw pleasure. 'Giacometti's *Elisabetta*,' I find him writing, 'fetched the house vastly. Poor Queen Elizabeth! And yet it was a little good.' And again, after a night of Salvini: 'I do not suppose any one with feelings could sit out *Othello*, if Iago and Desdemona were acted.' Salvini was, in his view, the greatest actor he had seen. We were all indeed moved and bettered by the visit of that wonderful man.—'I declare I feel as if I could pray!' cried one of us, on the return from *Hamlet*.—'That is prayer,' said Fleeming. W. B. Hole and I, in a fine enthusiasm of gratitude, determined to draw up an address to Salvini, did so, and carried it to Fleeming; and I shall never forget with what coldness he heard and deleted the eloquence of our draft, nor with what spirit (our vanities once properly mortified) he threw himself into the business of collecting signatures. It was his part, on the ground of his Italian, to see and arrange with the actor; it was mine to write in the *Academy* a notice of the first performance of *Macbeth*. Fleeming opened the paper, read so far, and flung it on the floor. 'No,' he cried 'that won't do. You were thinking of yourself, not of Salvini!' The criticism was shrewd as usual, but it was unfair through ignorance; it was not of myself that I was thinking, but of the difficulties of my trade which I had not well mastered. Another unalloyed dramatic pleasure which Fleeming and I shared the year of the Paris Exposition, was the *Marquis de Villemer*, that blameless play, performed by Madeleine Brohan, Delaunay, Worms, and Broisat—an actress, in such parts at least, to whom I have never seen full justice rendered. He had his fill of weeping on that occasion; and when the piece was at an end, in front of a café, in the mild, midnight air, we had our fill of talk about the art of acting.

But what gave the stage so strong a hold on Fleeming was an inheritance from Norwich, from Edward Barron, and from Enfield of the *Speaker*. The theatre was one of Edward Barron's elegant hobbies; he read plays, as became Enfield's son-in-law, with a good discretion; he wrote plays for his family, in which Eliza Barron used to shine in the chief parts; and later in life, after the Norwich home was broken up, his little granddaughter would sit behind him in a great armchair, and be introduced, with his stately elocution, to the world of dramatic literature. From this, in a direct line, we can deduce the charades at Clay-gate; and after money came, in the Edinburgh days, that private theatre which took up so much of Fleeming's energy and thought. The company—Mr. and Mrs. R. O. Carter of Colwall, W. B. Hole, Captain Charles Douglas, Mr. Kunz, Mr. Burnett, Professor Lewis Campbell, Mr. Charles Baxter, and many more—made a charming society for themselves and gave pleasure to their audience. Mr. Carter in Sir Toby Belch it would be hard to beat. Mr. Hole in broad farce, or as the herald in the *Trachiniæ* showed true stage talent. As for Mrs. Jenkin, it was for her the rest of us existed and were forgiven; her powers were an endless spring of pride and pleasure to her husband; he spent hours hearing and schooling her in private; and when it came to the performance, though there was perhaps no one in the audience more critical, none was more moved than Fleeming. The rest of us did not aspire so high. There were always five performances and weeks of busy rehearsal; and whether we came to sit and stifle as the prompter, to be the dumb (or rather the inarticulate) recipients of Carter's dog whip in the *Taming of the Shrew*, or having earned our spurs, to lose one more illusion in a leading part, we were always sure at least of a long and an exciting holiday in mirthful company.

In this laborious annual diversion, Fleeming's part was large. I never thought him an actor, but he was something of a mimic, which stood him in stead. Thus he had seen Got in Poirier; and his own Poirier, when he came to play it, breathed meritoriously of the model. The last part I saw him play was Triplet, and at first I thought it promised well. But alas! the boys went for a holiday, missed a train, and were not heard of at home till late at night. Poor Fleeming, the man who never hesitated to give his sons a chisel or a gun, or to send them abroad in a canoe or on a horse, toiled all day at his rehearsal, growing hourly paler, Triplet growing hourly less meritorious. And though the return of the children, none the worse for their little adventure, brought the colour back into his face, it could not restore him to his part. I remember finding him seated on the stairs in some rare moment of quiet during the subsequent performances. ' Hullo, Jenkin,' said I, ' you look down in the mouth.'—' My dear boy,' said he, ' haven't you heard me? I have not one decent intonation from beginning to end.'

But indeed he never supposed himself an actor ; took a part, when he took any, merely for convenience, as one takes a hand at whist ; and found his true service and pleasure in the more congenial business of the manager. Augier, Racine, Shakespeare, Aristophanes in Hookham Frere's translation, Sophocles and Æschylus in Lewis Campbell's, such were some of the authors whom he introduced to his public. In putting these upon the stage, he found a thousand exercises for his ingenuity and taste, a thousand problems arising which he delighted to study, a thousand opportunities to make those infinitesimal improvements which are so much in art and for the artist. Our first Greek play had been costumed by the professional costumier, with unforgettable results of comicality and indecorum : the second, the *Trachiniæ* of Sophocles, he took in hand himself, and a delightful task he made of it. His study was then in antiquarian books, where he found confusion, and on statues and bas-reliefs, where he at last found clearness ; after an hour or so at the British Museum, he was able to master ' the chitôn, sleeves and all ; ' and before the time was ripe, he had a theory of Greek tailoring at his fingers' ends, and had all the costumes made under his eye as a Greek tailor would have made them. ' The Greeks made the best plays and the best statues, and were the best architects ; of course, they were the best tailors, too,' said he; and was never weary, when he could find a tolerant listener, of dwelling on the simplicity, the economy, the elegance both of means and effect, which made their system so delightful.

But there is another side to the stage-manager's employment. The discipline of acting is detestable; the failures and triumphs of that business appeal too directly to the vanity; and even in the course of a careful amateur performance such as ours, much of the smaller side of man will be displayed. Fleeming, among conflicting vanities and levities, played his part to my admiration. He had his own view; he might be wrong; but the performances (he would remind us) were after all his, and he must decide. He was, in this as in all other things, an iron taskmaster, sparing not himself nor others. If you were going to do it at all, he would see that· it was done as well as you were able. I have known him to keep two culprits (and one of these his wife) repeating the same action and the same two or three words for a whole weary afternoon. And yet he gained and retained warm feelings from far the most of those who fell under his domination, and particularly (it is pleasant to remember) from the girls. After the slipshod training and the incomplete accomplishments of a girls' school, there was something at first annoying, at last exciting and bracing, in this high standard of accomplishment and perseverance.

III.

Sanitary associations.

It did not matter why he entered upon any study or employment, whether for amusement like the Greek tailoring or the Highland reels, whether from a desire to serve the public as with his sanitary work, or in the view of benefiting poorer men as with his labours for technical education, he ' pitched into it ' (as he would have said himself) with the same headlong zest. I give in the Appendix a letter from Colonel Fergusson, which tells fully the nature of the sanitary work and of Fleeming's part and success in it. It will be enough to say here that it was a scheme of protection against the blundering of builders and the dishonesty of plumbers. Started with an eye rather to the houses of the rich, Fleming hoped his Sanitary Associations would soon extend their sphere of usefulness and improve the dwellings of the poor. In this hope he was disappointed; but in all other ways the scheme exceedingly prospered, associations sprang up and continue to spring up in many quarters, and wherever tried they have been found of use.

Here, then, was a serious employment; it has proved highly useful to mankind; and it was begun besides, in a mood of bitterness, under the shock of what Fleming would so sensitively feel—the death of a whole family of children. Yet it was gone upon like a holiday jaunt. I read in Colonel Fergusson's letter that his schoolmates bantered him when he began to broach his scheme; so did I at first, and he took the banter as he always did with enjoyment, until he suddenly posed me with the question: 'And now do you see any other jokes to make? Well, then,' said he, 'that's all right. I wanted you to have your fun out first; now we can be serious.' And then with a glowing heat of pleasure, he laid his plans before me, revelling in the details, revelling in hope. It was as he wrote about the joy of electrical experiment: 'What shall I compare them to?— A new song? a Greek play?' Delight attended the exercise of all his powers; delight painted the future. Of these ideal visions, some (as I have said) failed of their fruition. And the illusion was characteristic. Fleming believed we had only to make a virtue cheap and easy, and then all would practise it; that for an end unquestionably good, men would not grudge a little trouble and a little money, though they might stumble at laborious pains and generous sacrifices. He could not believe in any resolute badness. 'I cannot quite say,' he wrote in his young manhood, 'that I think there is no sin or misery. This I can say: I do not remember one single malicious act done to myself. In fact it is rather awkward when I have to say the Lord's Prayer. I have nobody's trespasses to forgive.' And to the point, I remember one of our discussions. I said it was a dangerous error not to admit there were bad people; he, that it was only a confession of blindness on our part, and that we

probably called others bad only so far as we were wrapped in
ourselves and lacking in the transmigratory forces of imagination. I undertook to describe to him three persons irredeemably
bad and whom he should admit to be so. In the first case, he
denied my evidence : ' You cannot judge a man upon such testimony,' said he. For the second, he owned it made him sick to
hear the tale ; but then there was no spark of malice, it was
mere weakness I had described, and he had never denied nor
thought to set a limit to man's weakness. At my third gentleman, he struck his colours. ' Yes,' said he, ' I'm afraid that *is* a
bad man.' And then looking at me shrewdly : ' I wonder if it
isn't a very unfortunate thing for you to have met him.' I showed
him radiantly how it was the world we must know, the world as
it was, not a world expurgated and prettified with optimistic rain-
bows. ' Yes, yes,' said he ; ' but this badness is such an easy,
lazy explanation. Won't you be tempted to use it, instead of
trying to understand people ? '

The phono-
graph.

In the year 1878, he took a passionate fancy for the phonograph : it was a toy after his heart, a toy that touched the skirts
of life, art and science, a toy prolific of problems and theories.
Something fell to be done for a University Cricket Ground
Bazaar. ' And the thought struck him,' Mr. Ewing writes to
me, ' to exhibit Edison's phonograph, then the very newest scientific marvel. The instrument itself was not to be purchased
—I think no specimen had then crossed the Atlantic—but a
copy of the *Times* with an account of it was at hand, and by the
help of this we made a phonograph which to our great joy
talked, and talked, too, with the purest American accent. It was
so good that a second instrument was got ready forthwith. Both
were shown at the Bazaar : one by Mrs. Jenkin to people willing
to pay half a crown for a private view and the privilege of hearing their own voices, while Jenkin, perfervid as usual, gave half-
hourly lectures on the other in an adjoining room—I, as his
lieutenant, taking turns. The thing was in its way a little
triumph. A few of the visitors were deaf, and hugged the
belief that they were the victims of a new kind of fancy-fair
swindle. Of the others, many who came to scoff remained to

take raffle tickets ; and one of the phonographs was finally disposed of in this way.' The other remained in Fleeming's hands,
and was a source of infinite occupation. Once it was sent to
London, 'to bring back on the tinfoil the tones of a lady distinguished for clear vocalisation'; at another time 'Sir Robert
Christison was brought in to contribute his powerful bass ';
and there scarcely came a visitor about the house, but he was
made the subject of experiment. The visitors, I am afraid,
took their parts lightly : Mr. Hole and I, with unscientific
laughter, commemorating various shades of Scotch accent, or
proposing to 'teach the poor dumb animal to swear.' But
Fleeming and Mr. Ewing, when we butterflies were gone, were
laboriously ardent. Many thoughts that occupied the later years
of my friend were caught from the small utterance of that toy.
Thence came his inquiries into the roots of articulate language
and the foundations of literary art; his papers on vowel sounds,
his papers in the *Saturday Review* upon the laws of verse, and
many a strange approximation, many a just note, thrown out
in talk and now forgotten. I pass over dozens of his interests,
and dwell on this trifling matter of the phonograph, because
it seems to me that it depicts the man. So, for Fleeming,
one thing joined into another, the greater with the less. He
cared not where it was he scratched the surface of the ultimate
mystery—in the child's toy, in the great tragedy, in the laws
of the tempest, or in the properties of energy or mass—certain
that whatever he touched, it was a part of life—and however
he touched it, there would flow for his happy constitution interest and delight. 'All fables have their morals,' says Thoreau,
'but the innocent enjoy the story.' There is a truth represented for the imagination in these lines of a noble poem, where
we are told, that in our highest hours of visionary clearness, we
can but

> 'see the children sport upon the shore
> And hear the mighty waters rolling evermore.'

To this clearness Fleeming had attained; and although he
heard the voice of the eternal seas and weighed its message,

he was yet able, until the end of his life, to sport upon these shores of death and mystery with the gaiety and innocence of children.

IV.

It was as a student that I first knew Fleeming, as one of that modest number of young men who sat under his ministrations in a soul-chilling class-room at the top of the University buildings. His presence was against him as a professor: no one, least of all students, would have been moved to respect him at first sight: rather short in stature, markedly plain, boyishly young in manner, cocking his head like a terrier with every mark of the most engaging vivacity and readiness to be pleased, full of words, full of paradox, a stranger could scarcely fail to look at him twice, a man thrown with him in a train could scarcely fail to be engaged by him in talk, but a student would never regard him as academical. Yet he had that fibre in him that order always existed in his class-room. I do not remember that he ever addressed me in language; at the least sign of unrest, his eye would fall on me and I was quelled. Such a feat is comparatively easy in a small class; but I have misbehaved in smaller classes and under eyes more Olympian than Fleeming Jenkin's. He was simply a man from whose reproof one shrank; in manner the least buckrammed of mankind, he had, in serious moments, an extreme dignity of goodness. So it was that he obtained a power over the most insubordinate of students, but a power of which I was myself unconscious. I was inclined to regard any professor as a joke, and Fleeming as a particularly good joke, perhaps the broadest in the vast pleasantry of my curriculum. I was not able to follow his lectures; I somehow dared not misconduct myself, as was my customary solace; and I refrained from attending. This brought me at the end of the session into a relation with my contemned professor that completely opened my eyes. During the year, bad student as I was, he had shown a certain leaning to my society; I had been to his house, he had asked

me to take a humble part in his theatricals; I was a master in
the art of extracting a certificate even at the cannon's mouth;
and I was under no apprehension. But when I approached
Fleeming, I found myself in another world; he would have
naught of me. 'It is quite useless for *you* to come to me, Mr.
Stevenson. There may be doubtful cases, there is no doubt
about yours. You have simply *not* attended my class.' The
document was necessary to me for family considerations; and
presently I stooped to such pleadings and rose to such adjura-
tions, as made my ears burn to remember. He was quite un-
moved; he had no pity for me.—'You are no fool,' said he,
'and you chose your course.' I showed him that he had miscon-
ceived his duty, that certificates were things of form, attendance
a matter of taste. Two things, he replied, had been required
for graduation, a certain competency proved in the final trials
and a certain period of genuine training proved by certificate;
if he did as I desired, not less than if he gave me hints for an
examination, he was aiding me to steal a degree. 'You see,
Mr. Stevenson, these are the laws and I am here to apply
them,' said he. I could not say but that this view was tenable,
though it was new to me; I changed my attack: it was only
for my father's eye that I required his signature, it need never
go to the Senatus, I had already certificates enough to justify
my year's attendance. 'Bring them to me; I cannot take your
word for that,' said he. 'Then I will consider.' The next day
I came charged with my certificates, a humble assortment. And
when he had satisfied himself, 'Remember,' said he, 'that I can
promise nothing, but I will try to find a form of words.' He
did find one, and I am still ashamed when I think of his shame
in giving me that paper. He made no reproach in speech, but
his manner was the more eloquent; it told me plainly what a
dirty business we were on; and I went from his presence, with
my certificate indeed in my possession, but with no answerable
sense of triumph. That was the bitter beginning of my love
for Fleeming; I never thought lightly of him afterwards.

Once, and once only, after our friendship was truly founded,
did we come to a considerable difference. It was, by the rules

cxxxivMEMOIR

of poor humanity, my fault and his. I had been led to dabble in society journalism; and this coming to his ears, he felt it like a disgrace upon himself. So far he was exactly in the right; but he was scarce happily inspired when he broached the subject at his own table and before guests who were strangers to me. It was the sort of error he was always ready to repent, but always certain to repeat; and on this occasion he spoke so freely that I soon made an excuse and left the house with the firm purpose of returning no more. About a month later, I met him at dinner at a common friend's. 'Now,' said he, on the stairs, 'I engage you—like a lady to dance—for the end of the evening. You have no right to quarrel with me and not give me a chance.' I have often said and thought that Fleeming had no tact; he belied the opinion then. I remember perfectly how, so soon as we could get together, he began his attack: 'You may have grounds of quarrel with me; you have none against Mrs. Jenkin; and before I say another word, I want you to promise you will come to *her* house as usual.' An interview thus begun could have but one ending: if the quarrel were the fault of both, the merit of the reconciliation was entirely Fleeming's.

His late maturity of mind.

When our intimacy first began, coldly enough, accidentally enough on his part, he had still something of the Puritan, something of the inhuman narrowness of the good youth. It fell from him slowly, year by year, as he continued to ripen, and grow milder, and understand more generously the mingled characters of men. In the early days he once read me a bitter lecture; and I remember leaving his house in a fine spring afternoon, with the physical darkness of despair upon my eyesight. Long after he made me a formal retractation of the sermon and a formal apology for the pain he had inflicted; adding drolly, but truly, 'You see, at that time I was so much younger than you!' And yet even in those days there was much to learn from him; and above all his fine spirit of piety, bravely and trustfully accepting life, and his singular delight in the heroic.

His piety was, indeed, a thing of chief importance. His views (as they are called) upon religious matters varied much; and he

Religion
and
morality

could never be induced to think them more or less than views.
'All dogma is to me mere form,' he wrote; 'dogmas are mere
blind struggles to express the inexpressible. I cannot conceive
that any single proposition whatever in religion is true in the
scientific sense; and yet all the while I think the religious view
of the world is the most true view. Try to separate from the
mass of their statements that which is common to Socrates,
Isaiah, David, St. Bernard, the Jansenists, Luther, Mahomet,
Bunyan—yes, and George Eliot: of course you do not believe
that this something could be written down in a set of propositions like Euclid, neither will you deny that there is something
common and this something very valuable. . . . I shall be sorry
if the boys ever give a moment's thought to the question of what
community they belong to—I hope they will belong to the great
community.' I should observe that as time went on his conformity to the church in which he was born grew more complete, and
his views drew nearer the conventional. 'The longer I live, my
dear Louis,' he wrote but a few months before his death, ' the more
convinced I become of a direct care by God—which is reasonably
impossible—but there it is.' And in his last year he took the
communion.

But at the time when I fell under his influence, he stood
more aloof; and this made him the more impressive to a youthful atheist. He had a keen sense of language and its imperial
influence on men; language contained all the great and sound
metaphysics, he was wont to say; and a word once made and
generally understood, he thought a real victory of man and
reason. But he never dreamed it could be accurate, knowing
that words stand symbol for the indefinable. I came to him
once with a problem which had puzzled me out of measure:
what is a cause? why out of so many innumerable millions of
conditions, all necessary, should one be singled out and ticketed
' the cause?' ' You do not understand,' said he. ' A cause is
the answer to a question: it designates that condition which I
happen to know and you happen not to know.' It was thus,
with partial exception of the mathematical, that he thought of
all means of reasoning: they were in his eyes but means of

communication, so to be understood, so to be judged, and only so far to be credited. The mathematical he made, I say, exception of: number and measure he believed in to the extent of their significance, but that significance, he was never weary of reminding you, was slender to the verge of nonentity. Science was true, because it told us almost nothing. With a few abstractions it could deal, and deal correctly; conveying honestly faint truths. Apply its means to any concrete fact of life, and this high dialect of the wise became a childish jargon.

Thus the atheistic youth was met at every turn by a scepticism more complete than his own, so that the very weapons of the fight were changed in his grasp to swords of paper. Certainly the church is not right, he would argue, but certainly not the anti-church either. Men are not such fools as to be wholly in the wrong, nor yet are they so placed as to be ever wholly in the right. Somewhere, in mid air between the disputants, like hovering Victory in some design of a Greek battle, the truth hangs undiscerned. And in the meanwhile what matter these uncertainties? Right is very obvious; a great consent of the best of mankind, a loud voice within us (whether of God, or whether by inheritance, and in that case still from God) guide and command us in the path of duty. He saw life very simple; he did not love refinements; he was a friend to much conformity in unessentials. For (he would argue) it is in this life as it stands about us, that we are given our problem; the manners of the day are the colours of our palette; they condition, they constrain us; and a man must be very sure he is in the right, must (in a favourite phrase of his) be 'either very wise or very vain,' to break with any general consent in ethics. I remember taking his advice upon some point of conduct. 'Now,' he said, 'how do you suppose Christ would have advised you?' and when I had answered that he would not have counselled me anything unkind or cowardly, 'No,' he said, with one of his shrewd strokes at the weakness of his hearer, 'nor anything amusing.' Later in life, he made less certain in the field of ethics. 'The old story of the knowledge of good and evil is a very true one,' I find him writing; only (he goes on) 'the effect

of the original dose is much worn out, leaving Adam's descendants with the knowledge that there is such a thing—but uncertain where.' His growing sense of this ambiguity made him
less swift to condemn but no less stimulating in counsel. 'You
grant yourself certain freedoms. Very well,' he would say, 'I
want to see you pay for them some other way. You positively
cannot do this : then there positively must be something else
that you can do, and I want to see you find that out and do it.'
Fleeming would never suffer you to think that you were living,
if there were not, somewhere in your life, some touch of heroism,
to do or to endure.

This was his rarest quality. Far on in middle age, when His love
men begin to lie down with the bestial goddesses, Comfort and of hero-
Respectability, the strings of his nature still sounded as high a ism.
note as a young man's. He loved the harsh voice of duty like
a call to battle. He loved courage, enterprise, brave natures, a
brave word, an ugly virtue ; everything that lifts us above the
table where we eat or the bed we sleep upon. This with no
touch of the motive-monger or the ascetic. He loved his virtues
to be practical, his heroes to be great eaters of beef; he loved
the jovial Heracles, loved the astute Odysseus ; not the Robes-
pierres and Wesleys. A fine buoyant sense of life and of man's
unequal character ran through all his thoughts. He could not
tolerate the spirit of the pickthank ; being what we are, he
wished us to see others with a generous eye of admiration, not
with the smallness of the seeker after faults. If there shone
anywhere a virtue, no matter how incongruously set, it was
upon the virtue we must fix our eyes. I remember having
found much entertainment in Voltaire's *Saül*, and telling him
what seemed to me the drollest touches. He heard me out, as
usual when displeased, and then opened fire on me with red-hot
shot. To belittle a noble story was easy ; it was not literature,
it was not art, it was not morality ; there was no sustenance in
such a form of jesting, there was (in his favourite phrase) ' no
nitrogenous food' in such literature. And then he proceeded to
show what a fine fellow David was ; and what a hard knot he was
in about Bathsheba, so that (the initial wrong committed) honour

might well hesitate in the choice of conduct; and what owls those people were who marvelled because an Eastern tyrant had killed Uriah, instead of marvelling that he had not killed the prophet also. 'Now if Voltaire had helped me to feel that,' said he, 'I could have seen some fun in it.' He loved the comedy which shows a hero human, and yet leaves him a hero; and the laughter which does not lessen love.

Taste in literature. It was this taste for what is fine in humankind, that ruled his choice in books. These should all strike a high note, whether brave or tender, and smack of the open air. The noble and simple presentation of things noble and simple, that was the 'nitrogenous food' of which he spoke so much, which he sought so eagerly, enjoyed so royally. He wrote to an author, the first part of whose story he had seen with sympathy, hoping that it might continue in the same vein. 'That this may be so,' he wrote, 'I long with the longing of David for the water of Bethlehem. But no man need die for the water a poet can give, and all can drink it to the end of time, and their thirst be quenched and the pool never dry—and the thirst and the water are both blessed.' It was in the Greeks particularly that he found this blessed water; he loved 'a fresh air' which he found 'about the Greek things even in translations;' he loved their freedom from the mawkish and the rancid. The tale of David in the Bible, the *Odyssey*, Sophocles, Æschylus, Shakespeare, Scott; old Dumas in his chivalrous note; Dickens rather than Thackeray, and the *Tale of Two Cities* out of Dickens: such were some of his preferences. To Ariosto and Boccaccio he was always faithful; *Burnt Njal* was a late favourite; and he found at least a passing entertainment in the *Arcadia* and the *Grand Cyrus*. George Eliot he outgrew, finding her latterly only sawdust in the mouth; but her influence, while it lasted, was great, and must have gone some way to form his mind. He was easily set on edge, however, by didactic writing; and held that books should teach no other lesson but what 'real life would teach, were it as vividly presented.' Again, it was the thing made that took him, the drama in the book; to the book itself, to any merit of the making, he was long strangely blind. He would prefer the *Agamemnon* in the prose of Mr. Buckley, ay, to Keats. But he was his mother's son, learning to the last. He told me one day that literature was not a trade; that it was no craft; that the professed author was merely an amateur with a door-plate. 'Very well,' said I, 'the first time you get a proof, I will demonstrate that it is as much a trade as bricklaying, and that you do not know it.' By the very next post, a proof came. I opened it with fear; for he was indeed, as the reader will see by these volumes, a formidable amateur; always wrote brightly, because he always thought trenchantly; and sometimes wrote brilliantly, as the worst of whistlers may sometimes stumble on a perfect intonation. But it was all for the best in the interests of his education; and I was able, over that proof, to give him a quarter of an hour such as Fleeming loved both to give and to receive. His subsequent training passed out of my hands into those of our common friend, W. E. Henley. 'Henley and I,' he wrote, 'have fairly good times wigging one another for not doing better. I wig him because he won't try to write a real play, and he wigs me because I can't try to write English.' When I next saw him, he was full of his new acquisitions. 'And yet I have lost something too,' he said regretfully. 'Up to now Scott seemed to me quite perfect, he was all I wanted. Since I have been learning this confounded thing, I took up one of the novels, and a great deal of it is both careless and clumsy.'

V.

He spoke four languages with freedom, not even English His Talk. with any marked propriety. What he uttered was not so much well said, as excellently acted: so we may hear every day the inexpressive language of a poorly-written drama assume character and colour in the hands of a good player. No man had more of the *vis comica* in private life; he played no character on the stage, as he could play himself among his friends. It was one of his special charms; now when the voice is silent and the face still, it makes it impossible to do justice to his power in conversation. He was a delightful companion to such as can bear

bracing weather ; not to the very vain ; not to the owlishly wise,
who cannot have their dogmas canvassed ; not to the painfully
refined, whose sentiments become articles of faith. The spirit in
which he could write that he was ' much revived by having an
opportunity of abusing Whistler to a knot of his special admirers,'
is a spirit apt to be misconstrued. He was not a dogmatist,
even about Whistler. 'The house is full of pretty things,' he
wrote, when on a visit ; ' but Mrs. ——'s taste in pretty things
has one very bad fault : it is not my taste.' And that was the
true attitude of his mind ; but these eternal differences it was
his joy to thresh out and wrangle over by the hour. It was no
wonder if he loved the Greeks ; he was in many ways a Greek
himself ; he should have been a sophist and met Socrates ; he
would have loved Socrates, and done battle with him staunchly
and manfully owned his defeat ; and the dialogue, arranged by
Plato, would have shone even in Plato's gallery. He seemed in
talk aggressive, petulant, full of a singular energy ; as vain you
would have said as a peacock, until you trod on his toes, and
then you saw that he was at least clear of all the sicklier
elements of vanity. Soundly rang his laugh at any jest against
himself. He wished to be taken, as he took others, for what
was good in him without dissimulation of the evil, for what
was wise in him without concealment of the childish. He
hated a draped virtue, and despised a wit on its own defence.
And he drew (if I may so express myself) a human and
humorous portrait of himself with all his defects and qualities, as he thus enjoyed in talk the robust sports of the intelligence ; giving and taking manfully, always without pretence, always with paradox, always with exuberant pleasure ;
speaking wisely of what he knew, foolishly of what he knew
not ; a teacher, a learner, but still combative ; picking holes in
what was said even to the length of captiousness, yet aware
of all that was said rightly ; jubilant in victory, delighted by
defeat : a Greek sophist, a British schoolboy.

His late popularity. Among the legends of what was once a very pleasant spot,
the old Savile Club, not then divorced from Savile Row, there
are many memories of Fleeming. He was not popular at first,

being known simply as 'the man who dines here and goes up to Scotland'; but he grew at last, I think, the most generally liked of all the members. To those who truly knew and loved him, who had tasted the real sweetness of his nature, Fleeming's porcupine ways had always been a matter of keen regret. They introduced him to their own friends with fear; sometimes recalled the step with mortification. It was not possible to look on with patience while a man so loveable thwarted love at every step. But the course of time and the ripening of his nature brought a cure. It was at the Savile that he first remarked a change; it soon spread beyond the walls of the club. Presently I find him writing : ' Will you kindly explain what has happened to me ? All my life I have talked a good deal, with the almost unfailing result of making people sick of the sound of my tongue. It appeared to me that I had various things to say, and I had no malevolent feelings, but nevertheless the result was that expressed above. Well, lately some change has happened. If I talk to a person one day, they must have me the next. Faces light up when they see me.—" Ah, I say come here "—" come and dine with me." It's the most preposterous thing I ever experienced. It is curiously pleasant. You have enjoyed it all your life, and therefore cannot conceive how bewildering a burst of it is for the first time at forty-nine.' And this late sunshine of popularity still further softened him. He was a bit of a porcupine to the last, still shedding darts ; or rather he was to the end a bit of a schoolboy, and must still throw stones ; but the essential toleration that underlay his disputatiousness, and the kindness that made of him a tender sicknurse and a generous helper, shone more conspicuously through. A new pleasure had come to him ; and as with all sound natures, he was bettered by the pleasure.

I can best show Fleeming in this later stage by quoting from a vivid and interesting letter of M. Emile Trélat's. Here, admirably expressed, is how he appeared to a friend of another nation, whom he encountered only late in life. M. Trélat will pardon me if I correct, even before I quote him ; but what the Frenchman supposed to flow from some particular

Letter from M. Trélat.

bitterness against France, was only Fleeming's usual address.
Had M. Trélat been Italian, Italy would have fared as ill; and
yet Italy was Fleeming's favourite country.

Vous savez comment j'ai connu Fleeming Jenkin !   C'était en Mai
1878.   Nous étions tous deux membres du jury de l'Exposition Uni-
verselle.   On n'avait rien fait qui vaille à la première séance de notre
classe, qui avait eu lieu le matin.   Tout le monde avait parlé et re-
parlé pour ne rien dire.   Cela durait depuis huit heures ; il était midi.
Je demandai la parole pour une motion d'ordre, et je proposai que la
séance fut levée à la condition que chaque membre français *emportât*
à déjeuner un juré étranger.   Jenkin applaudit.   ' Je vous emmène
déjeuner,' lui criai-je.   ' Je veux bien.' . . . Nous partîmes ; en chemin
nous vous rencontrions ; il vous présente et nous allons déjeuner tous
trois auprès du Trocadéro.

Et, depuis ce temps, nous avons été de vieux amis.   Non seule-
ment nous passions nos journées au jury, où nous étions toujours
ensemble, côte-à-côte.   Mais nos habitudes s'étaient faites telles que,
non contents de déjeuner en face l'un de l'autre, je le ramenais dîner
presque tous les jours chez moi.   Cela dura une quinzaine : puis il
fut rappelé en Angleterre.   Mais il revint, et nous fîmes encore une
bonne étape de vie intellectuelle, morale et philosophique.   Je crois
qu'il me rendait déjà tout ce que j'éprouvais de sympathie et d'estime,
et que je ne fus pas pour rien dans son retour à Paris.

Chose singulière !   nous nous étions attachés l'un à l'autre par les
sous-entendus bien plus que par la matière de nos conversations.   À
vrai dire, nous étions presque toujours en discussion ; et il nous
arrivait de nous rire au nez l'un et l'autre pendant des heures, tant
nous nous étonnions réciproquement de la diversité de nos points de
vue.   Je le trouvais si Anglais, et il me trouvait si Français !   Il était
si franchement révolté de certaines choses qu'il voyait chez nous, et
je comprenais si mal certaines choses qui se passaient chez vous !   Rien
de plus intéressant que ces contacts qui étaient des contrastes, et que
ces rencontres d'idées qui étaient des choses ; rien de si attachant que
les échappées de cœur ou d'esprit auxquelles ces petits conflits donnaient
à tout moment cours.   C'est dans ces conditions que, pendant son séjour
à Paris en 1878, je conduisis un peu partout mon nouvel ami.   Nous
allâmes chez Madame Edmond Adam, où il vit passer beaucoup
d'hommes politiques avec lesquels il causa.   Mais c'est chez les minis-
tres qu'il fut intéressé.   Le moment était, d'ailleurs, curieux en France.
Je me rappelle que, lorsque je le présentai au Ministre du Commerce,

il fit cette spirituelle repartie : ' C'est la seconde fois que je viens en
France sous la République. La première fois, c'était en 1848, elle
s'était coiffée de travers : je suis bien heureux de saluer aujourd'hui
votre excellence, quand elle a mis son chapeau droit.' Une fois je le
menai voir couronner la Rosière de Nanterre. Il y suivit les céré-
monies civiles et religieuses ; il y assista au banquet donné par le
Maire ; il y vit notre de Lesseps, auquel il porta un toast. Le soir,
nous revînmes tard à Paris ; il faisait chaud ; nous étions un peu fati-
gués ; nous entrâmes dans un des rares cafés encore ouverts. Il
devint silencieux.—' N'êtes-vous pas content de votre journée ? ' lui
dis-je.—' O, si ! mais je réfléchis, et je me dis que vous êtes un peuple
gai—tous ces braves gens étaient gais aujourd'hui. C'est une vertu,
la gaieté, et vous l'avez en France, cette vertu ! ' Il me disait cela
mélancoliquement ; et c'était la première fois que je lui entendais
faire une louange adressée à la France. . . . Mais il ne faut pas que
vous voyiez là une plainte de ma part. Je serais un ingrat si je me
plaignais ; car il me disait souvent : ' Quel bon Français vous faites ! '
Et il m'aimait à cause de cela, quoiqu'il semblât n'aimer pas la France.
C'était là un trait de son originalité. Il est vrai qu'il s'en tirait en
disant que je ne ressemblai pas à mes compatriotes, ce à quoi il ne connaissait rien !—Tout cela était fort curieux ; car, moi-même, je l'aimais
quoiqu'il en eût à mon pays !

En 1879 il amena son fils Austin à Paris. J'attirai celui-ci. Il
déjeunait avec moi deux fois par semaine. Je lui montrai ce
qu'était l'intimité française en le tutoyant paternellement. Cela
reserra beaucoup nos liens d'intimité avec Jenkin. . . . Je fis in-
viter mon ami au congrès de l'*Association française pour l'avance-
ment des sciences*, qui se tenait à Rheims en 1880. Il y vint.
J'eus le plaisir de lui donner la parole dans la section du génie
civil et militaire, que je présidais. Il y fit une très intéressante communication, qui me montrait une fois de plus l'originalité de ses
vues et la sûreté de sa science. C'est à l'issue de ce congrès que je
passai lui faire visite à Rochefort, où je le trouvai installé en famille
et où je présentai pour la première fois mes hommages à son éminente
compagne. Je le vis là sous un jour nouveau et touchant pour moi.
Madame Jenkin, qu'il entourait si galamment, et ses deux jeunes
fils donnaient encore plus de relief à sa personne. J'emportai des
quelques heures que je passai à côté de lui dans ce charmant paysage
un souvenir ému.

J'étais allé en Angleterre en 1882 sans pouvoir gagner Edimbourg,
J'y retournai en 1883 avec la commission d'assainissement de la ville
de Paris, dont je faisais partie. Jenkin me rejoignit. J le fis

entendre par mes collègues ; car il était fondateur d'une société de
salubrité. Il eut un grand succès parmi nous. Mais ce voyage me
restera toujours en mémoire parce que c'est là que se fixa définitive-
ment notre forte amitié. Il m'invita un jour à dîner à son club et
au moment de me faire asseoir à côté de lui, il me retint et me dit :
' Je voudrais vous demander de m'accorder quelque chose. C'est
mon sentiment que nos relations ne peuvent pas se bien continuer si
vous ne me donnez pas la permission de vous tutoyer. Voulez-vous
que nous nous tutoyions ?' Je lui pris les mains et je lui dis qu'une
pareille proposition venant d'un Anglais, et d'un Anglais de sa haute
distinction, c'était une victoire, dont je serais fier toute ma vie. Et
nous commencions à user de cette nouvelle forme dans nos rapports.
Vous savez avec quelle finesse il parlait le francais : comme il en
connaissait tous les tours, comme il jouait avec ses difficultés, et même
avec ses petites gamineries. Je crois qu'il a été heureux de pratiquer
avec moi ce tutoiement, qui ne s'adapte pas à l'anglais, et qui est si
français. Je ne puis vous peindre l'étendue et la variété de nos
conversations de la soirée. Mais ce que je puis vous dire, c'est que,
sous la caresse du *tu*, nos idées se sont élevées. Nous avions toujours
beaucoup ri ensemble ; mais nous n'avions jamais laissé des banalités
s'introduire dans nos échanges de pensées. Ce soir-là, notre horizon
intellectuel s'est élargie, et nous y avons poussé des reconnaissances
profondes et lointaines. Après avoir vivement causé à table, nous
avons longuement causé au salon ; et nous nous séparions le soir à
Trafalgar Square, après avoir longé les trottoirs, stationné aux coins
des rues et deux fois rebroussé chemin en nous reconduisant l'un
l'autre. Il était près d'une heure du matin ! Mais quelle belle passe
d'argumentation, quels beaux échanges de sentiments, quelles fortes
confidences patriotiques nous avions fournies ! J'ai compris ce soir-
là que Jenkin ne détestait pas la France, et je lui serrai fort les mains
en l'embrassant. Nous nous quittions aussi amis qu'on puisse l'être ;
et notre affection s'était par lui étendue et comprise dans un *tu*
francais.

CHAPTER VII.

1875–1885.

Mrs. Jenkin's Illness—Captain Jenkin—The Golden Wedding—Death of Uncle
John—Death of Mr. and Mrs. Austin—Illness and Death of the Captain—
Death of Mrs. Jenkin—Effect on Fleeming—Telpherage—The End.

AND now I must resume my narrative for that melancholy
business that concludes all human histories. In January of the
year 1875, while Fleeming's sky was still unclouded, he was
reading Smiles. 'I read my engineers' lives steadily,' he writes,
'but find biographies depressing. I suspect one reason to be
that misfortunes and trials can be graphically described, but
happiness and the causes of happiness either cannot be or are
not. A grand new branch of literature opens to my view: a
drama in which people begin in a poor way and end, after
getting gradually happier, in an ecstasy of enjoyment. The
common novel is not the thing at all. It gives struggle followed
by relief. I want each act to close on a new and triumphant
happiness, which has been steadily growing all the while. This
is the real antithesis of tragedy, where things get blacker and
blacker and end in hopeless woe. Smiles has not grasped my
grand idea, and only shows a bitter struggle followed by a little
respite before death. Some feeble critic might say my new
idea was not true to nature. I'm sick of this old-fashioned
notion of art. Hold a mirror up, indeed! Let's paint a picture
of how things ought to be and hold that up to nature, and
perhaps the poor old woman may repent and mend her ways.'
The 'grand idea' might be possible in art; not even the ingenuity of nature could so round in the actual life of any man.
And yet it might almost seem to fancy that she had read the
letter and taken the hint; for to Fleeming the cruelties of fate

were strangely blended with tenderness, and when death came, it came harshly to others, to him not unkindly.

In the autumn of that same year 1875, Fleeming's father and mother were walking in the garden of their house at Merchiston, when the latter fell to the ground. It was thought at the time to be a stumble; it was in all likelihood a premonitory stroke of palsy. From that day, there fell upon her an abiding panic fear; that glib, superficial part of us that speaks and reasons could alledge no cause, science itself could find no mark of danger, a son's solicitude was laid at rest; but the eyes of the body saw the approach of a blow, and the consciousness of the body trembled at its coming. It came in a moment; the brilliant, spirited old lady leapt from her bed, raving. For about six months, this stage of her disease continued with many painful and many pathetic circumstances; her husband who tended her, her son who was unwearied in his visits, looked for no change in her condition but the change that comes to all. 'Poor mother,' I find Fleeming writing, 'I cannot get the tones of her voice out of my head. . . . I may have to bear this pain for a long time; and so I am bearing it and sparing myself whatever pain seems useless. Mercifully I do sleep, I am so weary that I must sleep.' And again later: 'I could do very well, if my mind did not revert to my poor mother's state whenever I stop attending to matters immediately before me.' And the next day: 'I can never feel a moment's pleasure without having my mother's suffering recalled by the very feeling of happiness. A pretty, young face recalls hers by contrast —a careworn face recalls it by association. I tell you, for I can speak to no one else; but do not suppose that I wilfully let my mind dwell on sorrow.'

In the summer of the next year, the frenzy left her; it left her stone deaf and almost entirely aphasic, but with some remains of her old sense and courage. Stoutly she set to work with dictionaries, to recover her lost tongues; and had already made notable progress, when a third stroke scattered her acquisitions. Thenceforth, for nearly ten years, stroke followed upon stroke, each still further jumbling the threads of her intelligence, but by degrees so gradual and with such partiality of loss and of survival, that her precise state was always and to the end a matter of dispute. She still remembered her friends; she still loved to learn news of them upon the slate; she still read and marked the list of the subscription library; she still took an interest in the choice of a play for the theatricals, and could remember and find parallel passages; but alongside of these surviving powers, were lapses as remarkable, she misbehaved like a child, and a servant had to sit with her at table. To see her so sitting, speaking with the tones of a deaf mute not always to the purpose, and to remember what she had been, was a moving appeal to all who knew her. Such was the pathos of these two old people in their affliction, that even the reserve of cities was melted and the neighbours vied in sympathy and kindness. Where so many were more than usually helpful, it is hard to draw distinctions; but I am directed and I delight to mention in particular the good Dr. Joseph Bell, Mr. Thomas and Mr. Archibald Constable with both their wives, the Rev. Mr. Belcombe (of whose good heart and taste I do not hear for the first time—the news had come to me by way of the Infirmary) and their next-door neighbour, unwearied in service, Miss Hannah Mayne. Nor should I omit to mention that John Ruffini continued to write to Mrs. Jenkin till his own death, and the clever lady known to the world as Vernon Lee until the end: a touching, a becoming attention to what was only the wreck and survival of their brilliant friend.

But he to whom this affliction brought the greatest change was the Captain himself. What was bitter in his lot, he bore with unshaken courage; only once, in these ten years of trial, has Mrs. Fleeming Jenkin seen him weep; for the rest of the time his wife—his commanding officer, now become his trying child—was served not with patience alone, but with a lovely happiness of temper. He had belonged all his life to the ancient, formal, speech-making compliment-presenting school of courtesy; the dictates of this code partook in his eyes of the nature of a duty; and he must now be courteous for two. Partly from a happy illusion, partly in a tender fraud, he kept

Captain Jenkin.

his wife before the world as a still active partner. When he paid a call, he would have her write 'with love' upon a card; or if that (at the moment) was too much, he would go armed with a bouquet and present it in her name. He even wrote letters for her to copy and sign: an innocent substitution, which may have caused surprise to Ruffini or to Vernon Lee, if they ever received, in the hand of Mrs. Jenkin, the very obvious reflections of her husband. He had always adored this wife whom he now tended and sought to represent in correspondence : it was now, if not before, her turn to repay the compliment; mind enough was left her to perceive his unwearied kindness; and as her moral qualities seemed to survive quite unimpaired, a childish love and gratitude were his reward. She would interrupt a conversation to cross the room and kiss him. If she grew excited (as she did too often) it was his habit to come behind her chair and pat her shoulder; and then she would turn round, and clasp his hand in hers, and look from him to her visitor with a face of pride and love; and it was at such moments only that the light of humanity revived in her eyes. It was hard for any stranger, it was impossible for any that loved them, to behold these mute scenes, to recall the past, and not to weep. But to the Captain, I think it was all happiness. After these so long years, he had found his wife again; perhaps kinder than ever before; perhaps now on a more equal footing; certainly, to his eyes, still beautiful. And the call made on his intelligence had not been made in vain. The merchants of Aux Cayes, who had seen him tried in some 'counter-revolution' in 1845, wrote to the consul of his 'able and decided measures,' 'his cool, steady judgment and discernment' with admiration ; and of himself, as 'a credit and an ornament to H.M. Naval Service.' It is plain he must have sunk in all his powers, during the years when he was only a figure, and often a dumb figure, in his wife's drawing-room; but with this new term of service, he brightened visibly. He showed tact and even invention in managing his wife, guiding or restraining her by the touch, holding family worship so arranged that she could follow and take part in it. He took (to the world's surprise) to reading—voyages, biographies, Blair's *Sermons*, even (for her letters' sake) a work of Vernon Lee's, which proved however more than he was quite prepared for. He shone more, in his remarkable way, in society ; and twice he had a little holiday to Glenmorven, where, as may be fancied, he was the delight of the Highlanders. One of his last pleasures was to arrange his dining-room. Many and many a room (in their wandering and thriftless existence) had he seen his wife furnish 'with exquisite taste' and perhaps with 'considerable luxury' : now it was his turn to be the decorator. On the wall he had an engraving of Lord Rodney's action, showing the *Prothée*, his father's ship, if the reader recollects ; on either side of this on brackets, his father's sword, and his father's telescope, a gift from Admiral Buckner who had used it himself during the engagement ; higher yet, the head of his grandson's first stag, portraits of his son and his son's wife, and a couple of old Windsor jugs from Mrs. Buckner's. But his simple trophy was not yet complete ; a device had to be worked and framed and hung below the engraving ; and for this he applied to his daughter-in-law : 'I want you to work me something, Annie. An anchor at each side—an anchor—stands for an old sailor, you know—stands for hope, you know—an anchor at each side, and in the middle THANKFUL.' It is not easy, on any system of punctuation, to represent the Captain's speech. Yet I hope there may shine out of these facts, even as there shone through his own troubled utterance, some of the charm of that delightful spirit.

In 1881, the time of the golden wedding came round for that sad and pretty household. It fell on a Good Friday, and its celebration can scarcely be recalled without both smiles and tears. The drawing-room was filled with presents and beautiful bouquets ; these, to Fleeming and his family, the golden bride and bridegroom displayed with unspeakable pride, she so painfully excited that the guests feared every moment to see her stricken afresh, he guiding and moderating her with his customary tact and understanding, and doing the honours of the day with more than his usual delight. Thence they were brought to the dining-room, where the Captain's idea of a feast awaited them : tea and champagne, fruit and toast and childish little

The golden wedding.

luxuries, set forth pell-mell and pressed at random on the guests. And here he must make a speech for himself and his wife, praising their destiny, their marriage, their son, their daughter-in-law, their grandchildren, their manifold causes of gratitude : surely the most innocent speech, the old, sharp contemner of his innocence now watching him with eyes of admiration. Then it was time for the guests to depart ; and they went away, bathed, even to the youngest child, in tears of inseparable sorrow and gladness, and leaving the golden bride and bridegroom to their own society and that of the hired nurse.

It was a great thing for Fleeming to make, even thus late, the acquaintance of his father ; but the harrowing pathos of such scenes consumed him. In a life of tense intellectual effort, a certain smoothness of emotional tenor were to be desired ; or we burn the candle at both ends. Dr. Bell perceived the evil that was being done ; he pressed Mrs. Jenkin to restrain her husband from too frequent visits ; but here was one of those clear-cut, indubitable duties for which Fleeming lived, and he could not pardon even the suggestion of neglect.

Death of Uncle John.
And now, after death had so long visibly but still innocuously hovered above the family, it began at last to strike and its blows fell thick and heavy. The first to go was uncle John Jenkin, taken at last from his Mexican dwelling and the lost tribes of Israel ; and nothing in this remarkable old gentleman's life, became him like the leaving of it. His sterling, jovial acquiescence in man's destiny was a delight to Fleeming. ' My visit to Stowting has been a very strange but not at all a painful one,' he wrote. ' In case you ever wish to make a person die as he ought to die in a novel,' he said to me, ' I must tell you all about my old uncle.' He was to see a nearer instance before long ; for this family of Jenkin, if they were not very aptly fitted to live, had the art of manly dying. Uncle John was but an outsider after all ; he had dropped out of hail of his nephew's way of life and station in society, and was more like some shrewd, old, humble friend who should have kept a lodge ; yet he led the procession of becoming deaths, and began in the mind of Fleeming that train of tender and grateful thought, which was like

a preparation for his own. Already I find him writing in the plural of ' these impending deaths '; already I find him in quest of consolation. ' There is little pain in store for these wayfarers,' he wrote, ' and we have hope—more than hope, trust.'

On May 19, 1884, Mr. Austin was taken. He was seventy-eight years of age, suffered sharply with all his old firmness, and died happy in the knowledge that he had left his wife well cared for. This had always been a bosom concern; for the Barrons were long-lived and he believed that she would long survive him. But their union had been so full and quiet that Mrs. Austin languished under the separation. In their last years, they would sit all evening in their own drawing-room hand in hand: two old people who, for all their fundamental differences, had yet grown together and become all the world in each other's eyes and hearts; and it was felt to be a kind release, when eight months after, on January 14, 1885, Eliza Barron followed Alfred Austin. ' I wish I could save you from all pain,' wrote Fleeming six days later to his sorrowing wife, ' I would if I could—but my way is not God's way; and of this be assured,—God's way is best.'

In the end of the same month, Captain Jenkin caught cold and was confined to bed. He was so unchanged in spirit that at first there seemed no ground of fear; but his great age began to tell, and presently it was plain he had a summons. The charm of his sailor's cheerfulness and ancient courtesy, as he lay dying, is not to be described. There he lay, singing his old sea songs; watching the poultry from the window with a child's delight; scribbling on the slate little messages to his wife, who lay bed-ridden in another room; glad to have Psalms read aloud to him, if they were of a pious strain—checking, with an ' I don't think we need read that, my dear,' any that were gloomy or bloody. Fleeming's wife coming to the house and asking one of the nurses for news of Mrs. Jenkin, ' Madam, I do not know,' said the nurse; ' for I am really so carried away by the Captain that I can think of nothing else.' One of the last messages scribbled to his wife and sent her with a glass of the champagne that had been ordered for himself, ran, in his most

finished vein of childish madrigal : 'The Captain bows to you,
my love, across the table.' When the end was near and it was
thought best that Fleeming should no longer go home but sleep
at Merchiston, he broke his news to the Captain with some trepidation, knowing that it carried sentence of death. 'Charming,
charming—charming arrangement,' was the Captain's only com-
mentary. It was the proper thing for a dying man, of Captain
Jenkin's school of manners, to make some expression of his
spiritual state ; nor did he neglect the observance. With his
usual abruptness, 'Fleeming,' said he, 'I suppose you and I
feel about all this as two Christian gentlemen should.' A last
pleasure was secured for him. He had been waiting with painful interest for news of Gordon and Khartoum ; and by great
good fortune, a false report reached him that the city was relieved,
and the men of Sussex (his old neighbours) had been the first
to enter. He sat up in bed and gave three cheers for the Sussex
regiment. The subsequent correction, if it came in time, was
prudently withheld from the dying man. An hour before midnight on the fifth of February, he passed away : aged eighty-four.

Death of Mrs. Jenkin.

Word of his death was kept from Mrs. Jenkin ; and she
survived him no more than nine and forty hours. On the day
before her death, she received a letter from her old friend Miss
Bell of Manchester, knew the hand, kissed the envelope and
laid it on her heart ; so that she too died upon a pleasure.
Half an hour after midnight, on the eighth of February, she fell
asleep : it is supposed in her seventy-eighth year.

Effect on Fleeming.

Thus, in the space of less than ten months, the four seniors
of this family were taken away ; but taken with such features
of opportunity in time or pleasant courage in the sufferer, that
grief was tempered with a kind of admiration. The effect on
Fleeming was profound. His pious optimism increased and
became touched with something mystic and filial. 'The grave
is not good, the approaches to it are terrible,' he had written
in the beginning of his mother's illness : he thought so no more,
when he had laid father and mother side by side at Stowting.
He had always loved life ; in the brief time that now remained
to him, he seemed to be half in love with death. 'Grief is no

duty,' he wrote to Miss Bell ; ' it was all too beautiful for grief,'
he said to me ; but the emotion, call it by what name we please,
shook him to his depths, his wife thought he would have broken
his heart when he must demolish the Captain's trophy in the
dining-room, and he seemed thenceforth scarcely the same
man.

These last years were indeed years of an excessive demand
upon his vitality ; he was not only worn out with sorrow, he
was worn out by hope. The singular invention to which he
gave the name of telpherage, had of late consumed his time,
overtaxed his strength and overheated his imagination. The
words in which he first mentioned his discovery to me—'I am
simply Alnaschar'—were not only descriptive of his state of
mind, they were in a sense prophetic; since whatever fortune
may await his idea in the future, it was not his to see it bring forth
fruit. Alnaschar he was indeed ; beholding about him a world
all changed, a world filled with telpherage wires; and seeing
not only himself and family but all his friends enriched. It was
his pleasure, when the company was floated, to endow those
whom he liked with stock ; one, at least, never knew that he
was a possible rich man until the grave had closed over his
stealthy benefactor. And however Fleeming chafed among material and business difficulties, this rainbow vision never faded ;
and he, like his father and his mother, may be said to have died
upon a pleasure. But the strain told, and he knew that it was
telling. 'I am becoming a fossil,' he had written five years
before, as a kind of plea for a holiday visit to his beloved Italy.
'Take care! If I am Mr. Fossil, you will be Mrs. Fossil, and
Jack will be Jack Fossil, and all the boys will be little fossils,
and then we shall be a collection.' There was no fear more
chimerical for Fleeming ; years brought him no repose; he was as
packed with energy, as fiery in hope, as at the first ; weariness,
to which he began to be no stranger, distressed, it did not quiet
him. He feared for himself, not without ground, the fate which
had overtaken his mother ; others shared the fear. In the
changed life now made for his family, the elders dead, the sons
going from home upon their education, even their tried domestic

Telpher-
age.

(Mrs. Alice Dunns) leaving the house after twenty-two years of service, it was not unnatural that he should return to dreams of Italy. He and his wife were to go (as he told me) on ' a real honeymoon tour.' He had not been alone with his wife 'to speak of,' he added, since the birth of his children. But now he was to enjoy the society of her to whom he wrote, in these last days, that she was his ' Heaven on earth.' Now he was to revisit Italy, and see all the pictures and the buildings and the scenes that he admired so warmly, and lay aside for a time the irritations of his strenuous activity. Nor was this all. A trifling operation was to restore his former lightness of foot; and it was a renovated youth that was to set forth upon this reënacted honeymoon.

The end. The operation was performed; it was of a trifling character, it seemed to go well, no fear was entertained; and his wife was reading aloud to him as he lay in bed, when she perceived him to wander in his mind. It is doubtful if he ever recovered a sure grasp upon the things of life; and he was still unconscious when he passed away, June the twelfth, 1885, in the fifty-third year of his age. He passed; but something in his gallant vitality had impressed itself upon his friends, and still impresses. Not from one or two only, but from many, I hear the same tale of how the imagination refuses to accept our loss and instinctively looks for his reappearing, and how memory retains his voice and image like things of yesterday. Others, the well-beloved too, die and are progressively forgotten : two years have passed since Fleeming was laid to rest beside his father, his mother and his Uncle John; and the thought and the look of our friend still haunts us.

APPENDIX.

———•◦•———

I.

*NOTE ON THE CONTRIBUTIONS OF FLEEMING JENKIN
TO ELECTRICAL AND ENGINEERING SCIENCE.*

By Sir WILLIAM THOMSON, F.R.S., LL.D.

IN the beginning of the year 1859 my former colleague (the first
British University Professor of Engineering), Lewis Gordon, at that
time deeply engaged in the then new work of cable making and
cable laying, came to Glasgow to see apparatus for testing submarine
cables and signalling through them, which I had been preparing for
practical use on the first Atlantic cable, and which had actually
done service upon it, during the six weeks of its successful working
between Valencia and Newfoundland. As soon as he had seen
something of what I had in hand, he said to me, 'I would like to
show this to a young man of remarkable ability, at present engaged
in our works at Birkenhead.' Fleeming Jenkin was accordingly
telegraphed for, and appeared next morning in Glasgow. He remained for a week, spending the whole day in my class room and
laboratory, and thus pleasantly began our lifelong acquaintance. I
was much struck, not only with his brightness and ability, but with
his resolution to understand everything spoken of, to see if possible
thoroughly through every difficult question, and (no *if* about this !)
to slur over nothing. I soon found that thoroughness of honesty
was as strongly engrained in the scientific as in the moral side of
his character.

In the first week of our acquaintance, the electric telegraph
and, particularly, submarine cables, and the methods, machines, and
instruments for laying, testing, and using them, formed naturally

the chief subject of our conversations and discussions ; as it was in
fact the practical object of Jenkin's visit to me in Glasgow ; but not
much of the week had passed before I found him remarkably interested in science generally, and full of intelligent eagerness on many
particular questions of dynamics and physics.   When he returned
from Glasgow to Birkenhead a correspondence commenced between
us, which was continued without intermission up to the last days of
his life.   It commenced with a well-sustained fire of letters on each
side about the physical qualities of submarine cables, and the practical results attainable in the way of rapid signalling through them.
Jenkin used excellently the valuable opportunities for experiment
allowed him by Newall, and his partner Lewis Gordon, at their
Birkenhead factory.   Thus he began definite scientific investigation
of the copper resistance of the conductor, and the insulating resistance and specific inductive capacity of its gutta-percha coating, in
the factory, in various stages of manufacture ; and he was the very
first to introduce systematically into practice the grand system of
absolute measurement founded in Germany by Gauss and Weber.
The immense value of this step, if only in respect to the electric
telegraph, is amply appreciated by all who remember or who have
read something of the history of submarine telegraphy ; but it can
scarcely be known generally how much it is due to Jenkin.

Looking to the article ' Telegraph (Electric) ' in the last volume
of the old edition of the ' Encyclopædia Britannica,' which was published about the year 1861, we find on record that Jenkin's measurements in absolute units of the specific resistance of pure gutta-percha,
and of the gutta-percha with Chatterton's compound constituting
the insulation of the Red Sea cable of 1859, are given as the only
results in the way of absolute measurements of the electric resistance
of an insulating material which had then been made.   These remarks
are prefaced in the ' Encyclopædia ' article by the following statement : ' No telegraphic testing ought in future to be accepted in
any department of telegraphic business which has not this definite
character ; although it is only within the last year that convenient
instruments for working, in absolute measure, have been introduced
at all, and the whole system of absolute measure is still almost
unknown to practical electricians.'

A particular result of great importance in respect to testing is
referred to as follows in the ' Encyclopædia ' article : ' The importance of having results thus stated in absolute measure is illustrated by the circumstance, that the writer has been able at once to

compare them, in the manner stated in a preceding paragraph, with his own previous deductions from the testings of the Atlantic cable during its manufacture in 1857, and with Weber's measurements of the specific resistance of copper.' It has now become universally adopted—first of all in England ; twenty-two years later by Germany, the country of its birth ; and by France and Italy, and all the other countries of Europe and America—practically the whole scientific world—at the Electrical Congress in Paris in the years 1882 and 1884.

An important paper of thirty quarto pages published in the ' Transactions of the Royal Society ' for June 19, 1862, under the title ' Experimental Researches on the Transmission of Electric Signals through submarine cables, Part I. Laws of Transmission through various lengths of one cable, by Fleeming Jenkin, Esq., communicated by C. Wheatstone, Esq., F.R.S.,' contains an account of a large part of Jenkin's experimental work in the Birkenhead factory during the years 1859 and 1860. This paper is called Part I. Part II. alas never appeared, but something that it would have included we can see from the following ominous statement which I find near the end of Part I. ' From this value, the electrostatical capacity per unit of length and the specific inductive capacity of the dielectric, could be determined. These points will, however, be more fully treated of in the second part of this paper.' Jenkin had in fact made a determination at Birkenhead of the specific inductive capacity of gutta-percha, or of the gutta-percha and Chatterton's compound constituting the insulation of the cable, on which he experimented. This was the very first true measurement of the specific inductive capacity of a dielectric which had been made after the discovery by Faraday of the existence of the property, and his primitive measurement of it for the three substances, glass, shellac, and sulphur ; and at the time when Jenkin made his measurements the existence of specific inductive capacity was either unknown, or ignored, or denied, by almost all the scientific authorities of the day.

The original determination of the microfarad, brought out under the auspices of the British Association Committee on Electrical Standards, is due to experimental work by Jenkin, described in a paper, ' Experiments on Capacity,' constituting No. IV. of the appendix to the Report presented by the Committee to the Dundee Meeting of 1867. No other determination, so far as I know, of this important element of electric measurement has hitherto been made ;

and it is no small thing to be proud of in respect to Jenkin's fame as a scientific and practical electrician that the microfarad which we now all use is his.

The British Association unit of electrical resistance, on which was founded the first practical approximation to absolute measurement on the system of Gauss and Weber, was largely due to Jenkin's zeal as one of the originators, and persevering energy as a working member, of the first Electrical Standards Committee. The experimental work of first making practical standards, founded on the absolute system, which led to the unit now known as the British Association ohm, was chiefly performed by Clerk Maxwell and Jenkin. The realisation of the great practical benefit which has resulted from the experimental and scientific work of the Committee is certainly in a large measure due to Jenkin's zeal and perseverance as secretary, and as editor of the volume of Collected Reports of the work of the Committee, which extended over eight years, from 1861 till 1869. The volume of Reports included Jenkin's Cantor Lectures of January 1866, 'On Submarine Telegraphy,' through which the practical applications of the scientific principles for which he had worked so devotedly for eight years became part of general knowledge in the engineering profession.

Jenkin's scientific activity continued without abatement to the end, as will be seen by the detailed account of his published papers contained in the present volume. For the last two years of his life he was much occupied with a new mode of electric locomotion, a very remarkable invention of his own, to which he gave the name of 'Telpherage.' He persevered with endless ingenuity in carrying out the numerous and difficult mechanical arrangements essential to the project, up to the very last days of his work in life. He had completed almost every detail of the realisation of the system which was recently opened for practical working at Glynde, in Sussex, four months after his death.

His book on 'Magnetism and Electricity,' published as one of Longman's elementary series in 1873, marked a new departure in the exposition of electricity, as the first text-book containing a systematic application of the quantitative methods inaugurated by the British Association Committee on Electrical Standards. In 1883 the seventh edition was published, after there had already appeared two foreign editions, one in Italian and the other in German.

His papers on purely engineering subjects, though not numerous, are interesting and valuable. Amongst these may be mentioned the article ' Bridges,' written by him for the ninth edition of the 'Encyclopædia Britannica,' and afterwards republished as a separate treatise in 1876 ; and a paper 'On the Practical Application of Reciprocal Figures to the Calculation of Strains in Framework,' read before the Royal Society of Edinburgh, and published in the ' Transactions' of that Society in 1869. But perhaps the most important of all is his paper 'On the Application of Graphic Methods to the Determination of the Efficiency of Machinery,' read before the Royal Society of Edinburgh, and published in the 'Transactions,' vol. xxviii. (1876–78), for which he was awarded the Keith Gold Medal. This paper was a continuation of the subject treated in ' Reulaux's Mechanism,' and, recognising the value of that work, supplied the elements required to constitute from Reulaux's kinematic system a full machine receiving energy and doing work.

II.

NOTE ON THE WORK OF FLEEMING JENKIN IN CONNECTION WITH SANITARY REFORM.

By Lieutenant-Colonel ALEXANDER FERGUSSON.

It was, I believe, during the autumn of 1877 that there came to Fleeming Jenkin the first inkling of an idea, not the least in importance of the many that emanated from that fertile brain, which, with singular rapidity, took root, and under his careful fostering expanded into a scheme the fruits of which have been of the utmost value to his fellow-citizens and others.

The phrase which afterwards suggested itself, and came into use, ' Healthy houses,' expresses very happily the drift of this scheme, and the ultimate object that Jenkin had in view.

In the summer of that year there had been much talk, and some newspaper correspondence, on the subject of the unsatisfactory condition of many of the best houses in Edinburgh as regards their sanitary state. One gentleman, for example, drew an appalling picture of a large and expensive house he had bought in the West-end of Edinburgh, fresh from the builder's hands. To ascertain precisely what was wrong, and the steps to be taken to remedy the evils, the effects

of which were but too apparent, obviously demanded the expendi-
ture of much time and careful study on the part of the intelligent
proprietor himself and the professional experts he had to call in,
and, it is needless to add, much money.   There came also, from the
poorer parts of the town, the cry that in many cases the houses of
our working people were built anyhow that the dictates of a narrow
economy suggested to the speculative and irresponsible builder.   The
horrors of what was called the 'Sandwich system,' amongst other
evils, were brought to light.   It is sufficient to say, generally, that
this particular practice of the builder consists in placing in a block
of workmen's houses, to save space and money, the water cisterns of
one flat, directly under the sanitary appliances of the other, and so
on to the top of a house of several storeys.   It is easy to conceive the
abominations that must ensue when the leakage of the upper floors
begins to penetrate to the drinking water below.   The picture was
a hideous one, apart from the well-known fact that a whole class of
diseases is habitually spread by contaminated water.

In October 1876, a brisk and interesting discussion had been
carried on in the columns of the *Times* at intervals during the
greater part of that month, in which the same subject, that of the
health and sewage of towns, had been dealt with by several writers
well informed in such matters.   Amongst others, Professor Jenkin
himself took part, as did Professor G. F. Armstrong, who now occupies
the chair of Civil Engineering in Edinburgh.   Many of the truths
then advanced had been recently discussed at a meeting of the British
Association.

It was while such topics were attracting attention that Fleeming Jenkin's family were shocked by the sad intelligence of the loss
that friends of theirs had sustained in the deaths of several of their
children from causes that could be traced up to the insanitary condition of their house.   Sympathy took the practical form of an
intense desire that something might be done to mitigate the chance
of such calamities ; and, I am permitted to say, the result of a home-
talk on this subject was an earnest appeal to the head of the house
to turn his scientific knowledge to account in some way that should
make people's homes more healthy, and their children's lives more
safe.   In answer to the call Jenkin turned his thoughts in this
direction.   And the scheme which I shall endeavour briefly to
sketch out was the result.

The obvious remedy for a faulty house is to call in a skilful
expert, architect or engineer, who will doubtless point out by means

of reports and plans what is wrong, and suggest a remedy; but, as remarked by Professor Jenkin, 'it has not been the practice for leading engineers to advise individuals about their house arrangements, except where large outlay is in contemplation.' A point of very considerable importance in such a case as that now supposed.

The problem was to ensure to the great body of the citizens sound professional advice concerning their houses, such as had hitherto been only obtainable at great cost—but 'with due regard to economical considerations.'

The advantages of co-operation are patent to all. Everyone can understand how, if a sufficient number of persons combine, there are few luxuries or advantages that are not within their reach, for a moderate payment. The advice of a first-rate engineer regarding a dwelling-house was a palpable advantage ; but within the reach of comparatively few. One has heard of a winter in Madeira being prescribed as the cure for a poor Infirmary sufferer.

Like most good plans Jenkin's scheme was simple in the extreme, and consisted in *combination* and a small subscription.

' Just,' he says ' as the leading physician of the day may give his services to great numbers of poor patients when these are gathered in a hospital, although he could not practically visit them in their own houses, so the simple fact of a number of clients gathered into a group will enable the leading engineer to give them the benefit of his advice.'

But it was his opinion that only ' continual supervision could secure the householder from danger due to defects in sanitary-appliances.' He had in his eye a case precisely similar. The following passage in one of his first lectures, afterwards repeated frequently, conveys the essence of Professor Jenkin's theory, as well as a graceful acknowledgment of the source from which this happy idea was derived : —

'An analogous case occurred to him,' he said, 'in the "Steam Users' Association," in Lancashire. So many boilers burst in that district for want of inspection that an association was formed for having the boilers under a continual course of inspection. Let a perfect boiler be bought from a first-rate maker, the owner has then an apparatus as perfect as it is now sought to make the sanitary appliances in his house. But in the course of time the boiler must decay. The prudent proprietor, therefore, joins the Steam-boiler Association, which, from time to time, examines his boiler, and by the tests they apply are able to give an absolute guarantee against accident. This idea of an inspection by an association was due,' the lecturer continued, 'to Sir William Fairbairn, under whom he had the honour of serving his apprenticeship.'[1] The steam users were thus absolutely protected from danger ; and the same idea it was sought to apply to the sanitary system of a house.

To bring together a sufficient number of persons, to form such a 'group' as had been contemplated, was the first step to be taken. No time was lost in taking it. The idea hitherto roughly blocked out was now given a more definite form. The original sketch, as dictated by Jenkin himself, is before me, and I cannot do better than transcribe it, seeing it is short and simple. Several important alterations were afterwards made by himself in consultation with one or two of his Provisional Council ; and as experience suggested :—

'The objects of this Association are twofold.

' 1. By taking advantage of the principle of co-operation, to provide its members at moderate cost with such advice and supervision as shall ensure the proper sanitary condition of their own dwellings.

' 2. By making use of specially qualified officers to support the inhabitants and local authorities in enforcing obedience to the provisions of those laws and by laws which affect the sanitary condition of the community.

'It is proposed that an Association with these objects be formed ; and that all residents within the municipal boundaries of Edinburgh be eligible as members. That each member of the Association shall subscribe *one guinea* annually. That in return for the annual subscription each member shall be entitled to the following advantages :—

'1. A report by the Engineer of the Association on the sanitary condition of his dwelling, with specific recommendations as to the improvement of drainage, ventilation, &c., should this be found necessary.

' 2. The supervision of any alterations in the sanitary fittings of his dwelling which may be carried out by the advice, or with the approval, of the officers of the Association.

' 3. An annual inspection of his premises by the Engineer of the Association, with a report as to their sanitary condition.

' 4. The right, in consideration of a payment of five shillings, of

[1] See paper read at the Congress of the Social Science Association, Edinburgh, October 8, 1880.

calling on the Engineer, and legal adviser [1] of the Association to inspect and report on the existence of any infraction or supposed infraction of any law affecting the sanitary condition of the community.

'It is proposed that the Association should be managed by an unpaid Council, to be selected by ballot from among its members.

'That the following salaried officers be engaged by the Association.

'1. One or more acting engineers, who should give their services exclusively to the Association.

'2. A consulting engineer, who should exercise a general supervision, and advise both on the general principles to be followed, and on difficult cases.

'3. A legal agent, to be engaged on such terms as the Council shall hereafter think fit.

'4. A permanent secretary.

'It is also proposed that the officers of the Association should, with the sanction of the Council, have power to take legal proceedings against persons who shall, in their opinion, be guilty of any infraction of sanitary regulations in force throughout the district ; and generally it is intended that the Association shall further and promote all undertakings which, in their opinion, are calculated to improve the sanitary condition of Edinburgh and its immediate neighbourhood.

'In one aspect this association will be analogous to the Steam Boiler Users' Association, who co-operate in the employment of skilled inspectors. In a second aspect it will be analogous to the Association for the Prevention of Cruelty to Animals, which assists the community in enforcing obedience to existing laws.'

Towards the end of November 1877 this paper was handed about among those who were thought most likely, from their position and public spirit, to forward such a scheme, so clearly for the good of the community. Nay more, a systematic 'canvass' was set on foot ; personal application the most direct was made use of. The thing was new, and its advantages not perfectly obvious to all, at a glance. Every one who knows with what enthusiastic earnestness Jenkin would take hold of, and insist upon, what he felt to be wholesome and right will understand how he persisted, how he patiently explained, and swept away objections that were raised. One could not choose but listen, and understand, and agree.

---

[1] It was ultimately agreed not to appoint an officer of this kind till occasion should arise for his services none has been appointed.

On the evening of 2nd January 1878, or, to be more correct, the morning of the 3rd, two old school-fellows of his at the Edinburgh Academy walked home with him from an annual dinner of their 'Class.' All the way in glowing language he expounded his views of house inspection, and the protection of health, asking for sympathy. It was most readily given, and they parted from him with pleasant words of banter regarding this vision of his of grafting 'cleanliness' upon another quality said to be a growth, in some sort, of this northern land of ours.

But they reckoned hardly sufficiently on the fact that when Jenkin took a thing of this kind in hand it must *be* ; if it lay within the scope of a clear head and boundless energy.

Having secured a nucleus of well-wishers, the next step was to enlist the sympathies of the general public. It was sought to effect this by a series of public lectures. The first of these (one of two) was given on 22nd January under the auspices of the Edinburgh Philosophical Institution. It was apparent to the shrewd lecturer that in bringing before the people a scheme like this, where there was much that was novel, it was necessary first of all that his audience should be aware of the evils to which they were exposed in their own houses, before unfolding a plan for a remedy. The correspondence already referred to as having been carried on in the summer of the previous year had shown how crude were the ideas of many persons well informed, or considered to be so, on this subject. For example, there are few now-a-days who are not aware that a drain, to be safe, must have at intervals along its course openings to the upper air, or that it must be ' ventilated' as the phrase goes. But at the time spoken of there were some who went so far as to question this principle ; even to argue against it ; calling forth this forcible reply— ' Here is a pretty farce. You pour out a poison and send it off on its way to the sea, and forget that on its way there its very essence will take wings and fly back into your house up the very pipes it but recently ran down.' A properly ' trapped' and ventilated drain was the cure for this.

And the lecturer proceeded to show that in Edinburgh, where for the most part house construction is good and solid, but, as in other towns, the bulk of the houses were built when the arrangements for internal sewerage and water supply were very little understood, many serious errors were made. ' But,' the lecturer went on to say, ' Sanitary Science was now established on a fairly sound basis, and the germ theory, or theory of septic ferments, had explained much which used to be obscure. This theory explained how it was that families might in certain cases live with fair health for many years in the midst of great filth, while the dwellers in large and apparently clean mansions were struck down by fever and diphtheria. The filth which was found compatible with health was always isolated filth, and until the germs of some specific disease were introduced, this dirt was merely injurious, not poisonous. The mansions which were apparently clean and yet fever-visited were found to be those in which arrangements had been made for the removal of offensive matter, which arrangements served also to distribute poison germs from one house to another, from one room to another. These mansions had long suckers extended from one to another through the common sewer. Through these suckers, commonly called "house drains," they imbibed every taint which any one house in the system could supply. In fact, arrangements were too often made which simply "laid on" poison to bedrooms just as gas or water was laid on. He had known an intelligent person declare that no harm could come up a certain pipe which ended in a bed-room, because nothing offensive went down. That person had never realised the fact that his pipe joined another pipe, which again joined a sewer, which again, whenever there was an epidemic in the neighbourhood, received innumerable poison germs ; and that, although nothing more serious than scented soap and water went down, the germs of typhoid fever might any day come up.'

Professor Jenkin then proceeded to show how a house might be absolutely cut off from all contamination from these sources of evil. Then by means of large diagrams he showed the several systems of pipes within a house. One system coloured *red* showed the pipes that received foul matter. A system marked in *blue* showed pipes used to ventilate this red system. The essential conditions of safety in the internal fittings of a house — it was inculcated—were that no air to be breathed, no water to be drunk, should ever be contaminated by connection with *red* or *blue* systems. Then in *yellow* were shown the pipes which received dirty water, which was not necessarily foul. Lastly a *white* system, which under no circumstances must ever touch the 'red,' 'blue' or 'yellow' systems. Such a diagram recalled the complicated anatomical drawings which illustrate the system of arteries and veins in the human frame. Little wonder, then, that one gentleman remarked, in perplexity, that he had not room in his house for such a mass of pipes ; but they were already there, with other pipes besides, all carefully hidden away, as in the human tenement

MEMOIR

with the inevitable result—as the preacher of cleanliness and health declared—'out of sight, out of mind.'

In plain and forcible language were demonstrated the ills this product of modern life is heir to ; and the drastic measures that most of them demand to secure the reputation of a healthy house. Lastly the formation of an Association to carry out the idea (already sketched) cheaply, was briefly introduced.

Next morning, January 23rd, was the moment chosen to lay the scheme formally before the public. In all the Edinburgh newspapers, along with lengthy reports of the lecture, appeared, in form of an advertisement, a statement [1] of the scheme and its objects, supported by an imposing array of ' Provisional Council.' In due course several of the Scots newspapers and others, such as the *Building News*, gave leading articles, all of them directing attention to this new thing, as ' an interesting experiment about to be tried in Edinburgh,' 'what promises to be a very useful sanitary movement, now being organised, and an example set that may be worthy of imitation elsewhere,' and so on.

Several of the writers waxed eloquent on the singular ingenuity of the scheme ; the cheap professional advice to its adherents, &c.; and the rare advantages to be gained by means of co-operation and the traditional ' one pound one.'

The Provisional Council was absolutely representative of the community, and included names more than sufficient to inspire confidence. It included the Lord-Lieutenant of the county, Lord Rosebery ; the Lord Justice Clerk, Lord Moncrieff; the Lord Advocate ; Sir Robert Christison ; several of the Judges of the Court of Session ;

---

[1] Briefly stated the points submitted in this prospectus were these :—

1. That the proposed Association was a Society for the benefit of its members and the community that cannot be used for any purposes of profit.

2. The privileges of members include the annual inspection of their premises, as well as a preliminary report on their condition, with an estima'e of the cost of any alterations recommended.

3. The skilled inspection from time to time of drains and all sanitary arrangements.

4. No obligation on the part of members to carry out any of the suggestions made by the engineers of the Association, who merely give skilled advice when such is desired.

5 The officers of the Association to have no interest in any outlay recommended.

6. The Association might be of great service to the poorer members of the community.

the Presidents of the Colleges of Physicians, and of Surgeons ; many of the Professors of the University ; the Bishop of Edinburgh, and the Dean ; several of the best known of the Clergy of the Church of Scotland, Established, Free, and of other branches ; one or two members of Parliament ; more than one lady (who should have been perhaps mentioned earlier on this list) well known for large views and public spirit ; several well-known country gentlemen ; one or two distinguished civil engineers and architects ; and many gentlemen of repute for intelligence and business qualities.

Very soon after the second of the promised lectures, the members of the new Society began to be numbered by hundreds. By the 28th of February, 500 subscribers having been enrolled, they were in a position to hold their first regular meeting under the presidency of Sir Robert Christison, when a permanent Council composed of many of those who had from the first shown an interest in the movement—for example, Professor (now Sir Douglas) Maclagan and Lord Dean of Guild (now Sir James) Gowans, Professor Jenkin himself undertaking the duties of Consulting Engineer—were appointed. And Jenkin was singularly fortunate in securing as Secretary the late Captain Charles Douglas, a worker as earnest as himself. It was the theory of the originator that the Council, composed of leading men not necessarily possessed of engineering knowledge, should 'give a guarantee to the members that the officials employed should have been carefully selected, and themselves work under supervision. Every householder in this town,' he adds, ' knows the names of the gentlemen composing our Council.'

The new Association was a success alike in town and country. Without going far into statistics it will be evident what scope there was, and is, for such operations when it is stated that last year (1885) 60 per cent. of the houses inspected in London and its neighbourhood were found to have foul air escaping direct into them, and 81 per cent. had their sanitary appliances in an unsatisfactory state. Here in Edinburgh things were little, if any, better ; as for the country houses, the descriptions of some were simply appalling. As the new Association continued its operations it became the *rôle* of the Consulting Engineer to note such objections, hypothetical or real, as were raised against the working of his scheme. Some of these were ingenious enough : but all were replied to in order, and satisfactorily resolved. It was shown for example, that ' you might have a dinner party in your house on the day of your inspection ; ' that the Association worked in the utmost harmony with the city

authorities, and with the tradesmen usually employed in such busi
ness ; and that the officials were as ' confidential' as regards the in-
firmities of a house as any physician consulted by a patient. The
strength of the engineering staff has been varied from time to time
as occasion required ; at the moment of writing employment is found
in Edinburgh and country districts in various parts of Scotland for
five engineers temporarily or permanently engaged.

The position Jenkin claimed for the Engineers was a high one,
but not too high : thus he well defined it :—

'In respect of Domestic Sanitation the business of the Engineer
and that of the medical man overlap ; for while it is the duty of the
engineer to learn from the doctor what conditions are necessary to
secure health, the engineer may, nevertheless, claim in his turn the
privilege of assisting in the warfare against disease by using his
professional skill to determine what mechanical and constructive
arrangements are best adapted to secure these conditions.' [1]

Flattery in the form of imitation followed in due course. A
branch was established at St. Andrews, and one of the earliest of
similar institutions was founded at Newport in the United States.
Another sprang up at Wolverhampton. In 1881 two such societies
were announced as having been set on foot in London. And the
Times of April 14th, in a leading article of some length, drew attention to the special features of the plan which it was stated had
followed close upon a paper read by Professor Fleeming Jenkin before
the Society of Arts in the preceding month of January. The adherents included such names as those of Sir William Gull, Professor
Huxley, Professor Burdon Sanderson, and Sir Joseph Fayrer. The
Saturday Review, in January, had already in a characteristic article
enforced the principles of the scheme, and shown how, for a small
annual payment, 'the helpless and hopeless condition of the householder at the mercy of the plumber' might be for ever changed.

The London Association, established on the lines of the parent
society, has been followed by many others year by year ; amongst
these are Bradford, Cheltenham, Glasgow, and Liverpool in 1882 ;
Bedford, Brighton, and Newcastle in 1883 ; Bath, Cambridge, Cardiff,
Dublin, and Dundee in 1884 ; and Swansea in 1885 ; and while we
write the first steps are being taken, with help from Edinburgh, to
establish an association at Montreal ; sixteen Associations.

Almost, it may be said, a bibliography has been achieved for
Fleeming Jenkin's movement.

    [1] *Healthy Houses*, by Professor Fleeming Jenkin, p. 54.

In 1878 was published *Healthy Houses* (Edin., David Douglas), being the substance of the two lectures already mentioned as having been delivered in Edinburgh with the intention of laying open the idea of the scheme then in contemplation, with a third addressed to the Medico-Chirurgical Society. This book has been long out of print, and, such has been the demand for it that the American edition [1] is understood to be also out of print, and unobtainable.

In 1880 was printed (London, Spottiswoode and Co.) a pamphlet entitled *What is the Best Mode of Amending the Present Laws with Reference to Existing Buildings, and also of Improving their Sanitary Condition with due Regard to Economical Considerations ?* the substance of a paper read by Professor Jenkin at the Congress of the Social Science Association at Edinburgh in October of that year.

The first item of *Health Lectures for the People* (Edin. 1881) consists of a discourse on the 'Care of the Body' delivered by Professor Jenkin in the Watt Institution at Edinburgh, in which the theories of house sanitation are dwelt on.

House Inspection, reprinted from the *Sanitary Record,* was issued in pamphlet form in 1882. And another small tract, *Houses of the Poor ; their Sanitary Arrangement,* in 1885.

In this connection it may be said that while the idea formulated by Jenkin has been carried out with a measure of success that could hardly have been foreseen, in one point only, it may be noted, has expectation been somewhat disappointed as regards the good that these Associations should have effected—and the fact was constantly deplored by the founder—namely, the comparative failure as a means of improving the condition of the dwellings of the poorer classes. It was ' hoped that charity and public spirit would have used the Association to obtain reports on poor tenements, and to remedy the most glaring evils.' [1]

The good that these associations have effected is not to be estimated by the numbers of their membership. They have educated the public on certain points. The fact that they exist has become generally known, and, by consequence, persons of all classes are induced to satisfy themselves of the reasons for the existence of such institutions, and thus they learn of the evils that have called them into being.

Builders, burgh engineers, and private individuals in any way

---

[1] It is perhaps worth mentioning as a curiosity of literature that the American publishers who produced this book in the States, without consulting the author, afterwards sent him a handsome cheque, of course unsolicited by him.

connected with the construction of dwellings in town or country have been put upon their mettle, and constrained to keep themselves abreast with the wholesome truths which the engineering staff of all these Sanitary Associations are the means of disseminating.

In this way, doubtless, some good may indirectly have been done to poorer tenements, though not exactly in the manner contemplated by the founder.[1]

Now, if it be true that Providence helps those who help themselves, surely a debt of gratitude is due to him who has placed (as has been attempted to be shown in this brief narrative) the means of self-help and the attainment of a palpable benefit within the reach of all through the working of a simple plan, whose motto well may be, ' Healthy Houses ' ; and device a strangled snake.

A. F.

[1] It is true, handsome tenements for working people have been built, such as the picturesque group of houses erected with this object by a member of the Council of the Edinburgh Sanitary Association, at Bell's Mills, so well seen from the Dean Bridge, where every appliance that science can suggest has been made use of. But for the ordinary houses of the poor the advice of the Association's engineers has been but rarely taken advantage of.

PAPERS

BY

FLEEMING JENKIN

PREFATORY NOTE.

By Sidney Colvin, M.A.

MOST of the papers which follow in this Volume, and in the first section of Volume II., are republished, by the kind permission of the proprietors, from one or other of the following periodicals :— the *Edinburgh Review*, the *North British Review*, the *Nineteenth Century*, *Macmillan's Magazine*, the *Saturday Review*, the *Art Journal*.

The reader will not fail to bear in mind that the contents of these sections do not represent their author's main occupations, but are the πάρεργα or by-labours of a life busily occupied in scientific and professional pursuits. They touch on many matters; but such were the keenness and loyalty of the writer's intelligence that on whatever subject engaged his attention he was almost sure to find something to say that was well worth hearing. When masters in such divers fields as Darwin and Munro have acknowledged the value of his arguments, weaker testimony is needless. Neither does it seem necessary, in a memorial collection such as this, to apologise for including some pieces in which farther reflection might have modified our friend's conclusions, or others (e.g. the paper on 'Greek Dress') in which his observations have lost some of their freshness through lapse of time or the labours of others. The essay 'On Rhythm' I have put together out of three separate articles, and in this instance alone have ventured on some slight abridgment and correction of the text. The fragments

'On the Life of George Eliot' and 'On Truth' in the section Literature and Drama, and those on the 'Time Labour System' and 'Is one Man's Gain another Man's Loss?' in the Political Economy section, are now printed for the first time. The play of 'Griselda,' in which the writer essayed to bring to the test of practice his own principles of dramatic composition, was printed during his life, but not published.

LITERATURE AND DRAMA

THE 'AGAMEMNON' AND 'TRACHINIÆ.' [1]

No one, without comparing some portion of Mr. Browning's rendering of the 'Agamemnon' with the text of Æschylus, can have any idea of the closeness with which the original has been followed. The translator has not been content with the ordinary Liddell and Scott version, but has endeavoured, often with success, to give even the etymological force of the original words. Careful toil marks every line, and to some few this transcript will be a source of real enjoyment, but these few must possess the rare power of putting life and poetry into such literal translations as are published in Bohn's series. This is said not as a sneer, for we have known men of true poetic temperament who, ignorant of Greek, read the dramatists by preference in Bohn's series, wishing to feel certain that 'where they were gaping for Æschylus they did not get Theognis.' Well, such men as these may read Mr. Browning's 'Agamemnon' with more implicit faith than they could give to many prose literal translations, but they must also possess the qualification, happily not rare nowadays, of being able to understand Mr. Browning's manner of expressing himself. Moreover, it would be well that they should have at their elbow some fairly literal translation such as Mr. Conington's to help them when, after reading a passage two or three times, they quite fail to see what the poet means to say.

To the general British public Mr. Browning's version will be almost as completely a sealed book as the original of

[1] A review of *The Agamemnon*, a transcript by Robert Browning: *The Agamemnon of Æschylus*, translated into English verse by E. O. A. Morshead, M.A.: and *Three Plays of Sophocles*, translated into English verse by Lewis Campbell, M.A., LL.D. From the *Edinburgh Review*, April, 1878.

Æschylus; a fact much to be deplored, for a version of a great poet by a true poet should have been a boon to the mass of readers, not merely to a select few. In 'Balaustion's Adventure' Mr. Browning gave this boon. The translation of the 'Alcestis' can be read aloud, so as in most places to be followed by the hearer, which cannot be done with the present play.

We do not ask Mr. Browning to write in such perspicuous sort that he who runs may read, but we think it not unreasonable to ask him to write so that a man who has mastered the meaning of any passage may be able when reading aloud to convey that meaning to his hearers, with the aid of due emphasis and inflection. The translator obviously felt that his version was obscure, and therefore himself qualifies his effort as perhaps a fruitless adventure. He argues that Æschylus is hard reading, and therefore that, to resemble Æschylus, the translation must be hard reading also. But surely Æschylus must have been intelligible, the phrases if not the thoughts of the man, to the thirty thousand hearers in the theatre at Athens. Can a dramatist be popular on the stage if unintelligible to the masses? Surely we must think that the obscurity of Æschylus arises chiefly from our ignorance of the language, of popular well-understood allusions and customs, and from corruptions in the text. Mr. Browning, in his version of passages usually accepted as corrupt, has maintained an obscurity that corresponds with the original in a way which is almost humorous. This he may justify, but it is difficult to excuse him when the obscurity is really due to his own style and not to Æschylus at all. But let his version speak for itself. After the prologue by the Warder, the chorus enters and speaks as follows:

> The tenth year this, since Priamos' great match,
> King Menelaos, Agamemnon King,
> —The strenuous yoke-pair of the Atreidai's honor,
> Two-throned, two-sceptred, whereof Zeus was donor—
> Did from this land the aid, the armament dispatch,
> The thousand-sailored force of Argives clamouring
> ' Ares ' from out the indignant breast, as fling
> Passion forth vultures which, because of grief
> Away,—as are their young ones,—with the thief,

Lofty above their brood-nests wheel in ring,
Now round and round with oar of either wing,
Lament the bedded chicks, lost labour that was love
Which hearing, one above
—Whether Apollon, Pan or Zeus—that wail,
Sharp-piercing bird-shriek of the guests who fare
Housemates with gods in air—
Suchanone sends, against who these assail,
What, late-sent, shall not fail
Of punishing—Erinus.'

The man who tries to read this aloud must first master the fact
that 'match' means 'antagonist;' he must also secure an
audience able to understand the expression 'clamouring Ares,'
and he will then with some difficulty make the first eight lines
intelligible. 'This is the tenth year since Menelaus and Agamemnon started with an army shouting a warlike cry.' Then
comes the simile of the vultures, which, in both Greek and
English, is hard to construe, the passage being possibly corrupt,
the English as much so as the Greek; but the passage which
begins 'Which hearing, one above,' &c., and ends with 'Erinus,'
could not be made intelligible to any hearer, and it owes its
obscurity to Mr. Browning. The words of Æschylus may be
construed as follows: 'Some one above, whether Apollo, Pan,
or Zeus, hearing the sharp-piercing bird-shriek of those who
are his guests, sends against the transgressors the sure but
tardy Erinus.' The construction is even more straightforward
than this English version, because a single Greek word expresses
what requires several in English. The involution of this broken
sentence is Browning, not Æschylus, and is a defect, not a
beauty. The Greek has no double construction answering to
the 'which hearing that wail;' the Greek says plainly and
simply that some one sends Erinus. The long phrase, 'what,
late-sent, shall not fail of punishing,' is an obscure way of
rendering a single adjective, and is as remote from the Greek
construction as Johnsonian magniloquence would be. 'Who
these assail' represents a single noun, so that in fine the whole
passage sins against the simplicity of the Greek as much as
Potter's old version, although in quite a different way. Those
who know the Greek will recognise a close adherence to the

original, and will readily admit that 'the guests who fare
housemates with gods in air' renders the true meaning of τῶνδε
μετοίκων.

We much wish that Mr. Browning had set a different aim
before him, for in truth we do not give the best idea of a foreign
author by using in English the very turn of each foreign phrase.
On the contrary, this practice is a cheap and common method
of raising a laugh. Thackeray began it or practised it with
French, making his Frenchmen speak a literal translation of
French phrases. The thing was droll, and is now copied in every
comic publication, but we should not get a good translation of
Racine by following this method. His graceful lines would
become grotesque, and the matter would not be much mended
if for each noun or adjective we substituted a periphrasis giving
the force of the word as indicated by its etymology.

Perfect, or even nearly perfect, translation is of course
impossible, but good work has been done from time to time
when a poet has felt the beauty of some foreign poem strongly,
and has written in his own language another poem giving the
beauty which he saw. It almost seems as if Mr. Browning did
not very much admire the 'Agamemnon.' Now and then his
version suggests the almost incredible suspicion that he wished
to show his friends how inferior Æschylus was as a writer to
Euripides. Surely he must have had a sense of fun when he
made the chorus (of reputed sonority and magniloquence) speak
as follows:

> For there's no bulwark in man's wealth to him
> Who, through a surfeit, kicks—into the dim
> And disappearing—Right's great altar.

Could a more ludicrous image be presented to us than that of
a man who, in consequence of overeating, kicks a great altar
into the dim? Hermann, who is followed by many scholars,
connects εἰς ἀφάνειαν with ἔπαλξις, and so obtains the rational
meaning that wealth affords no bulwark behind which the guilty
man can hide.

Our author has given us no setting to the play such as the
adventure of Balaustion, a story which enabled him by the
comments of the lyric girl to show us what he himself saw in

the 'Alcestis.' Mr. Browning acted that play for us, creating, as the French would say, the part of Heracles. And the creation has been very successful. Some may think that Euripides never intended his Heracles to be acted in that fashion, but no one will deny that Mr. Browning's Heracles is a fine conception. Mr. Browning often writes as if he were acting. He generally receives the title of dramatic from the public, but he is not a dramatic author in the old-fashioned sense. The words he puts into the mouths of his characters are not such as actors would like to use, but he is dramatic in the sense that he seems himself to act each part in succession, so that every character appears as if acted by Browning, and he has a large range of characters which he can act well. Of course we all recognise the actor and his mannerisms in every dress, but so it must be with all actors. Now, Mr. Browning has not acted Agamemnon, nor Clytemnestra, nor Cassandra, more's the pity, and therefore we say with some fear and trembling perhaps he did not see how these parts should be acted. If he had ever thought of acting Cassandra himself, he could never have made her say, speaking of Apollo,

> He was athlete to me—huge grace breathing ;

nor would he have liked in the part of Clytemnestra to announce the capture of Troy in these words:

> I think a noise—no mixture—reigns i' the city.

'No mixture' is a simple adjective in the original, and Miss Swanwick translates the two words βοὴν ἄμικτον by 'ill-blending clamour.'

The following is a sample of Mr. Browning's work where he seems to have been more in sympathy with Æschylus:

> For Ares, gold-exchanger for the dead,
> And balance-holder in the fight o' the spear,
> Due-weight from Ilion sends—
> What moves the tear on tear—
> A charred scrap to the friends :
> Filling with well-packed ashes every urn,
> For man that was the sole return.

> And they groan—praising much, the while,
> Now this man as experienced in the strife,
> Now that, fallen nobly on a slaughtered pile
> Because of—not his own—another's wife.

The absence of tawdry additions to the original is certainly a great comfort, going far to counterbalance the oddity of some expressions.   One cannot help regretting, however, that a natural reaction against smooth commonplace should lately in all branches of art have led to affected harshness.   After all, to be quaint is a small merit, while to be queer is, sooner or later, to be damned.   We will not quarrel with the new style of spelling.   There is really no right or wrong in the matter; the effect of a word as seen or pronounced is a matter of association; to those for whom more and nobler associations gather round Klutaimnestra and Kikero than round Clytemnestra and Cicero, the modern antiques are best.   Heracles shall at once displace Hercules, since even now he is the stronger.   When we turn to other translators, we see well enough why Mr. Browning was tempted to sacrifice everything to fidelity.   Mr. Morshead, the latest adventurer, gives us a flowing version made with care, but, not being a poet, he worries his reader by a frequent use of stock expressions, such as 'rapine fell' and 'presage fair,' often reducing Æschylus to the level of Scott's lays.   We think he has been most successful in rendering the difficult scene with Cassandra, and he shows everywhere a keen and just appreciation of the beauty of the original work.   Mr. Fitzgerald has more poetical *verve* than Mr. Morshead, and here and there rises to a high level, but he misses out all that does not come home to him, and it is not a little amusing to find the translator or writer of the 'Rubaiyat' omitting all the simple straightforward religion preached by the chorus.   A plain man reading the 'Agamemnon' of Æschylus would come to the conclusion that the old men of the chorus believed in a supreme God, Zeus, by no means unlike Jehovah, that they regarded moral conduct as pleasing to God and the punishment of crime as inflicted by God soon or late.   The reader of Mr. Fitzgerald's chorus would imagine that Æschylus had forestalled the nineteenth century in mild pessimistic mooning, and that Menelaus was a Scandinavian sentimentalist such as Mr. Morris loves to paint. We shall not on this account quarrel with Mr. Fitzgerald, who frankly warns us in his preface that he has poured away some of the wine of Æschylus and mixed some water with what is left. On the contrary, we are grateful to him for some well-turned phrases, and for the part of Clytemnestra, which is well translated. We do quarrel with him because his work is unequal and slovenly even in respect of grammar.

An amusing comparison of the various translations may be made by the help of two words. There is a refrain in the hymn concerning Iphigenia,

αἴλινον αἴλινον εἰπὲ, τὸ δ᾿ εὖ νικάτω,

of which the first three words may be baldly translated as 'say alas, alas.'

Potter paraphrases the whole thus:

> Sound high the strain, the according notes prolong,
> Till conquest listens to the raptured song.

The proportion of Potter to Æschylus here is really overwhelming.

Milman:

> Ring out the dolorous hymn, yet triumph still the good.

We see that the translator felt bound to elevate the style of Æschylus.

Conington:

> Sing Sorrow! sing Sorrow! but triumph the good—

conscientious, but not poetical.

Swanwick:

> Chant the dirge, uplift the wail, but may the right prevail—

a fair paraphrase.

Morshead:

> Ah, woe and well-a-day! but be the issue fair.

A scrap of old ballad is here made to do new duty.

Fitzgerald: Leaves out the refrain.

Browning:

> Ah, Linos say—Ah, Linos, song of wail,
> But may the good prevail.

The reader who does not know the passage would in this last
version be puzzled by what seems to be an invocation to some
one called Linos, but a peep at the Greek will show him that
' Ah, Linos ' is merely an exclamation, while Liddell and Scott
will prove that the received etymology of αἴλινον is correctly
indicated by Mr. Browning. Who Linos was does not much
matter, but the first syllable of the name was short. Once all
these facts have been mastered, we may perhaps think that
Mr. Browning has made the best transcript ; it is certain that
any one knowing the Greek will, after trying other versions,
come back to this one with a sense of relief.

By the way, no translator of the above refrain has adopted
the rendering taught by the late James Riddell, that the old
men wished the note of rejoicing in the song to prevail over
the note of woe, not that good generally should prevail over
evil. Perhaps the words really have the double meaning, and
to get their full force we ought to imagine dispirited trebles
piping their wail in a minor key, followed by a burst of san-
guine baritones with a grand swell in the major, closing on the
triumphant νικάτω.

A very slight acquaintance with ancient mythology and
ancient customs would be required to enable a spectator to
enjoy a great part of the ' Agamemnon ' and many other Greek
plays if he saw them acted, but unfortunately a drama when
simply read, not seen, makes such large demands on the imagination and intelligence of the reader, that great plays even in our
own language remain unread and unknown. For this very reason we should have valued highly comments such as those of
Balaustion on the demeanour and thoughts of the personages
in the ' Agamemnon,' which, however, is hardly a play in the
sense in which we use the word now. It is not a realistic
representation of a series of incidents. By far the greater part
consists either of poems recited by the chorus—who are not,
properly speaking, actors at all, and were not on the stage—or
of long speeches addressed by a single personage from the stage
to the crowd below who formed the chorus. Even when two
people happen to be on the stage, dialogue is almost wholly
avoided, and it may well be that part of the popularity of the
play is owing to this undramatic form which makes it almost a

poem, although it is also true that this poem contains dramatic
scenes of extraordinary power.

The 'Agamemnon' may be analysed into three parts, each
of which is extremely beautiful, even if considered separately
from the rest. We have a complete poem recited by the chorus
describing the sacrifice of Iphigenia, and ending with her death ;
another complete poem describing the flight of Helen, and ending with her reception in Troy, seemingly a blessing, really a
curse ; and lastly we have a drama showing the arrival and
murder of Agamemnon. The three parts are all harmonious,
conspiring to produce one general effect, but the mere written
copy without stage directions often leaves us in doubt as to how
these parts were blended together. It is highly probable that
by change of attitude, position, and demeanour, the chorus
marked in the clearest possible way the separation between
those periods during which they were the singers of a sacred
hymn, and those in which they represented, with some approach
to realism, personages taking part in the action of the drama.
When these stage directions are wholly omitted, as in Mr.
Browning's transcript, the reader will often be startled by a
sudden descent from passages of great lyrical grandeur to others
spoken by the same men, but so worded as to indicate plainly
that the speakers were commonplace people, incapable of inventing the words they had previously delivered. The sacred hymns
are appropriate to the old men who sing them, being such as
they would love ; but the old people who prattle about tottering
along on three legs could not in their proper characters have
used the language describing the sacrifice of Iphigenia or the
flight of Helen. It would be a mistake to suppose that the
portion headed KOMMATIKA included all those parts which
are spoken or sung by the choreutai in character.

The poem of Iphigenia begins with the entrance of the
chorus after a prologue has been spoken by the Warder, and
ends before the first dramatic interlude. The course of this
poem is interrupted by a passage in which the old men describe
themselves, and ask Clytemnestra the meaning of the sacrificial
fires ; the Queen does not seem to be on the stage, and they
wait for no answer, but abruptly resume the story of Iphigenia.
When the play is simply read, no reason appears why the poem

should be interrupted by a passage which is almost humorous ; but if we think of the necessary action accompanying the words we see that the first verses were chanted as the chorus marched in; that the passage spoken in character was probably delivered with a complete change of demeanour while the crowd was taking up its place round the central altar; and that when their final solemn station had been reached on or round this thymele, it would be quite appropriate that they should resume the solemn hymn with dignity. It is strange, however, that they should address Clytemnestra and get no response. Another break in the poem is formed by a passage expressing a kind of protest against the conduct of Artemis, to the effect that, if men would simply trust to the supreme god and his divine laws, all would be well. This protest might be so acted on the stage as to be no interruption. A change of melody, even a change of attitude on the part of the chorus, would suffice. The hymn closes by a strophe in which the chorus give their own reflections on the story, which are rather commonplace. A similar return from the past to the present, from the elevated hymn to the customary reflections of respectable people, will be found on examination to precede each dramatic interlude in the play. The name dramatic interlude is given advisedly ; for while the plays of Sophocles are dramas with choral interludes, the ' Agamemnon ' is a lyrical performance with dramatic interludes.

When the poem was ended and the old men had moralised thereon, Clytemnestra entered, and standing on a raised platform, our modern stage, was addressed by the chorus, grouped below in front, where the pit of a modern theatre is. She answers their questions as a modern orator answers questions from the hustings, tells them Troy is taken, and describes how the signal came from Ida by beacons successively fired ; in another speech she draws a picture of the sack of Troy. These speeches serve the dramatic purpose of introducing Clytemnestra to us with great splendour.

The first interlude is now over, Clytemnestra retires, and the chorus, after a few words of farewell, ' prepare to address the gods rightly.' They cease to be mere commonplace elders of the city, and with an invocation to Zeus begin the second solemn hymn, which describes the flight of Helen. This poem is interrupted by the scene in which a Herald brings confirmation of the news that Troy had fallen. In preparation for this scene the old men at line 440 revert from the past to the present, from the flight of Helen to their own position and feelings, which are described in two strophes.

Mr. Browning has followed the usual arrangement in giving Clytemnestra the speech announcing the Herald's approach, but it would be more consistent with the general scheme if, as Scaliger thought, this speech were allotted to the chorus; we should then, as usual, have some lines spoken by them after the conclusion of their ode, and telling the spectators the name of the coming actor. This is no work for a queen, and the Herald, when he comes, addresses himself exclusively to the chorus, which he could hardly have done if Clytemnestra had been present. He greets his country and his countrymen, expresses his own joy, tells the great news, and describes the sufferings of the army before Troy. Clytemnestra certainly hears some part of the Herald's speech, for when he ends she speaks, boasting of the accuracy of the news she had long before announced. She then despatches the Herald to Agamemnon with a message, and leaves the stage. There is nothing realistic in the dialogue; her short appearance seems designed simply to give the Herald breathing space. He proceeds to describe the tempest which separated Menelaus from Agamemnon while they were returning to Greece, and then this interlude ends abruptly. He takes no farewell, nor do the chorus bid him godspeed. Without a line of preparation they return to the story of Helen, taking it up at line 662, exactly where they left off at line 440. The description of Helen's flight and her reception in Troy closes at line 724, with the declaration that this lovely bride was really an avenging fury sent by Jove. The chorus now become mere old men full of ancient saws; as usual the concluding strophes are full of moralising, and, when the poem ends, the chorus as usual announce the approach of the personage who is to open the next dramatic scene. Agamemnon arrives, and the main action of the play begins.

We are probably justified in regarding this earlier part of the 'Agamemnon' as an example of the old-fashioned Thespian drama, which never attempted to represent any other incident than such as could be indicated by the coming or going of one actor. Unity of time and place has nothing to do with such a performance as this. Of course, the old men did not stand singing from the time Troy was taken until Agamemnon arrived. Whether Clytemnestra was or was not present when the Herald told his news would not matter. The form of the art was familiar to the spectators, and they knew what they were to take for granted as well as the audience listening to the Italian 'Maggi' know this to-day. The chorus took care to tell them what personage the actor was going to represent as he came upon the stage, and this was quite sufficient. They gave the actor a little rest from time to time by addressing him, and he gave them a long rest as he recited his speeches. When he went off, the chorus reverted to the main business of the day, the lyric song, and might take it up exactly at the place where they left off.

Extraordinary art is shown throughout the ' Agamemnon ' in so arranging the incidents that each actor may speak at the full pitch of his voice with truth to nature. The presence of the chorus forming a crowd who might be addressed collectively enabled this to be done, and this use of the chorus had an important influence on the earlier forms of Greek tragedy.

When Agamemnon entered, the performance became much more like our own stage play, but even then Æschylus seems to have avoided dialogue between two actors *on* the stage, feeling, perhaps, that the shouting necessary to make the spectators hear would seem unnatural in a mere conversation. Agamemnon arrived in a chariot, with Cassandra beside him or following him in another car, and was probably accompanied by a retinue of soldiers. He did not dismount at once, but remained standing in his chariot in the orchestra before the stage, as if at the front door of his palace. Clytemnestra and he did not converse. He harangued his people from the chariot, and then she harangued them also; some lines in her speech are addressed to Agamemnon, but in the main her speech, though in honour of the king, was directed to the chorus. She was surrounded by mute attendants, who, at her command, laid down splendid garments for Agamemnon to tread on as he left the chariot to ascend the stage by the central steps. He did not quite like being kept standing so long, and compared the length of Clytemnestra's speech to the length of his own absence. He was averse to exhibit pride by treading on these purple trappings, but was over-persuaded, taking his boots off first, so as not to spoil the robes. No sooner had Agamemnon reached the stage than he was silent, and apparently passed with Clytemnestra into the palace. The chorus, knowing the queen's real character, were awe-struck, not triumphant. Cassandra meanwhile stood motionless in her chariot. Agamemnon's attendants, and the slaves, who laid the trappings down, went off while the chorus were singing their prophetic fears, so as to prepare the audience for what was to come. When the stage and orchestra were ready, the queen returned to tell Cassandra she must enter the palace. Observe that she spoke to her at a distance. Cassandra was mute, and Clytemnestra left her with some show of contempt. The chorus expressed pity for the captive, who probably left the chariot and ascended the stage while she cried on Apollo to tell her to what roof she was come. Here begins a scene worthy of the greatest actress the world ever saw. In a dialogue with the chorus Cassandra gives her own story, the fall of Troy, the legend of Thyestes, and a description of the murder of Agamemnon, instantly about to happen. Sometimes she speaks in her own mind, and sometimes possessed by the prophetic frenzy. We have no more long set speeches suitable simply for declamation, but an unparalleled dramatic scene. The words of the prophetess of truth, cursed in the fated unbelief of all who hear her, would rouse the spectators almost to the frenzy of the seer herself. What? Agamemnon is in there. The woman, beautiful, miserable, wise, warns you—screams to you—tells you the whole story, which will even now be irrevocable fact; and you, crowd of old wiseacres, stand, and listen, and admire, and ask questions, and all the while the fatal net is closing round the king, and Clytemnestra, unseen within, grasps the fatal axe. Would that Mr. Browning had given us one phrase that Cassandra might use! If ever inspired poet wrote, Æschylus was inspired when he wrote this scene. The old Thespian mummery was gone—gone for ever. Warm flesh and blood had spoken on the stage—spoken with beauty and with power—and in the whirl of emotion which the audience felt they hailed the birth of a new art.

Mr. Browning lets Cassandra quit the scene using these words :

> But I will go—even in the household wailing
> My fate and Agamemnon's. Life suffice me !
> Ah, strangers !
> I cry not 'ah '—as bird at bush—through terror
> Idly ! To me, the dead, bear witness this much :
> When, for me—woman, there shall die a woman,
> And, for a man ill-wived, a man shall perish !
> This hospitality I ask as dying.

CHOROS.

> O sufferer, thee— thy foretold fate I pity.

KASSANDRA.

> Yet once for all, to speak a speech, I fain am:
> No dirge, mine for myself ! The sun I pray to,
> Fronting his last light !—to my own avengers—
> That from my hateful slayers they exact too
> Pay for the dead slave— easy-managed hand's work :

A poet should have felt what Cassandra felt, and put such words into her mouth as would have enabled an actress to show those feelings and carry her audience with her. This is what Æschylus did. Mr. Browning has preferred to give us a mosaic copy where every beautiful tint in the original is represented by half a dozen coarse broken bits ill patched together. He is faithful to word arrangement and etymology, false to art and feeling. He is clearly right in giving the next four lines to the chorus, who maunder about the frail state of mortals in a style which Cassandra at this supreme moment could never adopt.

Mr. Symonds, whose description of the ' Agamemnon ' is very beautiful, mentions as especially dramatic the moment of suspense which follows, filled by this moralising of the old men and ended by the cry of the king as he is murdered. There is no doubt that the audience has been by this time roused into intense excitement, and that the cry adds to the feeling, but the climax can hardly come when the stage is empty. Mere incident will not affect an audience to the utmost; for this the presence of a great actor is required. After the cry the hearers would be in a state of horror and suspense, very well represented by the trembling, undecided crowd of old men. The real climax comes when, doors thrown wide, Clytemnestra advances, defiant, axe in hand, and glories in her deed.

Here again is a situation which might make a Rachel or a Siddons rise from the dead if they could have a stage to act it on :

> πολλῶν πάροιθεν καιρίως εἰρημένων,
> τἀναντί' εἰπεῖν οὐκ ἐπαισχυνθήσομαι.

Every word strong, every word capable of receiving a true and forcible intonation. Browning puts it:

> Much having been before to purpose spoken,
> The opposite to say I shall not shamed be.

From this the reader can gather that Æschylus put into Clytemnestra's lips words by which she defiantly defends herself against the charge of cowardly deceit. She anticipates accusation, and answers not the charge of murder, but the charge of lying. What the reader cannot gather is the forcible and splendid burst of language in which this is expressed. In English the suspended sense of the first clause is a fault which in a schoolboy's exercise would be corrected by his master. 'Having been spoken,' with its weak little words all scattered, is no representative of the Greek εἰρημένων. 'To purpose' is obscure and affected where καιρίως is clear. 'I shall not shamed be' is neither English nor Greek. The Greek future is a single word corresponding in no way to the awkward inversion and unusual use of ' shamed.' No actress could produce her effect with such a speech, and yet Mr. Browning is called a dramatic poet. Assuredly he could have done much better had he not deliberately chosen to do the wrong thing.

Clytemnestra and the chorus wrangle. Ægisthus comes on and bullies, and finally the play subsides without solution. The old men are cowed, and Clytemnestra in a grand way forgives them. The play is a prologue to the great act of the Oresteia related in the 'Choephoroi.'

In the scenes with Cassandra before the murder, and with Clytemnestra afterwards, the poet was swept away by his dramatic feelings, and in writing these scenes he invented the real Greek drama, not by plan aforethought, but by the inspiration of his subject. In form he adheres to an address from one actor to the chorus, but the spirit is changed. The arrival of Agamemnon, the prophecy of Cassandra, the murder of the king, and the boast of Clytemnestra form a real dramatic representation of a fact happening then and there. The chorus changed its character, and the words assigned to it might have been spoken by a few persons on the stage. They became actors, whereas before they had been alternately singers of a sacred hymn and listeners to set speeches.

The proposition that Æschylus invented a new art while writing the ' Agamemnon' is not a mere figure of speech. The ' Choephoroi' which follows is a complete drama from beginning to end. The chorus takes part in the action throughout, and, when the stage was empty, recited only such short poems as might serve to divide acts. In its arrangement the ' Choephoroi' might have been planned by Sophocles. As usual when we pass from one artistic form to that next evolved, something was gained, something lost. As a dramatic entertainment far more was gained than lost; and if even now the ' Agamemnon' and ' Choephoroi' were successively *acted*, the spectators would, we venture to say, prefer the later play. The long hymns of the ' Agamemnon,' so beautiful to read, would be a trifle dull recited by bands of performers. The declamation of the single actor about the taking of Troy or the shipwreck of Menelaus, magnificent poetry as it is, would be somewhat like a reading of Milton : we should admire but remain cold. The play would not begin until Agamemnon arrived, and it would be over by the time Clytemnestra had finished her great speech after Agamemnon's death. In the ' Choephoroi,' on the contrary, the interest is dramatic from first to last. The return of Orestes, the present woe of Electra, the recognition of the brother and sister, the invocation of Agamemnon, whose hidden shade listens to son and daughter, the meeting of Clytemnestra and her son, the death of Ægisthus, the pleading for life or death between mother and son, with the final frenzy of Orestes, form one unbroken chain of dramatic scenes of the most perfect kind, ending in a climax far finer than that of the 'Agamemnon.' Yet the translations of the 'Agamemnon' outnumber those of the 'Choephoroi' perhaps by ten to one, precisely because the 'Agamemnon' is as much a poem as a drama, while the 'Choephoroi' is above all things a play.

One of Sophocles' great plays, 'The Trachinian Virgins,' has met with injustice owing to the same cause. F. W. Schlegel treats this tragedy, recounting the death of Deianira and Heracles, with positive contempt, and the general impression seems to be that as a work of art it is inferior to the other plays of Sophocles. A translation by Professor Lewis Campbell is obviously intended to enable the English reader to compare the three great heroines of Sophocles, and it may be said that no previous translator has shown so keen a sympathy with the womanly qualities of Antigone, Electra, and Deianira. It may be granted that many readers will find the story of Deianira and Heracles somewhat dull, whereas the histories of Antigone and Electra cannot fail to move all who are not insensible. If, however, the three tragedies were put upon the stage, the verdict as to their relative merits might not improbably be reversed; it would certainly be much modified. The noble figure of Antigone would, as was proved by Helen Faucit, command our deepest reverence and admiration; the devotion of Electra to her father, and, above all, her love for Orestes, would perhaps touch the spectator even more than the heroism of Antigone; but seen on the stage the gentle and noble Deianira, wrecked by her very love and simplicity, would sway our hearts with sweeter and surer touch than either the stern devotee or the vengeful daughter plotting her mother's murder. Antigone is almost above the earth, and the object for which she sacrifices her life and her love is one which to us nowadays savours of superstition. Electra, if she loved her father and brother dearly, yet hated her mother with a rancour which seen on the stage would repel—not only would her entrance, squalid and full of hate as of grief, be unprepossessing, but her final appearance would be most horrible as she listened to the murder of her mother, and urged the striking of another blow. Yet almost every educated English man or woman has

some acquaintance with Antigone and Electra, while very few
know anything of Deianira or the death of Heracles as Sophocles
conceived it.   For this very reason the attempt shall be made to
give, with Mr. Campbell's help, such comment on this play as
may in some feeble way supply the elements which actual representation affords.

The story of Iphitus is the background from which the
action of the play stands out.   Heracles, upon a trifling quarrel,
killed both Iphitus and his father Eurytus, king of Œchalia,
sacked the unoffending town, and sent the dead chief's daughter,
Iole, home to supplant his own true wife.   What wonder if the
ancient oracles, promising him rest at this very hour, mean not
happy rest, but swift death ?   Deianira, in mere simplicity of
guileless love, sends him a poisoned robe, and, learning the
result of her action, takes her own life before the hero is brought
in agony by his son Hyllus to their home in Trachis.   Heracles
and Deianira do not meet.   Dialogue between them would have
distracted our mind from greater tragic issues.   If they met
after the poison began to work, we should for a while be chiefly
interested in watching whether the woman would or would not
be able to exculpate herself, whether the man would or would
not believe her.   This situation belongs to melodrama, being
pathetic, not tragic.   Whether this woman does or does not win
credence is a mere accident, and after all does not much concern
the world at large.   Pain, failure, hate where love should be,
our crimes, our follies, and their bitter fruits, the irony of the
gods—these are the tragic points which pierce men's souls for
ever, and from these Sophocles allows no distraction.   Our
author here, as in the 'Philoctetes,' accepts physical pain as
thoroughly tragic, and, we think, rightly.   Surely those who
call pain a small thing can never have felt it, and it lies with
the actor to prevent the ugliness of all suffering from alienating
our sympathy.   When we meet with the written Greek cries
arranged so as not to break the metre, we must remember that
in Greek every sob is indicated.   In our freer form of art the
actor puts in exclamations where he feels they are required.   If
all Mr. Irving's cries were found in Shakespeare's text, unthink-
ing readers might call Hamlet chicken-hearted.   Heracles
would have met simple death with defiance ; but in the fangs

of pain, in the grip of disease, even the mightiest are crushed. Pain then we have, not continuous, for this would exclude all other emotions, but, with truth to nature, in mighty spasms, with intervals of rest.

When the tragedy of Deianira—to which we will presently return—has wrought the audience to such a point that they are prepared for any scene, however terrible, Heracles is brought upon the stage in fevered sleep. Hyllus, who now knows that his mother had wrought innocently, stands near his father, and by heedless speech wakes him to conscious pain.

Heracles' first impulse is to complain of the ingratitude and injustice of the gods. For a moment the hope that even yet Zeus may send a healer deludes him, only to cause a bitterer agony. The attendants would fain help him, as in the eagerness of prayer and hope he strives to raise himself; but their touch brings on a paroxysm. Then comes the thought of failure, of how much he had done for men; and this was the end of all :

> Where are ye, men, whom over Hellas wide
> This arm hath freed, and o'er the ocean tide,
> And through rough brakes, from every monstrous thing ?
> But now in my misfortune none will bring
> A sword to aid, a fire to quell this fire.

The attendants flinch in terror, and the son takes their place, when, in an agony of still more terrible pain, the hero calls upon his son to kill him, and then thinks upon the cause of all —the mother. Observe with what perfect nature Deianira is brought to his mind. Looking on the son he thinks of the mother, and to think and to curse are one. The boy is taken too much unawares to be ready with exculpation, and Heracles continues wildly :

> Many hot toils and hard beyond report
> With hands and struggling shoulders have I borne,
> But no such labour has the Thunderer's wife
> Or sour Eurystheus ever given, as this,
> Which Œneus' daughter of the treacherous eye
> Hath fastened on my back, this amply woven
> Net of the Furies that is breaking me.

He describes his agony in words too terrible to be quoted on cold paper :

> Yet me nor Lapiths, nor Earth's giant brood,
> Nor Centaur's monstrous violence could subdue,
> Nor Hellas, nor the stranger, nor all lands
> Where I have gone, cleansing the world from harms ;
> But a soft woman, without manhood's strain,
> Alone and weaponless hath conquered me.

And so it still is. Our mighty ones fall before a little grain, a poison-germ. Our hearts are broken when our wives betray us. All this is true now in England as then in Attica. But no more curses ; bring out the woman, says the hero, and I will slay her righteously. The son shrinks at the mention of his mother, dead even now, but Heracles supposes that he shrinks through simple pity, and claims that pity for himself, not her. Then suddenly he shudders at the thought that he, Heracles, has asked for pity ; but he will justify even this. He bares his breast.

> O see !
> Ye people, gaze on this poor quivering flesh,
> Look with compassion on my misery.

Another spasm follows, when, looking on his bare body, he thus addresses it :

> O breast and back,
> O hands and arms of mine, ye are the same
> That crushed the dweller of the Neméan wild.

And then a little comforted, even in death, to think of all that he has done, his mind runs over those great triumphs.

> But now
> Jointless and riven to tatters, I am wrecked
> Thus utterly by imperceptible woe ;

one thing only is left. Bring Deianira hither. Then Hyllus tells her fatal error and her death. The hero's mind is dull with pain and sickness, but at last he understands. No word of pity comes from him for Deianira. It is easy to explain this by saying that hero and author were mere pagans, but if

Sophocles had thought the sentiment artistically right we should have found it here. The revulsion of feeling in favour of Deianira would give rise to a sort of *attendrissement* wholly out of keeping with the situation. To make this softening effective, much love must have been shown by Heracles to Deianira previously, whereas the hero fell by his ungoverned passions, and we have no hint that he was romantically attached to his wife. Hyllus, in giving his explanation, names Nessus as author of the charm; then a great awe falls on Heracles, he remembers the oracle and accepts his fate. Now we see the hero once more noble and strong, resolved to face death with dignity. On the summit of Mount Œta, on a funeral pyre, he will depart, while the flames which take his life quench the worst agony of the poison. Hyllus is made to promise that this shall be so, and the pangs abate; the moment pain no longer heightens the situation the artist lets the fire die out. And here follows an incident which a little relieves our hearts. The dying hero cannot forget the maiden Iole, the love for whom has been his ruin; she at least shall be well cared for, and shall raise him seed. He commands Hyllus to marry her. The poor boy says :

> How can I do it, when my mother's death
> And thy sad state sprang solely from this girl ?

Mark the art with which Sophocles lets the audience see what really killed Heracles. Not Nessus, not Deianira, not Zeus, but Iole! Hyllus saw it quite well; but his great father will be obeyed. Heracles' last hope is that he may reach the funeral pyre undaunted—

> Seeming to do gladly still.

Hyllus, as Heracles is borne off, rebels against heaven, as any son would do in like case. As a representation of the extremity of a hero's suffering, this scene stands pre-eminent among all tragedies. Let Salvini act the hero, and its power would instantly be recognised; not only the power of the actor, but the fact that it gave him greater scope than any other part. There is no such sequence of long-drawn agony, which is yet the agony of a demigod, in the 'Agamemnon,' the 'Choephoroi,' or the 'Eumenides.' There is nothing to compare with it in the 'Antigone' or 'Electra.' Even the suffering of the blind Œdipus falls short of it. He lived years after the climax of his tragedy had come and gone. In the 'Prometheus Bound' the hero is too much of a god. His sufferings are too unlike our own to touch us much. The misery of Philoctetes is less; his pain was not fatal, and his grief was chiefly due to his abandonment by the army. The grief of Ajax is great, and with good cause; madness is a heavy curse, yet his suffering under a sense of disgrace is a small thing compared with this prostration of the mighty Heracles. Some may think the situation in 'Œdipus King' equally tragic, but many of the incidents which fill the mind of the son of Laius with horror leave our minds untouched, whereas we feel every suffering of Heracles to be pain *now*. Modern tragedies are purposely omitted from comparison, not because they are too strong.

It is clear that before Heracles can be brought on the stage, dying, the audience must be strongly moved. The play could not begin with this scene. Sophocles does not as a preparation show us Heracles putting on the poisoned vest; this would look like a juggling trick. Heracles' actions immediately before meeting his fate are not such as to awaken sympathy or any tragic emotion, so that our author is well advised in not presenting his hero until he is an object of compassion. Deianira's death is made to serve as a fit prelude, and, with the scenes leading to it, constitutes a great tragedy, ending where the other begins. As Heracles' tragedy is that of the great men of the world, a typical tragedy, so Deianira's fate shows one of the typical tragedies of women—one which, when seen, will move all women till humanity ends.

The hero's wife is a true woman and true wife, commanding all our sympathy from first to last. She perishes through her simplicity and love; yet her simplicity is queenly, and her love nobly placed. Some might say she fell through jealousy; but this feeling as shown by Sophocles is so refined, so free from all anger, so just, that her jealousy is almost sweeter than the love of other women. Heracles sends Iole, a young and beautiful captive, to his home, proposing to make her

a second and more loved wife, displacing Deianira, the mother
of his children. It would not be noble in any woman to submit to this. No special misfortunes these of Deianira. The
great man's work gives him no leisure for home life. The woman
in middle age finds herself supplanted. Yet such is the charm
of sweet heroic simplicity in Deianira, that she actually wins
our sympathy for Heracles, since we derive our first impression
of this demigod from the woman he had loved, and who loved
him ever.

The play begins when Deianira is alone at Trachis musing
over her past life. Musing in the porch, white-robed, fair-
armed, she tells us how her hero freed her long years ago from
peril worse than death, and won her as his bride, how she has
lived much alone, and ever anxious, and how even now she
has most cause of all for fear; and the very prophecy which
lends a bitter irony to Heracles' fate strikes the key-note of
Deianira's present dread; otherwise unreasonable, he her lord
being so mighty. In this way we learn quite simply the spot
where the tragedy takes place, the names of the principal persons of the drama, the absence of Heracles, his wife's character, and the impending doom. A short scene follows, in which
Deianira, at the suggestion of an aged matron, sends Hyllus,
her son, to seek for Heracles. There is here a touch of the
skilful playwright. Hyllus and the nurse are required later,
reappearing at the very crisis of Deianira's fate. If they had
not been previously introduced, the audience would have spent
some time in wondering who these were, and thus their attention would have been distracted. Sophocles, however, makes
skilful use of this little piece of scaffolding to confirm our
opinion of Deianira's character by showing the mutual love of
son and mother and her gentleness to the old nurse. When
Hyllus leaves, the chorus enters, a band of pleasant girls,
very sorry for their noble guest, and anxious to comfort her.
How? First by pointing out that Heracles has never come
to any harm yet, then by observing that things cannot always
remain at the worst, and lastly by that exhortation to trust in
Providence so natural to the young girl. The queen listens
very kindly, and thinks of quite other things as they speak—
thinks how young they are:

I see you have been told of my distress,
And that hath brought you. But my inward woe
Be it ever as unknown to you as now !
So free the garden of unruffled ease
Where the young life grows safely ; no fierce heat,
No rain, no wind, disturbs it ; but unharmed
It rises amid airs of peaceful joy,
Till maiden turns to matron, and a day
Brings years of care for husband or for child.
Then, imaged through her own calamity,
Some one may guess the burden of my life.[1]

Then she tells her feelings, hopes, and fears, so that by the end
of her speech the audience knows the whole situation, and
knows, too, that Deianira can still call Heracles

The best husband in the world of men.

How much better is this straightforward method than the plan
of introducing an underplot with secondary characters, whose
chief business is to tell us where the action takes place, and
what the main characters are ! The audience always sees
through the shallow artifice. Sophocles' method, too, has the
great advantage of putting the explanation—*exposition* the
French call it—into the best actor's mouth.

To the gentle Deianira telling her fears arrives a breathless
messenger, garlanded, and laden with good news. Heracles is
living and returning. Another moment and Heracles' own
herald will be here. When Deianira comes to believe this true,
with glorious sweep of outstretched arms and smooth strength
of voice, she turns to heaven :

O Zeus that rulest Œta's virgin wold,
At last, though late, thou hast vouchsafed us joy.

And so she calls on her friends and her whole household to
rejoice, and a sweet rapture seizes the maidens, who break into
innocent song and dance of womanly triumph, while as they
sing enters the herald Lichas with the captives of Heracles'
spear. What more pathetic can be seen than the swift turn of
the eager wife, joy in every feature, on every limb, with just

---

[1] These lines will not be found in Mr. Campbell's published translation.

that trembling eagerness for certainty which yet shows no
doubt? Noble and queenly, she omits no greeting:

> Herald, I bid thee hail, although so late
> Appearing, if thou bringest health with thee.

And then, gathering courage from Lichas' confident bearing
and cheerful salutation, with eyes in which the new-born joy
is beaming, she asks:

> Kind friend, first tell me what I first would know,
> Shall I receive my Heracles alive?

The herald had not the courage all at once to destroy this
beautiful happiness. He was a Greek to whom a lie was a small
thing, and so he lied with good intent, not telling that which
was false, but suppressing the main truth; and thus the
measure of Deianira's cup is full—full of joy to overflowing.
Terrible irony this. The audience know that she is doomed,
and yet listen to her sweet rejoicing:

> Yea, now I learn this triumph of my lord,
> Joy reigns without a rival in my breast.

But the mild womanliness of her checks all pride. The very
excess of joy humbles her great nature.

> Yet wise consideration even of good
> Is flecked with fear of what reverse may come ;
> And I, dear friends, when I behold these maids,
> Am visited with sadness deep and strange.
> Poor helpless beings, in a foreign land
> Wandering forlorn in homeless orphanhood ;
> Once sure of gentle parentage and free,
> Now snared in strong captivity for life.
> O Zeus of battles, breaker of the war,
> Ne'er may I see thee turn against my seed
> So cruelly ; or if thou meanest so,
> Let me be spared that sorrow by my death.

How can we but love this gentle wife, too noble to rejoice
loudly in the presence of others' grief? Surely this picture
of the perfect matron should take place in the heart of mankind
beside the portrait of the perfect girl Nausicaa.

Sophocles does not allow Iole to speak. Had she spoken
we must have been led into the sorrows of a new group of people
in whom we have no interest. So Deianira with kind words
ushers the silent captive across the threshold, and as she follows
pauses at a word from that same garlanded messenger, his
garland tossed aside now. Cunningly, being a poor creature,
he had brought the good news, leaving Lichas to tell the
evil truth; but now that he has seen the sweet queen deceived
by another, even his heart goes out to her in pity, and he
blurts out all the facts. Iole is no slave, nor unregarded, but
sent by Heracles to reign at home, for all his heart is kindled
with desire. The full measure of her misery cannot break on
Deianira in an instant. Stunned, she sinks into her chair,
bewildered, turns even to the band of inexperienced girls for
counsel. What counsel can there be but to question Lichas,
who returns, hastening to quit the palace before the inevitable blow falls? Unable to rise, she sits and questions,
and he who reads the words must try to hear the hardened
voice as it comes from the tightened throat—must try to see the
rhythmic spasms of pain that shake the body as Deianira sits
striving to be as she had been, but will never be again. Lichas
can but lie and lie again, till the old messenger fiercely taxes
him with treachery to his queen. The two men wrangle, and
she listens, and as she listens learns that there is hardly room
for doubt; yet, knowing that Lichas lies, she still hopes against
hope, rises, comes to the unhappy man, and questions him herself, no longer striving to conceal her misery, but with terrible
appeal begging for the truth, even though the truth be a very
sword that slays. Solemnly, almost calmly, she first adjures the
man in the name of the highest god to speak the truth. She
is no weak woman, but one who knows the ways of man; she
too knows love. She will not blame her lord—no, nor the
woman. Only not a lie—a lie can be no kindness; and as
the herald, all unmanned, trembles in his grief and doubt, she
towers for a moment in fierce indignation and contempt:

> To one free born
> The name of liar is a hateful lot,
> And thou canst not be hid.

But quickly catching herself she bids him not to fear, for—Ah, me! the truth, she says, will not hurt her.

> For doubtfulness is pain,
> But to know all what hurts it?

She almost believes this as she says it. But the man is silent still. Then she thinks he fears for Iole, and says with voice of utter nobleness:

> Many a love
> Hath fettered him ere now, and none hath borne
> Reproach or evil word from me. She shall not,
> Though he be drownèd in affection's spell;
> Since most mine eye hath pitied her, because
> Her beauty was the ruin of her life.

The man falters, and she concludes:

> Well, this must pass, as Heaven hath willed, but thou,
> If false to others, still be true to me.

All her argument issues at last in the direct illogical appeal of nature. She throws herself upon his mercy, and he yields, telling her all the fatal truth, and giving the counsel, little needed, of submission; and Iole must take no harm:

> For he whose might is in all else supreme,
> Is solely overmastered by her love.

Deianira bends; she will enter on no bootless strife with Heaven. But the gentle voice sounds bitterly now, as, turning to the herald, she says:

> Come, go we in, that thou mayest bear from me,
> Such message as is meet; and also carry
> Gifts, such as are befitting to return
> For gifts new given. Thou ought'st not to depart
> Unladen, having brought so much with thee.

Here an act ends, for, with the exception of the first ode, the singing of the chorus in this play fills the interval between separate acts. The art with which the antecedents of the story are told is altogether admirable. In the course of the play we learn Deianira's present situation, her wooing, the nature of her married life, the recent acts of Heracles, the quarrel with Iphitus, his slaying, the slavery of Heracles under Omphale and the terrible revenge he took on Œchalia, and finally the story of Heracles' love for Iole, the love which is the cause of all his woe. Yet these facts are so cunningly interwoven that we learn them unconsciously, and seem merely to have been watching the development of the action which, now and here, brings Deianira to this misery. We may analyse and admire this art. What passes analysis is the gentleness and nobility of Deianira's character, the truth and pathos in her appeal to Lichas. We see the art with which the situation is chosen, but the art with which the situation is used is too like nature to be fathomed.

The chorus take up their song. What strikes these maidens is the terrible, wonderful power of love. They recur to the day when young Heracles saved the maiden Deianira from her monster suitor, the river god and bull Achelous—saved her and bore her off far from her mother's care. Now this which we have just seen is her lot. Terrible subject of contemplation truly for these maidens, the power of Aphrodite!

By Deianira's sad mention of gifts, the spectator has been prepared a little for what is next to happen. Perhaps the thought of what she would do had already occurred to her.

See! no longer with mere misery in her face, but with trembling, eager, excited step, she comes, and in her hand a casket. Her friends are there, and she is come in part to tell the craft her hand had mastered, and in part to crave their sympathy. The fatal casket is set down, while Deianira tells her thoughts. Probably Iole, even when welcomed, was no maiden any more, but married. It was ungenerous in her lord, and yet, often as he has sinned, she knows not how to harbour indignation against *him*.

> But who that is a woman could endure
> To dwell with *her*, both married to one man ?
> One bloom is still advancing, one doth fade ;
> The budding flower is plucked ; the full-blown head
> Is left to wither, while love passeth by
> On the other side.

With tight throat, swimming eyes, and sadly shaken head, the fatal truth is spoken—no remedy in her own youth and beauty

any longer, but possibly in this. When she went forth with
Heracles a new-made bride, the centaur Nessus had laid wanton
touch on her, and as she speaks the spirit of the scene comes
back. Once more with glorious reminiscence of the power in
that bow and archer, she, for the last time, triumphs in the
thought of her hero :

> And I cried out ; and he,
> Zeus' son, turned suddenly, and from his bow
> Sent a winged shaft that whizzed into his chest
> To the lungs.

Then Nessus, dying, told her that his blood was a charm, a
charm of soul for Heracles—that never through the eye he
should receive another love than hers. And she, guileless
woman, too fondly had treasured up this fatal gift, had learned
by heart most carefully each treacherous word of the dying
monster, and now she has applied the charm ; the robe is ready.
Some doubt appears in her friends' faces, and the pride of
Deianira's virtue is infinitely touching when, with some slight
trace of haughtiness, she says :

> No criminal attempts
> Could e'er be mine. Far be they from my thought,
> As I abhor the woman who conceives them.
> But if by any means through gentle spells . . .

(here all her face lightens, and the casket is in her hand again,
advanced almost with joy)

> And bonds on Heracles' affection, we
> May triumph o'er this maiden in his heart,
> My scheme is perfected.

Alas ! the eagerness, the faith of the queenly woman are all to
bring such utter woe. The girls are not swept away by the
torrent of hope ; for an instant their faces check her.

> Unless you deem
> My action wild. If so, I will desist.

The girls timidly say they think she has no proof that the charm
will succeed. The check makes even this gracious queen a little
angry. True woman, she says :

> My confidence is grounded on belief.

Again the maidens speak their doubt, but when Lichas enters
her one idea is to put the charm to proof instantly, and he receives the casket with her instructions, anxious, poor man,
in all his best to undo the sad work done.   Take her last speech
to him, and think of the mingled woe and hope that blend
in it.

> What more is there to tell ?   But rash I fear
> Were thy report of longing on my part,
> Till we can learn if we are longed for there.

Another act is over, and another lyrical interlude gives the
audience time to breathe before the end.   For end it is of the
woman's tragedy.   The young things sing of Heracles' triumph,
ending with a doubtful hymn of hope.   Then Deianira returns.

Oh, what a change !   No more beauty now—no more youth
—no more hope.   Fever-stricken, death-marked, yet queenly,
graceful, noble still, even as she totters, seeing but dimly.   A
little flock of wool, with which the fatal robe had been mois-
tened, when the sun fell upon it, shrivelled away out of sight
before Deianira's eyes, and a clot of blood came where it lay.
She has seen this, and now comes partly to tell and partly to
clear her swimming thoughts.   Conviction is really burnt in
upon her ; you see with the first glance that the woman's life
is at an end.   Yet in this last agony she writhes against conviction, though every thought, every fact, turns upon her and
convicts her.

> For wherefore should the Centaur, for what end,
> Show kindness to the cause for whom he died ?
> This cannot be.   But seeking to destroy
> His slayer, he cajoled me.   This I learn
> Too late by sad experience, for no good.
> And if I err not now, my hapless fate
> Is all alone to be his murderess.

Some of our readers may have heard Rachel tell the dream in
the fourth act of Ponsard's ' Lucrèce.'   These may know how
Deianira looks when telling of this simple bit of wool plucked
from the household flock.

The iron proofs brand Deianira as a murderess, and she has
but one word.   Her life must ' follow at a bound.'   Hearing

this, the awed women mutter something of hope, but the dying
queen scorns all hope now.   Then they urge that she has acted
all unwittingly.   If any think that they will comfort themselves
with good intentions when the fruits of folly come to harvest
let them hear how Deianira's voice sounds as she says :

> So speaks not he who hath a share of sin,
> But who is clear of all offence at home.

We touch the goal ; for the full tragedy nothing remains but to
let her know the fatal issue of her credulity.   Her son brings
the tidings—her son, who saw his father in agony, and who
believes his mother to be a murderess.   She has to listen to all
the terrible details : to hear how great the hero looked, during
his mighty festival of triumph, when Lichas came, to hear that
her lord knows her guilt and thinks her wholly guilty ; to hear
that the murder of the innocent Lichas, killed in her hero's
fury, lies at her door ; to hear that her husband cursed her, that
he is coming, and will be here anon ; and last to hear a solemn
curse from her son's own lips.   What can she answer ? Nothing.
How much more terrible than any speech is her silence as she
slowly passes through the door !   The poor shaken, terror-
stricken women make one feeble call to her—'justify yourself
—plead,' they say ; and she turns round and looks—what a
look !—and sees her son, and then—silence all, she sinks into
the night.

The wretched youth begins to feel what he has done, but
repents not yet.   Why should he repent ? 'Her acts are all
unmotherly.'   And so he goes, leaving the maidens to fall
back on Fate and Doom for some little comfort.   So it was to
be, and so it is, the feeble things say.   The oracles promised
rest, but it is the rest of the grave.   And now a cry is heard
within—not the cry of Deianira, but the wailing of her maids—
and on the trembling girls bursts out the aged nurse to tell
them that the queen is even now departed.   Tender to the last,
'taking in her touch each household thing she formerly had
used, she wept o'er all ;' then, having prayed and taken
leave, she cast upon her bridal bed the finest sheets, undid her
robe where the brooch lay before her heart, and pierced her side.
To this household comes the dying Heracles.

Mark with what skill our interest is conciliated for him before he comes. His misdeeds lie in the background, although the very cause of the tragedy. We hear of him as a young man winning Deianira, as her protector against the centaur and for many years her kind husband. The chorus sings of him as the mighty conqueror. In their youthful minds his exploits are all glorious. Even if in this matter of Œchalia and Iole he be to blame, the power of that unmatched deity, Cypris, excuses him both to the herald and to the bevy of girls. Now, when he comes he is in agony, and we feel that he is more sinned against than sinning. It is left to Heracles himself to make the claim that his labours had the conscious end of freeing Hellas from every monstrous thing, that living he 'gave punishment to wrong,' that he had 'cleansed the world from harms.'

Assuredly this is one of the greatest tragedies of the world. If this be not universally acknowledged, the cause lies in the fact that it is more a play and less a poem than such works as the 'Agamemnon' or the 'Œdipus at Colonus.' In reading a play we are all apt to miss the proper point of view. If we read the speech of Hyllus as intended to exhibit his suffering in consequence of his father's death, we shall perhaps think it frigid. If, as we read it, we think of Deianira listening, we shall see that no more terrible torture could be inflicted than this slow speech, missing no detail of fact. Sophocles cares little what we think of Hyllus, but through Hyllus' speech he wrings our soul for Deianira, the most lovable woman of Greece.

One object of this article has been to draw attention to the extraordinary merits of some Greek plays as dramas fit for representation on the stage, and to insist that while read they should be conceived as actions occurring before us. Surely the object of a translator should be the same as that of his author. Æschylus and Sophocles meant, above all, to write good plays fit to be heard and understood by multitudes. Style and philosophy, religious teaching and lyric art were all means to this one end. Their object was to depict by these means heroic men and women, so as to move a large audience. We think that no English translator of Æschylus has as yet given a version fit for this purpose, also that the task is a worthy one and not impossible.

ON THE ANTIQUE GREEK DRESS FOR WOMEN.[1]

SOME ladies about to take part in private theatricals, required antique Greek dresses, and the task of providing these, which seemed easy at first, grew in difficulty as the ambition arose among us to produce a dress which should not be a mere piece of theatrical millinery, but should represent, as nearly as possible, the actual dress worn two thousand years ago by women in Athens.

Smith's Dictionary gives little more than the names χιτών, διπλοΐς, and πέπλος, applied to the tunic or under-garment, the mantle covering the shoulders, and the loose shawl or scarf arranged as suited the taste of the wearer. 'Hope's Costumes' offers some information as to the mantle, which he calls a bib, but none as to the make of the tunic. The limited circle of scholars, artists, critics, and costumiers known to me had little accurate knowledge to impart; but every difficulty met with in getting accurate information turned the hunt for a real Greek tunic into a hobby. Experiments with sheets and lay-figures, at first, and, later, with shawls and real women, led at last to so satisfactory a result that the whole party thought it would be well worth while to write down what we had discovered with some labour. No doubt the facts must be known to some artists and sculptors, but they are not readily accessible, and

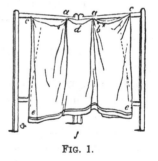

FIG. 1.

many artists are either ignorant of how Greek drapery is to be produced, or do not care to produce it. This was shown by more than one picture in the last exhibition of the Royal Academy.

No one who saw the ladies, as finally dressed, hesitated to affirm that the dress really was the dress they were familiar with in statues, and no one denied that it was eminently becoming. The sketches by which I shall illustrate my description must be judged leniently, as the work of a mere amateur, and they give little idea of the beauty of the real soft folds hanging on a graceful woman; they have only the merit of having been carefully drawn from the dress itself, as it actually hung in each case, so that each fold shown had its counterpart in reality: by doing them myself I could best illustrate the points which I wished to bring prominently forward. · The following is a description of the dress, as finally made and shown in the drawings. The tunic is simply a folded white Indian shawl (called a Chudder, in the language of the shopman), four yards long and two yards wide. This does not require to be cut anywhere; and, indeed, the result of the experiments was to convince us that the dress was invented before scissors came into use. The shawl is folded in half, lengthways, so as to form a square, the ends being sewn together, with the little fringe turned in.

The shawl thus doubled is shown in Fig. 1, hanging from a clothes-horse. Let us call the half of the shawl next us the front, and the other half next the clothes-horse the back; d is at the centre of the front. At two points, a a, equidistant from d, the shawl in the front is gathered together in a kind of bunch of folds and a large gold button sewn on; the distance between a and a must be somewhat less than the width of the shoulders of the person who is to wear the dress. Opposite a and a, but on the back, two loops are sewn, intended to slip over the buttons and hold the dress round the throat. The distance between the two loops must be considerably less than that between the buttons, otherwise the dress will not hang prettily, with that peculiar fulness always observed in statues; moreover, the loops should not be sewn to the edge of the stuff, but a little piece should be turned in, so as to give a double hold to the loop (capital engineers those old Greek tire-women); this gives an appearance to the dress at the back of being double. The dress is now complete, if no sleeves are wanted; but if sleeves are desirable—and in private life ladies seem to prefer them—sew two or more buttons of smaller size on the front at *b*, on each side of *a*; observe the same rules in sewing on these and the corresponding loops which have been given for the chief buttons at *a*—that is to say, turn in the edge and gather a little handful of stuff, on to which the button is to be secured. The dress is now finished, with the exception of such trimming or embroidery as may be put along the bottom—the only place where embroidery or any pattern (except stars and spots) was allowed.

The make of the dress is a trifle; how to put it on is the difficulty. Button the left-hand button at *a* and leave the other loose; slip the gown over the head, and, in the attitude so commonly represented, button the second button over the right shoulder;ʼ a petticoat and another garment or two may be allowed under the tunic, but no stays. If the sleeves are not buttoned, the garment will hang from the bust, as shown in Fig. 2; the gathered folds at *a* will spread as they fall and join in front, so as to make a series of curves which fall over the bosom somewhat like a series of chains, each longer than the preceding one. There

Fig. 2.

is ample fulness over the bust, but this fulness all comes from the two points at the shoulders. In nine out of ten statues this will be found to be the case; in nine out of ten modern pictures the fulness will be found produced in modern millinery fashion, either by the expedient of a tape running in a seam (true millinery terms are not known to the writer) or by a series of plaits sewn on to a stiff flat border—in both cases the folds run straight up and down, like the plaits in a night-gown, and are to my eye intolerably ungraceful. Examples could easily be given from modern pictures and photographs where the stuff is thin, the faces and figures pretty, but in which the busts are not in the least Greek, because the artist has not studied how the Greeks arranged the fulness of the dress. They had one plan, distinct from that described—of which more anon, when treating of the mantle, for which this second plan was more generally used than for the tunic.

Next fasten the buttons, $b\ b$, taking care to turn in the edge between c and a (Fig. 1). The dress will now hang as shown in Fig. 3, with a kind of pocket at c; this pocket is formed independently of any buttons at $b\ b$, and might have been shown in Fig. 2. The pocket, as it shall be called hereafter, plays an important part in the dress. It is easily recognised in several statues.

The Greek ladies wore both loose hanging sleeves and short close sleeves. If hanging sleeves are wanted, the zone alone is required to complete the dress. This zone should be a stiff metal band, fulfilling some of the functions of stays; the pocket should be lifted up a little and the zone put on underneath it, and as high up on the figure as possible. The clasping of the zone is a very delicate matter, because the dress is gathered together by it, and on the neatness of the folds at the zone depends the elegance of the dress. The folds are, in Greek examples, symmetrically arranged, of nearly equal size, turning outward towards the arms, as shown in Fig. 4. When the zone has been put on and the folds neatly made under it, the dress must be pulled up, without disturbing the folds, so far through the zone as to hang of even length all round the feet. Now observe that to do this it must be much more pulled up at the sides than in the front, the length from a to e (Fig. 1) being considerably greater than the mere width of the shawl measured vertically from a to the feet. The sleeve is quite complete when the zone has been clasped. The commonest statue of Diana affords an example of this sleeve, and Mr. Leslie has shown it in his picture of 'The Fountain.' Pretty as it is, this sleeve is

FIG. 3.

not very suitable for common wear, not so much because the body can occasionally be seen through it from one side (a defect easily remedied by a stitch or two out of sight) as because the whole figure looks rather topheavy when the dress is not kilted up to the knee. This kilting, through a second zone or rope over the hips, by enlarging the hips takes off the topheavy appearance, and the hanging sleeve is most commonly found in statues and pictures of Diana, who wanted to get superfluous drapery out of the way of her legs. The close sleeve is differently formed, and is more suited for indoor life, with long skirts; it is made by passing a cord round the front of each

FIG. 4.   FIG. 5.

shoulder, crossing at the back, and is shown in Figs. 5 and 7. The set of the sleeve on the arm, when pulled slightly through this cord, is excellent; scissors seldom succeed as well. Figs. 5 and 7 show this sleeve, made before the lady is zoned, and Fig. 8 shows the same sleeve when the zone has been applied and the dress pulled up through the zone so as to hang evenly round the feet at the front and sides. The series of folds or bags hanging low on the hips and barely covering the zone in front, form an arch which will be found more or less pronounced in every Greek statue or Greek drawing when the dress is at all drawn up through the zone. This arch is graceful, and no doubt was adhered to by the Greeks on account of its beauty,

adding as it does to the width of the hips; but it must have
been produced, in the first instance, from the necessity of getting
the dress off the ground equally at the front and sides; because,

Fig. 6.

as has been already explained, the length from *a* to *e* (Fig. 1) is greater than the width of the shawl at the centre, where the folds barely cover the zone. Mr. Leslie, in his graceful figures at the Fountain, has failed to seize this peculiarity, and each figure has a large bag over the stomach which, if seen in real life, would be grotesque. The folds in Fig. 8 are hardly sufficiently numerous, and this is especially visible when the arm is raised. This arises partly from the fact that even an Indian Chudder is thicker than the stuff used by the high-born Greeks, and partly from the difficulty of folding the dress by male fingers; the Greek tire-women folded materials to perfection, having practised the art all their lives; the writer never

FIG. 7.   FIG. 8.

could fold anything until practice taught him to fold this tunic. After a few attempts at folding, the ladies found that they could stitch the plaits where the stitching was concealed under the zone. This little row of plaits is parallel to the bottom of the gown, although the result in the overhanging folds is the arch drawn.

Let us recapitulate the results arrived at :

1. The Chudder shawls made now in India are exactly of the right size to make a Greek tunic, and fold very much as Greek drapery does. It is not impossible that the material and size of loom may have remained in the East unaltered for the last two or three thousand years.

2. The dress can be made without being cut from the simple woven parallelogram.

3. The set over the bust of most statues is exactly reproduced with the series of curved folds converging at the shoulders.

4. The mass of drapery at the hips, forming the arch as viewed from the front, results, not from design, but from the necessity of having the dress of equal length at the sides and in front.

5. Both varieties of sleeve can be produced at will by the cord or zone.

These are the arguments which induce me to believe that the dress produced is really made as Greek women made theirs.

Hitherto nothing has been said of the back. The backs of statues are much less perfectly finished than the fronts; moreover, it is awkward to arouse the suspicions of the guardians of galleries by creeping behind statues too much. Speaking from experience of the living model, I may say that the back view looks best to modern eyes when the dress is, at the back, not drawn up through the zone; this results in giving a small train. A few notes will complete what is to be said of the tunic. The pocket *c* may be shortened by folding in, or cutting off, the shawl along the dotted lines in Fig. 1 : I suspect that this was done in some examples of very full drapery, which might be produced by putting two shorter shawls together, instead of folding one longer shawl. For dancing girls and Spartan virgins the ends of the shawl were not sewn together, and so showed the leg from one side. In modern pictures I observe a cord crossed over the breast; this, no doubt, is founded on some examples with which I have not happened to meet.

Let us pass to the *diplots*, or mantle. Hope has correctly, though incompletely, described this : it is a mere scarf, twelve or fourteen feet long, and of different widths in different examples. It is folded in exactly the same way as the tunic, but the ends are not sewn together. The result is shown in Fig. 9. The pocket hangs down at one side, and balances the two ends hanging from the other shoulder. The zigzag folds are produced by the bunch gathered at the clasp on the shoulders; the same arched effect is produced as is given by the tunic drawn through the zone ; but there is a straight line crossing the body in a way not altogether graceful. Numerous examples of this may be seen in Hope's book. In the later dresses the bib was artificially arched, the width in the centre of the body being considerably reduced. In earlier examples the scarf was often doubled in front, as Hope suggests, probably to make it narrow enough ; but I think the scarf was always spread to its full width behind, otherwise I have by experiment found that the narrow *diploïs* produced an awkward hunchbacked appearance. There are many drawings in which the *diploïs* is really single, though it looks double ; this effect being produced merely by the habit of turning in the edge to give firm hold for the clasps.

FIG. 9.

In the *diploïs* the folds over the bosom are generally vertical, and in the simpler examples there is just one fold in the centre ; this is produced by catching the stuff together some three or four inches below the top edge, and more than one fold can be produced in this way if desired. The abomination of a tape running in a seam produces nothing like it, nor does simple looseness, the effect of which is quite correctly given in Mr. Leslie's dark green *diploïs*, folded double according to Hope ; and singularly awkward it is, forming a kind of cup projecting under the chin. For simplicity's sake, Fig. 9 shows the *diploïs* alone, when the dress is not gathered up through the zone ; but the Greeks generally wear it so narrow as to show the folds of the tunic hanging below the zone. Let any one buy a photograph of Mrs. Kendal as Galatea, and compare this description with the millinery mantle cut low at the back with a stiff top and vertical plaits all along, and then, whichever he prefers, he will allow that the difference is as great as between a Greek and a Gothic column.

The *péplos* was a mere shawl, exactly like the one of which the tunic was formed. In Fig. 9 we have an outline from life of the tunic and *péplos*; the tunic has slipped a little aside off one shoulder, as we see it frequently represented.

Now my hobby has been ridden, and I must dismount: I would fain persuade myself that it is a good useful hobby—that it may serve to robe many graceful figures in a singularly becoming dress—that it may amuse modern ladies to see how a Greek lady, merely by undoing a button here, or tying a cord there, could have a low gown or one tight round the throat; a gown with hanging sleeves or close sleeves; a gown with long sweeping folds, suited to indoor lounging on a couch, or a gown kilted to the knee suited to the muddiest roads or thickest heather; a gown too that would wash and wear for years; a gown that would fit all figures and required no trying on; a gown in which there was no hem and only one seam, which could be omitted if the habits of the country allowed it. I should like to fancy, moreover, that an artist here and there might be tempted to draw the folds which to my eye have a peculiar grace. But, relinquishing such aspirations, I will beg the reader to pardon this lengthy dissertation on the ground that some pains really were taken to ascertain the facts concerning Greek dress, however unimportant these facts may now be.

MRS. SIDDONS AS LADY MACBETH.[1]

WHEN any great work of art perishes from among us, we not only grieve, but we rebel against the decree of fate. The wars, the traffic, the mechanical arts of old, nay even the men and women, wither into an oblivion which is not painful but kindly. We sigh and smile and acquiesce—better so, for here was nothing fitted to endure for ever. They had their time, as we have ours, and who would wish that the strife, the bustle, the men of to-day should last for ever? But the destruction of any beautiful thing, whether it be the work of art or nature, fills us, on the contrary, with sickening regret. The temple, statue, picture gone imply a loss of joy to uncounted generations. We suffer real pain when we think of lost tragedies by Sophocles, and our whole classical system of education is a protest that though kingdoms, peoples, tongues may die, their works of·beauty shall endure.

If this be our feeling as to the more durable works of art, what shall we say of those triumphs which by their very nature last no longer than the action which creates them—the triumphs of the orator, the singer, or the actor? There is an anodyne in the words 'must be so,' 'inevitable,' and there is even some absurdity in longing for the impossible. This anodyne and our sense of humour temper the unhappiness we feel when, after hearing some great performance, we leave the theatre and think, 'Well, this great thing has been, and all that is now left of it is the feeble print upon my brain, the little thrill which memory will send along my nerves, mine and my neighbours'; as we live longer the print and thrill must grow feebler, and when we pass away the impress of the great artist will vanish from the world.' The regret that a great art should in its nature be transitory explains the lively interest which many feel in reading anecdotes or de-

[1] From the *Nineteenth Century*, 1878.

scriptions of a great actor, and it is this feeling which prompts
the publication of the following notes on Mrs. Siddons' acting
made by an eye-witness of ability and true artistic feeling.

The public of to-day are perhaps hardly aware of the height
to which the art of acting may rise. Yet those who have been
familiar with the creations of Rachel and Salvini will not only
credit the assertion that the genius of Mrs. Siddons in representing the characters of Murphy, Lillo, Southerne, and Otway,
was greatly superior to that of the writers, but that, even when
representing Shakespeare, she supplied much which enriched the
conceptions of the poet. To-day we often speak of an actor as
the mere interpreter of Shakespeare. We are apt to imagine
that there is some one Hamlet or Lady Macbeth, a creature of
Shakespeare's brain, an *eidolon* which the actor must of necessity
endeavour to represent, his success being measured by the approach which he makes to this unattainable ideal. Those, however, who have seen the acting of the last thirty years in Paris
will know that this view of the actor's province is far from true
when he interprets even the best modern authors. They know
that an actor, when he receives the manuscript, has to create
his part in the sense of conceiving a complete human being
who, under the given circumstances, employs the words which
the author has supplied. They know that no critic could, by
reading a play, evolve a portrait of the man whom an original
actor will represent as the embodiment of some new part. They
know that each new actor of real merit recreates the persons
of the older drama, sending traditions to the winds and producing a new person on the stage using the old words, but with
marvellous differences of manner, voice, gesture, and intention.
They know that there is not merely one good way of representing
a great part, but as many ways as there are great actors. Each
actor is bound so to fashion his conception that his own physical
attributes and mental powers will lend themselves to its execution, and thus the great parts on the French stage have bound
up with them a long series of portraits each representing the
creation of a separate actor—all the creations good and to be
judged of on their own merits, not by reference simply to the
mind of the author.

In small parts, and in the lower walks of the art, the English public will admit this truth readily. No one can suppose that the writer of ' Rip van Winkle ' conceived his man with the tones and gestures which we find so admirable in Mr. Jefferson ; but the majesty of Shakespeare's name overawes us when we hear that a Mrs. Siddons created a part which Shakespeare wrote— when we are told that an actor's first business is not to think how Shakespeare conceived his character as standing or looking, but how he, the actor, can make a real human being stand and look while speaking Shakespeare's words. Yet the words of the part do not by themselves supply the actor with one-hundredth part of the actions he has to perform. Every single word has to be spoken with just intonation and emphasis, while not a single intonation or emphasis is indicated by the printed copy. The actor must find the expression of face, the attitude of body, the action of the limbs, the pauses, the hurries—the life, in fact. There is no logical process by which all these things can be evolved out of the mere words of a part. The actor must go direct to nature and his own heart for the tones and action by which he is to move his audience ; these his author cannot give him, and in creating these, if he be a great actor, his art may be supremely great.

The distinction between the mechanical arts and what are commonly called the fine arts lies in the creation or invention required by the artist as compared with the skill or dexterity which are alone required by the craftsman. The one copies or executes ; the other creates, invents, or *finds* the treasure which he gives to the world. Arts are great or small as the thing created is noble or petty ; the artist is true or false as he possesses more or less of this creative power, for the exercise of which he in all cases requires skill more or less mechanical, which technical skill is often called ' art ' as if there were no other. This technical skill can be taught and must be learned by every artist. The poetic creative power can never be taught, though in a sense it is learned from every sight, sound, and feeling ; but this greater art is learned unconsciously, and few have the power to learn the lesson.

Judged by this canon, the art of the actor may claim high rank whenever its scope is the presentment of the highest human types. To truly great actors, the words they have to

speak are but opportunities of creating these types—opportunities
in the sense that a beautiful model, a fine landscape, are opportunities to the painter.   In these he finds his picture, in those the
actor finds his person ; but the dramatist does less for the actor
than nature for the painter.   It is the involuntary unconscious
perception of this truth which makes men accord a generous recognition to artists such as Mrs. Siddons while treating, not
without justice, the rank and file of the profession as mere
skilled workmen.

It is probable, nay certain, that in writing the words to be
uttered by each character, a great author has vividly present to
his mind an ideal man or woman speaking these with natural
and effective tones and gestures—perhaps in Shakespeare's case,
though not in others, the best tones and gestures possible; perhaps, however, with tones and gestures so old-fashioned that
they would not move us now; what is certain is that we have
no means of discovering these ; indeed, he could not himself
have imparted them to a fellow-actor.   Moreover, when writing
the words of Macbeth, he cannot have had present to his mind
all the gestures and expressions of Lady Macbeth as she listened.
Yet this by-play of the great actress was such that the audience,
looking at her, forgot to listen to Macbeth.   Corneille never
thought of how Camille would listen to the account of the death
of her lover in 'Les Horaces,' or, if he thought of it, his conception
must have been a mere sketch as compared with the long and
marvellous scene which Rachel, playing the part, showed to the
astonished audience.

In truth, the spectators do not know the marvellous study
which a great actor applies to every word of a speech.   Some
think that the study consists in finding out what the author
meant the hero to say or express by given words.   Sometimes
this demands study ; more often with great writers it is as plain
as can be, requiring no study.   When the meaning is understood, next comes the consideration of the feeling which the
speech implies or requires in the speaker.   The conception of
this is far more difficult than the simple interpretation of the
words, and will alter with each new actor; not differing *toto
cœlo*, but differing in shade, colour, and intensity.   Any one of
us can understand the reasoning in 'To be or not to be.'   Very

few of us can form any vivid conception of the state of Hamlet's mind, sentence by sentence, word by word, as he utters them. Of the few who can form any conception beyond a mere colourless, shadeless, pointless impression of gloom or bitterness, each one must of necessity form a distinct and new conception. In order that such a speech may sway a house, it must represent a series of emotions, each intense, natural, and noble—each succeeding the other in a natural sequence. After the speech has been understood and the feelings to which it corresponds conceived, comes a task of ineffable difficulty—that of finding tones, look, and action, which shall represent those feelings. The author gives an outline, which the actor must fill up with colour, light, and shade, so as to show a concrete fact; and no two actors can or ought to do this in one and the same way. Let any reader who doubts this—who thinks, for instance, that there is some one Hamlet, Shakespeare's Hamlet, who could only speak the speech in one attitude, with one set of tones—open the book, and in the solitude of his chamber try first to find out the emotions which Shakespeare meant his Hamlet to feel, and then try to express those emotions in tones which would indicate them to others. If honest and clever, he will find out after half an hour's study how little the author has done for the actor, how much the actor is called upon to do for the author.

These views will find their illustration in the remarkable notes by Professor G. J. Bell on Mrs. Siddons' acting, which are now published for the first time, having been kindly placed at the disposal of the writer by his surviving son, Mr. John Bell, of the Calcutta bar. Written apparently on the spot, and during the red-hot glow of appreciation, they bring the great actress before us in a way which no laboured criticism or description could do. They show how noble an art she practised, and might almost inspire some young and generous mind with the power once more to create heroic men and women on the stage.

Professor G. J. Bell, brother of the great surgeon Sir Charles Bell, was Professor of Scottish Law in the University of Edinburgh, and author of 'Commentaries on the Law of Scotland,' a standard work still in high repute. He was well known by his friends to be a man of fine taste and keen sensibility, as is

indeed proved by these notes. They were made in 1809, or
about that time, and are contained in three volumes, lettered
' Siddons,' which of themselves prove the great interest taken
in Mrs. Siddons' acting. They contain acting editions of
the plays in which she appeared, edited by Mrs. Inchbald.
Professor Bell was himself in the habit of reading aloud, and,
besides critical remarks, he has noted in many places the rise or
fall of Mrs. Siddons' voice, putting a mark / for a rise, and \
for a fall. The words on which the emphasis fell are underlined.
The following is an introductory note on ' Macbeth : '

> Of Lady Macbeth there is not a great deal in this play, but the
> wonderful genius of Mrs. Siddons makes it the whole. She makes
> it tell the whole story of the ambitious project, the disappointment,
> the remorse, the sickness and despair of guilty ambition, the attain-
> ment of whose object is no cure for the wounds of the spirit. Mac-
> beth in Kemble's hand is only a co-operating part. I can conceive
> Garrick to have sunk Lady Macbeth as much as Mrs. Siddons does
> Macbeth, yet when you see Mrs. Siddons play this part you scarcely
> can believe that any acting could make her part subordinate. Her
> turbulent and inhuman strength of spirit does all. She turns Mac-
> beth to her purpose, makes him her mere instrument, guides, directs,
> and inspires the whole plot. Like Macbeth's evil genius she hurries
> him on in the mad career of ambition and cruelty from which his
> nature would have shrunk. The flagging of her spirit, the melan-
> choly and dismal blank beginning to steal upon her, is one of the
> finest lessons of the drama. The moral is complete in the despair
> of Macbeth, the fond regret of both for that state of innocence from
> which their wild ambition has hurried them to their undoing.

The writer of this note obviously, like Milton, considered a
tragedy the moralest of poems, as indeed it is; but special
attention may be paid to two points. First, Mrs. Siddons did
not herself conceive Shakespeare's Lady Macbeth as turbulent
and with inhuman strength; she represented her as a woman
of this type because this conception suited her physical powers
and appearance. But in her own memoranda, published in her
life by Campbell, she speaks thus of Lady Macbeth's beauty :

> According to my notion it is of that character which I believe
> is generally allowed to be most captivating to the other sex—fair,
> feminine, nay perhaps even fragile—

> Fair as the forms that, wove in fancy's loom,
> Float in light visions round the poet's head.

Such a combination only, respectable in energy and strength of mind, and captivating in feminine loveliness, could have composed a charm of such potency as to fascinate the mind of a hero so dauntless, a character so amiable, so honourable as Macbeth.

There is something to be said for Mrs. Siddons' argument that an overbearing woman could never have guided Macbeth; but this point is for the moment of secondary interest, compared with the light which her remark throws on the statement made above, that there is not one conception which alone the actor must form of a given part. Here we have a great actress forming two distinct conceptions: for no one can believe that if Mrs. Siddons had been able to appear the fair and fragile beauty she conceived, she would have used a single gesture or one inflection similar to those employed when she was representing turbulent inhuman strength.

The second point of interest in Professor Bell's note is, that the melancholy and dismal blank beginning to steal on Lady Macbeth is more the creation of Siddons than of Shakespeare. There is nothing in the text to contradict it, but little to indicate it. This will become more apparent when we reach the detailed notes.

A second notice in another copy of 'Macbeth' appears as follows:

Mrs. Siddons is not before an audience. Her mind wrought up in high conception of her part, her eye never wandering, never for a moment idle, passion and sentiment continually betraying themselves. Her words are the accompaniments of her thoughts, scarcely necessary, you would imagine, to the expression, but highly raising it, and giving the full force of poetical effect.

What a tribute! Shakespeare's words hardly necessary! And this was written by a man who idolised Shakespeare.

Professor Bell elsewhere remarks:

A just observation that it is unhappy when the part of Lady Macbeth is in the hands of a Siddons, and Macbeth (with ?) an inferior actor. She then becomes not the affectionate aider of her

husband's ambition, but the fell monster who tempts him to trans-
gress, making him the mere instrument of her wild and uncontrollable
ambition.

The notes on this play will now be given, only so much of
each scene being quoted as is necessary to render the notes
intelligible. The text of Shakespeare is given as found in the
edition annotated by Professor Bell.

ACT I.

SCENE 5.—*Macbeth's Castle at Inverness.*

Enter LADY MACBETH,[2] *reading a letter.*

Lady. 'They met me in the day of success: and I have learned
by the *perfectest* report, they have *more* in them than mortal knowledge. When I burned in desire to question them further, they
made themselves air, into which they vanished. Whiles I stood
rapt in the *wonder of* it, came missives from the king, who all-hailed
me "Thane of Cawdor;" by which title, before, these weird sisters
saluted me, and referred me to the coming on of time,' with "Hail,
king that shalt be!" This have I thought good to deliver thee, my
dearest partner of greatness, that thou mightest not lose the dues
of rejoicing, by being ignorant of what greatness is promised thee.
Lay it to thy heart, and farewell.'

<blockquote>
Glamis thou art, and Cawdor, <i>and shalt be</i> [3]
What thou art promised : [4] yet do I fear thy nature ;
It is too full o' the milk of human kindness
To catch the nearest way : thou wouldst be great :
Art not without ambition, but without
The <i>illness</i> should attend it : [5] what thou wouldst highly,
That wouldst thou holily` ; wouldst not <i>play</i> false,
And yet wouldst wrongly win : thou'd'st have, great Glamis,
That which cries ' Thus thou must do, if thou have it ; '
And that which rather thou dost fear to do
Than wishest should be undone. Hie thee hither,
That I may pour <i>my</i> [6] spirits in thine ear ;
And chastise with the valour of my tongue
</blockquote>

---

[2] Mrs. Siddons.

[3] Exalted prophetic tone, as if the whole future were present to her soul.

[4] A slight tincture of contempt throughout.

[5] Here and in the night scenes it is plain that he had imparted to her his
ambitious thoughts and wishes.

[6] Starts into higher animation.

All that impedes thee from the golden round
Which *fate* [7] and metaphysical aid doth seem
To have thee crown'd withal.

Enter SEYTON.

        What is your tidings ?
 Seyton. The king comes here to-night.
 Lady.   *Thou'rt mad to say it* ; [8]
[9] Is not thy master with him ? who, were't so,
Would have inform'd for preparation.[9]
 Seyton. So please you, it is true : our thane is coming :
One of my fellows had the speed of him,
Who, almost dead for breath, had scarcely more
Than would make up his message.
 Lady.   Give him tending ;
He brings great news.   [*Exit* SEYTON.
     [10] The raven himself is hoarse
That croaks the *fatal* entrance of Duncan
Under my battlements. [11] Come, all you spirits
That tend on mortal thoughts, unsex me here,
And fill me from the crown to the toe top-full
Of direst cruelty ! make thick my blood ;
Stop up the access and passage to remorse,
That no compunctious visitings of nature
Shake my fell purpose, nor keep pace between
The effect and it ! [11] [12] Come to my woman's breasts,
And take my milk for gall, you murdering ministers,
Wherever in your sightless substances
You wait on nature's mischief ! Come, thick night,
And pall thee in the dunnest smoke of hell,
That my keen knife see not the wound it makes,

---

 [7] The whole of this following scene a picture of this highest working of
the soul. Kemble plays it not well, yet some things good. Much of the
effect depends on the fire which she strikes into him, and which the player
must make out.

 [8] ·Loud.

 [9] Soft, as if correcting herself, and under the tone of reasoning concealing
sentiments almost disclosed.

 [10] After a long pause when the messenger has retired. Indicates her fell
purpose settled and about to be accomplished.

 [11] In a low voice—a whisper of horrid determination.

 [12] Voice quite supernatural, as in a horrible dream. Chilled with horror by
the slow hollow whisper of this wonderful creature.

Nor heaven peep through the blanket of the dark,
To cry ' Hold, hold ! ' [12]

Enter MACBETH.

[13] Great Glamis ! worthy Cawdor !
Greater than both, by *the all-hail hereafter !*
Thy letters have transported me beyond
This ignorant present, and I feel now
The future in the instant.
    Macbeth.   My dearest love,
Duncan comes here to-night.
    Lady.[14]   And when goes hence ?
    Macbeth. To-morrow, as he purposes.
    Lady.   O, never` - - -
(never) Shall sun that morrow see` ! [15]
    [16] Your face, my thane, is as a book where men
May read strange matters.   To beguile the time,
Look like the time ; [16] bear welcome in your eye,
Your hand, your tongue : look like the innocent flower,
[17] *But be the serpent under 't.* He that's coming
Must be provided for : and you shall put
This night's great business into my dispatch ;
[18] Which shall to all our nights and days to come
Give solely sovereign sway and masterdom.[18]
    Macbeth. We will speak further.
    Lady.   Only look up clear ;
[19] To alter favour ever is to fear :
Leave all the rest to me.[19]   [*Exeunt.*

----

[13] Loud, triumphant and wild in her air.

[14] High purpose working in her mind.

[15] O, never`. A long pause, turned from him, her eye steadfast. Strong dwelling emphasis on ' never,' with deep downward inflection, ' never shall sun that morrow see ! ' Low, very slow sustained voice, her eye and her mind occupied steadfastly in the contemplation of her horrible purpose, pronunciation almost syllabic, note unvaried. Her self-collected solemn .energy, her fixed posture, her determined eye and full deep voice of fixed resolve never should be forgot, cannot be conceived nor described.

[16] Observing the effect of what she has said on him, now first turning her eye upon his face.

[17] Very slow, severe and cruel expression, her gesture impressive.

[18] Voice changes to assurance and gratulation.

[19] Leading him out, cajoling him, her hand on his shoulder clapping him. This vulgar—gives a mean conception of Macbeth, unlike the high mental working by which he is turned to her ambitious purpose.

Does not the reader feel that in these close personal observations is to be found a far better conception of what the genius of Siddons could do than is given in the long lives by Campbell and Boaden? Mrs. Siddons appears to have repeated the word ' never ' before ' shall sun that morrow see.' This appears not only from note (15), but from a manuscript insertion of a second ' never ' after the pause indicated above. The next notes are on the sixth scene, where Lady Macbeth addresses Duncan.

> *Lady.* [20] All our service
> In every point twice done and then done double
> Were poor and single business, to contend
> Against those honours deep and broad, wherewith
> Your majesty loads our house : for those of old,
> And the late dignities heap'd up to them,
> We rest your hermits.[20]

At her exit comes this note :

Bows gracefully to the king, when she gives him the *pas* in entering. Then graciously and sweetly to the nobles before she follows the king.

On Macbeth's speech, Scene 7, beginning

> If it were done, when 'tis done, then 'twere well
> It were done quickly,

there is the following :

Kemble speaks this, as if he had never seen his sister, like a *speech* to be recited. None of that hesitation and working of the mind which in Mrs. Siddons seems to inspire the words as the natural expression of the emotion.

After the entrance of Lady Macbeth the notes continue :

Lady. He has almost supp'd : why have you left the chamber ? [21]
Macbeth. Hath he ask'd for me ?
Lady. *Know* you not he has ?
Macbeth. We will proceed no further in this business :
[22] He hath honour'd me of late ; and I have bought

[20] Dignified and simple. Beautifully spoken; quite musical in her tones and in the pronunciation, soothing and satisfying the ear.
[21] Eager whisper of anger and surprise.
[22] Here again Mrs. Siddons appears with all her inimitable expression of emotion. The sudden change from animated hope and surprise to disappoint-

Golden opinions from all sorts of people,[22]
Which would be worn now in their newest gloss,
Not cast aside so soon.

    Lady.   [23] Was the hope drunk
Wherein you dressed yourself ? hath it slept since ?
And wakes it now, to look so green and pale
At what it did so freely ?[24]   From this time
Such I account thy love.[5] Art thou afeard
To be the same in thine own act and valour
As thou art in desire ?   Wouldst thou have that
Which thou esteem'st the ornament of life,
And live a coward in thine own esteem,
Letting ' I dare not ' wait upon ' I would,'
Like the poor cat i' the adage ?

    Macbeth.   [25] Prithee, peace :
I dare do all that may become a man ;
Who dares do more is none.[25]

    Lady.   What beast was it then
That made you break this enterprise to me ?
When you durst do it, then you were a man ;
And, to be more than what you were, you would
Be so much more than man.   Nor time nor place
Did then adhere, and yet you would make both :
They have made themselves, and that their fitness now
Does unmake you.[26]   [27] I have given suck, and know
How tender 'tis to love the babe that milks me :
I would, while it was smiling in my face,
Have plucked my nipple from his boneless gums
And dashed the brains out, had I but so sworn as *you*
Have done to this.

    Macbeth.   If we should fail ?
    Lady.   [28] We fail ! `

ment, depression, contempt, and rekindling resentment, is beyond any powers but hers.

  [23] Very cold, distant, and contemptuous.

  [24] Determined air and voice. Then a tone of cold contemptuous reasoning.

  [25] Kemble speaks this well.

  [26] Cold, still, and distant; slow with remarkable distinctness and great earnestness.

  [27] She has been at a distant part of the stage. She now comes close to him—an entire change of manner, looks for some time in his face, then speaks.

  [28] *We fail*. Not surprise, strong downward inflection, bowing with her hands down, the palms upward. Then voice of strong assurance, ' When

But screw your courage to the sticking-place,
And we'll *not* fail.²⁸ When Duncan is asleep—
Whereto the rather shall his day's hard journey
Soundly invite him—his two chamberlains
Will I with wine and wassail so convince
That memory, the warder of the brain,
Shall be a fume, and the receipt of reason
A limbec only ; ²⁹ when in swinish sleep
Their drenched natures lie as in a death,
What cannot you and I perform upon
The *unguarded Duncan '* ? *what* not put upon
His spongy officers, who shall bear the guilt
Of our great quell ?
    Macbeth.   Bring forth men-children only ;
For thy undaunted mettle should compose
Nothing but males. Will it not be received,
When we have marked with blood those sleepy two
Of his own chamber and used their very daggers,
That they have done't ?
    Lady.   Who *dares* receive it other,³⁰
As we shall make our griefs and clamour roar
Upon his death ?
    Macbeth.   I am settled, and bend up
Each corporal agent to this terrible feat.
Away, and mock the time with fairest show :
False face must hide what the false heart doth know.
                            [Exeunt.

The next note refers to Macbeth's dagger scene, and is very interesting, although referring more immediately to Kemble than to his sister. Professor Bell says :

There is much stage trick and very cold in this scene of Kemble —walks across the stage, his eyes on the ground, starts at the sight

Duncan,' &c. This spoken near to him, and in a low earnest whisper of discovery she discloses her plan.

²⁹ Pauses as if trying the effect on him. Then renews her plan more earnestly, low still, but with increasing confidence. Throughout this scene she feels her way, observes the wavering of his mind ; suits her earnestness and whole manner to it. With contempt, affection, reason, the conviction of her well-concerted plan, the assurance of success which her wonderful tones inspire, she turns him to her purpose with an art in which the player shares largely in the poet's praise.

³⁰ Pause. Look of great confidence, much dignity of mien. In 'dares' great and imperial dignity.

of the servant, whom he forgets for the purpose, renews his walk,
throws up his face, sick, sighs, then a start theatric and then the
dagger.   Why can't he learn from his sister ?

Charles Bell thinks (and justly) that he should stand or sit musing, his eye fixed on vacancy, then a more piercing look to seem to
see what still is in the mind's eye only, characterised by the bewildered look which accompanies the want of a fixed object of vision ;
yet the eye should not roll or start.   N.B.: Mrs. Siddons in reading
'Hamlet' showed how inimitably she could by a mere look, while sitting in a chair, paint to the spectators a horrible shadow in her
mind.

At the point where Macbeth says 'there's no such thing,'
Professor Bell continues :

Kemble here hides his eyes with his hand, then fearfully looks
up, and peeping first over then under his hand, as if for an insect whose
buzzing had disturbed him, he removes his hand, looks more abroad,
and then recovers—very poor—the recovery should be by an effort
of the mind.   It is not the absence of a physical corporeal dagger,
but the returning tone of a disordered fancy.   A change in the look,
the clearing of a bewildered imagination, a more steadfast and natural aspect, the hand drawn across the eyes or forehead, with something of a bitter smile.

These remarks illustrate well what was said before as to the
nature of an actor's study.   Professor Bell had reached the
second stage, and knew well what the actor should feel.   The
third stage, how to show it, can only be acted, not described.
With the entrance of Lady Macbeth the notes become detailed.

Scene 2.

Enter Lady Macbeth.

Lady.[31]   That which hath made them drunk hath made me bold ;
What hath quenched them hath given me fire.   *Hark! Peace!* [32]
It was the owl that shrieked, the fatal bellman,
Which gives the stern'st good-night.   *He is about it:* [33]
The doors are open, and the surfeited grooms
Do mock their charge with snores : I have drugg'd their possets,

[31] With a ghastly horrid smile.
[32] Hsh ! Hsh !   Whisper.
[33] Breathes with difficulty, hearkens towards the door.   Whisper horrible.

That death and nature do contend about them,
Whether they live or die.
    Macbeth. [*Within.*]   Who's there ? what, ho !
    Lady. 34 Alack, I am afraid they have awaked,
And 'tis not done. The attempt and not the deed
Confounds us. Hark ! I laid their daggers ready ;
He could not miss them. Had he not resembled
My father as he slept, I had done't.35—*My husband !*

Enter MACBETH.

    Macbeth. 36 I have done the deed. Didst thou not hear a noise ?
    Lady. I heard the owl scream and the crickets cry.
Did not you speak ? 35
    Macbeth.   When ? 36
    Lady.   Now.
    Macbeth.   As I descended ? 37
    Lady. Ay.
    Macbeth. Hark !
Who lies i' the second chamber ?
    Lady.   Donalbain.
    Macbeth. This is a sorry sight.   [*Looking on his hands.*
    Lady. A foolish thought, to say a sorry sight.
    Macbeth. 38 There's one did laugh in his sleep, and one cried
    ' Murder ! '
That they did wake each other : I stood and heard them :
But they did say their prayers, and address'd them
Again to sleep.
    Lady. There are two lodged together.
    Macbeth. One cried ' God bless us ! ' and ' Amen ' the other ;
As they had seen me with these hangman's hands :
Listening their fear, I could not say ' Amen,'
When they did say ' God bless us ! '
    Lady. Consider it not so deeply.
    Macbeth. But wherefore could not I pronounce ' Amen ' ?
I had most need of blessing, and ' Amen '
Stuck in my throat.

    34 The finest agony ; tossing of the arms.
    35 Agonised suspense, as if speechless with uncertainty whether discovered.
    36 Macbeth speaks all this like some horrid secret—a whisper in the dark.
    37 Very well spoken ; horrid whisper.
    38 Mrs. Siddons here displays her wonderful power and knowledge of nature.
As if her inhuman strength of spirit overcome by the contagion of his remorse
and terror. Her arms about her neck and bosom, shuddering.

Lady.   These deeds must not be thought
After these ways ; *so, it will make us mad.*38
    Macbeth.   39 Methought I heard a voice cry ' Sleep no more ! '
                                  . . . to all the house
' Glamis hath murder'd sleep, and therefore Cawdor
Shall sleep no more ; Macbeth shall sleep no more.' 39
    Lady.   Who was it that thus cried ?   Why, worthy thane,
40 You do unbend your noble strength, to think
So brainsickly of things.   Go get some water,
And wash this filthy witness from your hand.40
41 *Why did you bring these daggers from the place ?*
They must lie there : go carry them, and smear
The sleepy grooms with blood.
    Macbeth.   I'll go no more :
I am afraid to think what I have done ;
Look on't again I dare not.
    Lady.   Infirm of purpose !
42 Give me the daggers : the sleeping and the dead
Are but as pictures : 'tis the eye of childhood
That fears a painted devil.42   *If he do bleed,*43
I'll gild the faces of the grooms withal ;
For it must seem their guilt.   [*Exit. Knocking within*
    Macbeth.   Whence is that knocking ?
How is't with me, when every noise appals me ?
What hands are here ? ha ! they pluck out mine eyes.
Will all great Neptune's ocean wash this blood
Clean from my hand ?   No ; this my hand will rather
The multitudinous seas incarnadine,
Making the green one red.

Re-enter LADY MACBETH.

Lady.   44 My *hands* are of your colour ; but I shame
To wear a *heart* so white. [*Knocking within.*]   I hear a knocking

    39 Her horror changes to agony and alarm at his derangement, uncertain
what to do ; calling up the resources of her spirit.
    40 She comes near him, attempts to call back his wandering thoughts to
ideas of common life.   Strong emphasis on *who*.   Speaks forcibly into his
ear, looks at him steadfastly.   ' Why, worthy thane,' &c. : fine remonstrance,
tone fit to work on his mind.
    41 Now only at leisure to observe the daggers.
    42 Seizing the daggers very contemptuously.
    43 As stealing out she turns towards him stooping, and with the finger
pointed to him with malignant energy says, ' If he do bleed,' &c.
    44 Contempt.   Kemble plays well here ; stands motionless ; his bloody hands

At the south entry : retire we to our chamber :
A little water clears us of this deed :
How easy is it, then !   Your constancy
Hath left you unattended. [*Knocking within.*] Hark, more knocking,
Get on your nightgown, lest occasion call us
And show us to be watchers.   Be not lost
So poorly in your thoughts.
   Macbeth.   To know my deed, 'twere best not know myself.
[Knocking within.
Wake Duncan with this knocking !   Oh, would thou couldst ! 44
[Exeunt.

The notes are resumed where Lady Macbeth enters as queen.

ACT III.

SCENE 2.—*The Palace.*

Enter LADY MACBETH, *as Queen, and* SEYTON.

   Lady.   45 Is Banquo gone from court ?
   Seyton.   Ay, madam, but returns again to-night.
   Lady.   Say to the king, I would attend his leisure
For a few words.
   Seyton.   Madam, I will.   [*Exit.*
   Lady.   46 Nought's had, all's spent,
Where our desire is got without content :
'Tis safer to be that which we destroy
Than by destruction dwell in doubtful joy.46

Enter MACBETH.

47 How now, my lord ! why do you keep alone,
Of sorriest fancies your companions making ;
Using those thoughts which should indeed have died
With them they think on ?   Things without all remedy
Should be without regard : what's done is done.47

near his face ; his eye fixed, agony in his brow ; quite rooted to the spot. She
at first directs him with an assured and confident air.   Then alarm steals
on her, increasing to agony lest his reason be quite gone and discovery be
inevitable.   Strikes him on the shoulder, pulls him from his fixed posture,
forces him away, he talking as he goes.
   45 Great dignity and solemnity of voice ; nothing of the joy of gratified
ambition.   46 Very mournful.
   47 Still her accents very plaintive.   This is one of the passages in which her

Macbeth.   We have scotch'd the snake, not kill'd it :
She'll close and be herself, whilst our poor malice
Remains in danger of her former tooth.
But let the frame of things disjoint, both the worlds suffer,
Ere we will eat our meal in fear, and sleep
In the affliction of these terrible dreams
That shake us nightly ; better be with the dead,
Whom we, to gain our place, have sent to peace,
Than on the torture of the mind to lie
In restless ecstasy.   Duncan is in his grave ;
After life's fitful fever he sleeps well ;
Treason has done his worst : nor steel, nor poison,
Malice domestic, foreign levy, nothing,
Can touch him further.
Lady.   Come on ;
Gentle my lord, sleek o'er your rugged looks ;
Be bright and jovial 48 among your guests to-night.
Macbeth.   O, full of scorpions is my mind, dear wife !
Thou know'st that Banquo, and his Fleance, live.
Lady.   49 But in them nature's copy's not eterne.

There are no further remarks on this scene.

In Scene 4, where at the banquet Macbeth speaks to the murderers, the remark is written : 'During all this a growing uneasiness in her; at last she rises and speaks.' Full notes are resumed towards the end of the scene, as follows :

[*The Ghost of* BANQUO *enters, and sits in* MACBETH's *place.*

Macbeth.   Here had we now our country's honour roof'd,
Were the graced person of our Banquo present ;
Who may I rather challenge for unkindness
Than pity for mischance !
Ross.   His absence, sir,
Lays blame upon his promise.   Please 't your highness
To grace us with your royal company.
Macbeth.   50 The table's full.
Lennox.   Here is a place reserved, sir.

---

intense love of her husband should animate every word. It should not be contemptuous reproach, but deep sorrow and sympathy with his melancholy.

⁴⁸ Mournful : a forced cheerfulness breaking through it.

⁴⁹ A flash of her former spirit and energy.

⁵⁰ Her secret uneasiness very fine. Suppressed, but agitating her whole frame.

Macbeth. Where?

Lennox. Here, my good lord. What is't that moves your highness? 50

Macbeth. Which of you have done this?

Lennox. What, my good lord?

Macbeth. Thou canst not say, I did it : never shake
Thy gory locks at me.

Ross. Gentlemen, rise : his highness is not well.

Lady. 51 Sit, worthy friends :—my lord is often thus,
And hath been from his youth : pray you, keep seat ;
The fit is momentary ; upon a thought
He will again be well : if you much note him,
You shall offend him, and extend his passion ;
Feed, and regard him not. *Are you a man?* 52

Macbeth. Ay, and a bold one, that dare look on that
Which might appal the devil.

Lady. 53 O, proper stuff!
This is the very painting of your fear :
This is the air-drawn dagger, which, you said,
Led you to Duncan. O, these flaws and starts
(Impostors to true fear) would well become
A woman's story, at a winter's fire,53
Authorised by her grandam. Shame itself!
54 Why do you make such faces? When all's done,
You look but on a stool.54

Macbeth. Prithee, see there! behold! look! lo! how say you?
Why, what care I? If thou canst nod, speak too.
If charnel-houses and our graves must send
Those that we bury back, our monuments
Shall be the maws of kites.   [*Ghost vanishes.*

Lady.   What, quite unmann'd in folly?

Macbeth. If I stand here, I saw him.

Lady.   *Fie, for shame!* 55

Re-enter Ghost.

Macbeth. 56 Avaunt! and quit my sight! let the earth hide thee!
Thy bones are marrowless, thy blood is cold ;

---

51 Descends.   52 Comes up to him and catches his hand. Voice suppressed.
53 Peevish and scornful.
54 In his ear, as if to bring him back to objects of common life. Her
anxiety makes you creep with apprehension : uncertain how to act. Her emotion
keeps you breathless.   55 Returns to her seat ; this whispered.
56 Her secret agony again agitates her.

Thou hast no speculation in those eyes
Which thou dost glare with.
    Lady.[57]   Think of this, good peers,
But as a thing of custom : 'tis no other ;
Only it spoils the pleasure of the time.
    Macbeth. [58] What man dare, I dare :
Approach thou like the rugged Russian bear,
The arm'd rhinoceros, or the Hyrcan tiger ;
Take any shape but that, and my firm nerves
Shall never tremble : or be alive again,
And dare me to the desert with thy sword ;
If trembling I inhibit, then protest me
The baby of a girl.   Hence, horrible shadow !
Unreal mockery, hence ! [58]   *[Ghost vanishes.*
                Why, so : being gone,
I am a man again.
    Lady.   You have displaced the mirth, broke the good meeting,
With most admired disorder.
    Macbeth.   Can such things be,
And overcome us like a summer's cloud,
Without our special wonder ?   You make me strange
Even to the disposition that I owe,
When now I think you can behold such sights,
And keep the natural ruby of your cheeks,
When mine is blanch'd with fear.
    Ross.   What sights, my lord ?
    Lady.   [59] I pray you, speak not; he grows worse and worse ;
Question enrages him.   At once, good night :
Stand not upon the order of your going,
But go at once.
    Lennox.   Good night ; and better health
Attend his majesty !

[57] Rises and speaks sweetly to the company.

[58] Macready plays this well. Even Kemble chid and scolded the ghost out ! and rose in vehemence and courage as he went on.   Macready began in the vehemence of despair, but, overcome by terror as he continued to gaze on the apparition, dropped his voice lower and lower till he became tremulous and inarticulate, and at last uttering a subdued cry of mortal agony and horror, he suddenly cast his mantle over his face, and sank back almost lifeless on his seat.

[59] Descends in great eagerness; voice almost choked with anxiety to prevent their questioning; alarm, hurry, rapid and convulsive as if afraid he should tell of the murder of Duncan.

Lady.   A kind good night to all !

       [*Exeunt all but* MACBETH *and* LADY MACBETH.

Macbeth. It will have blood : they say blood will have blood :
Stones have been known to move and trees to speak ;
Augurs and understood relations have
By maggot-pies and choughs and rooks brought forth
The secret'st man of blood. What is the night ?

    Lady. 60 Almost at odds with morning, which is which.

    Macbeth. How say'st thou, that Macduff denies his person
At our great bidding ?

    Lady.   Did you send to him, sir ?

    Macbeth. I hear it by the way, but I will send :
There's not a one of them but in his house
I keep a servant fee'd. I will to-morrow,
And betimes I will, unto the weird sisters :
More shall they speak, for now I am bent to know,
By the worst means, the worst. For mine own good
All causes shall give way : I am in blood
Stepp'd in so far that, should I wade no more,
Returning were as tedious as go o'er.

    *Lady.*61 You lack the season of all natures, sleep.

    Macbeth. Come, we'll to sleep. My strange and self-abuse
Is the initiate fear that wants hard use :
We are yet but young in deed.   [*Exeunt.*

It is curious to see by these last two notes, as by the introductory remarks, that Mrs. Siddons conveyed by her demeanour
the impression of being already almost broken down, and quite as
much in need of sleep as Macbeth. This preparation for the sleeping scene is a very fine idea, and hardly seems to be suggested in
the insignificant remarks given by Shakespeare to Lady Macbeth
at the close of this scene. We now come to the fifth act.

    Gentlewoman. Lo you, here she comes ! This is her very guise ;
and, upon my life, fast asleep. Observe her ; stand close.

      Enter LADY MACBETH, *with a taper.*62

Physician. How came she by that light ?

  ⁶⁰ Very sorrowful. Quite exhausted.

  ⁶¹ Feeble now, and as if preparing for her last sickness and final doom.

  ⁶² I should like her to enter less suddenly. A slower and more interrupted
step more natural. She advances rapidly to the table, sets down the light
and rubs her hand, making the action of lifting up water in one hand at
intervals.

Gentlewoman. Why, it stood by her : she has light by her continually ; 'tis her command.

Physician. You see, her eyes are open.

Gentlewoman. Ay, but their sense is shut.

Physician. What is it she does now ? Look, how she rubs her hands.

Gentlewoman. It is an accustomed action with her, to seem thus washing her hands : I have known her continue in this a quarter of an hour.

Lady. Yet here's a *spot.*

Physician. Hark ! she speaks.

Lady. Out, damned spot ! out, I say !—*One* : [63] *two* : why, then 'tis time to do't.[64]—Hell is murky !—Fie, my lord, fie ! a soldier, and afeard ? What need we fear who knows it, when none can call our power to account ?—Yet who would have thought the old man to have had so much blood in him ?

Physician. Do you mark that ?

Lady. *The thane of Fife had a wife* : [65] where is she now ?—*What, will these hands ne'er* [66] *be clean* ?—No more o' that, my lord, *no more o' that* : [67] you mar all *with this* starting.

Physician. Go to, go to ; you have known what you should not.

Gentlewoman. She has spoke what she should not, I am sure of that : heaven knows what she has known.

Lady. Here's the smell of the blood still : all the perfumes of Arabia will not sweeten this little hand. *Oh, oh, oh !* [68]

Physician. What a sigh is there ! The heart is sorely charged.

This is the last of these notes by which we have been able to follow the great actress from the exalted prophetic tone of her entrance to the sigh of imbecility at the end.

---

[63] Listening eagerly.   [64] A strange unnatural whisper.
[65] Very melancholy tone.   [66] Melancholy peevishness.
[67] Eager whisper.
[68] This not a sigh. A convulsive shudder —very horrible. A tone of imbecility audible in the sigh.

MRS. SIDDONS AS QUEEN KATHARINE, MRS. BEVERLEY, AND LADY RANDOLPH.[1]

THE late Professor Bell's notes on Mrs. Siddons as Lady Macbeth were received with an interest which more than justifies the publication of his remarks on the part of Katharine, as played by the great actress. No other part played by Mrs. Siddons was annotated by Professor Bell in the thorough manner adopted by him when witnessing her Lady Macbeth and Queen Katharine. He left, however, some notes on her Mrs. Beverley and Lady Randolph, concerning which a few words may be said before speaking of Shakespeare's play.

Home's 'Douglas,' though known to all by name, is so little read that a sketch of the plot is necessary to make Professor Bell's remarks intelligible to the general reader. Lady Randolph was secretly married in early youth to one of a family at feud with her own, a Douglas, who was killed in battle three weeks after the marriage. The widow bore a son, but this infant, whose birth had been concealed, disappeared with his nurse, and his mother believes him to be dead. He, young Norval of the Grampian Hills, was however saved, and has been brought up in ignorance of his birth. Lady Randolph did not inform her second husband, Lord Randolph, of her first marriage, and explained her continual melancholy by attributing it to grief for the death of a brother. At the period when the play begins, young Norval is fortunate enough to save the life of his stepfather, Lord Randolph, who introduces him to his unknown mother and promotes him to an honourable command. In the course of the play the mother recognises her son, and makes herself known to him. The intimacy which results enables

---

[1] From *Macmillan's Magazine*, April 1882.

a villain, Glenalvon, so to poison the mind of Lord Randolph
with jealousy as to cause him to attempt the youth's life.
Young Norval or Douglas, while defending himself against
Lord Randolph, is wounded to death by the villain, and dies in
his mother's presence. She in despair commits suicide. In accordance with the taste of the day, neither combat nor suicide
takes place before the audience.

Although much of the sentiment in this play is expressed in
language which nowadays provokes a smile, an actress may find
great scope for her art in presenting the feelings of the mother,
who gradually acquires the certainty that her child still lives, and
is the gallant youth who has already shown himself worthy of her
love.

Professor Bell's notes, while sufficient to convince us that
Mrs. Siddons could express great tenderness and strong affection, no less than the sterner emotions with which her name is
more commonly connected, lack the precision by which, in
writing of Shakespeare's plays, he enables us in some measure
to understand the means she employed. Referring to the wish
expressed by the lady that every soldier of the two opposing
armies might return in 'peace and safety to his pleasant home,'
he writes :

The most musical sound I ever heard, and on the conclusion a
melancholy recollection seemed to fill her whole soul of the strength
of that wish in former times, and of its first disappointment.

Again, where Lady Randolph addresses Sincerity as the first
of virtues, the note says :

Fine apostrophe. Her fine eyes raised in tears to heaven, her
hands stretched out and elevated.

At the close of the well-known speech beginning, 'My name
is Norval,' the following remark is appended :

The idea of her own child seems to have been growing, and at
this point overwhelms her and fills her eyes with tears. Beautiful
acting of this sweet feeling throughout these speeches. The interest
she takes in the youth—her manifest retrospection.

The by-play of Lady Randolph throughout the long speeches
of her husband and son was obviously the centre of interest to
the spectator, and ended in what is called—

A great and affecting burst of affection and interest, as if she had already almost identified him with her son, or adopted him to supply the loss.

Answering Norval, who assures her that he will never be unworthy of the favour shown him, Lady Randolph says:

I will be sworn thou wilt not. Thou shalt be *my knight.*

The words printed in italics were underlined by Professor Bell.

Lady Randolph explains to her confidante that while Norval spoke she thought that, had the son of Douglas lived, he might have resembled this young gallant stranger.

Professor Bell writes:

It is this she has been acting during the preceding scene.

There are no further notes on this play, nothing to guide us as to the manner in which Mrs. Siddons said the famous ' Was he alive ? ' when a certain old man describes the finding of her infant son, who turns out to be Norval.

When we read Home's 'Douglas' we may feel a certain interest in our ancestors who liked it, but Moore's 'Gamester' awakens a feeling of loathing which extends even to the audience which can endure the degrading spectacle. The character of Lady Randolph is far from noble : this woman, who deceives her parents and husband, who lost her child and held her tongue, who has maundered through life for twenty years nursing her melancholy and despising all good things present, because they are not better things past, belongs to no heroic type.

We cannot admire her indifference to the excellent husband who after twenty years of married life still sues in vain for

Decent affection and complacent kindness.

But Lady Randolph's well-bred coldness is preferable to Mrs. Beverley's form of love. Says Mrs. Beverley : ' All may be well yet. When he has nothing to lose, I shall fetter him in these arms again ; and *then* what is it to be poor ? ' Professor Bell adds :

Such a speech as this the wonderful voice of Mrs. Siddons and her speaking eye make very affecting.

Surely no one but a Mrs. Siddons could do so.

An old servant offers to sacrifice his little fortune to the much-loved gamester who has been out all night for the first time : he proposes to go to him and if possible to bring him home. Mrs. Beverley says, ' Do so, then; but take care how you upbraid him—I have never upbraided him.' There is a note here :

Follows him to the door; then laying her hand on his arm detains him with an earnest look, and then speaks solemnly.

The lady uses much the same language to her husband's sister Charlotte, and Professor Bell notes :

She repeats an injunction she had given to Jarvis, more famili- arly but with equal earnestness, with more sorrow and less of dignity ; then crossing the stage to go out, she bows kindly to Charlotte ; then, with her finger up and a fine look of determination, leaves her.

In a subsequent scene the husband has come home, and his honest friend Jewson tries to open his eyes to the machinations of the villain Stukeley by telling what a bad boy he had been at school. Mrs. Siddons, who listens, is described thus :

She stands with riveted attention. She is behind at a little distance. The earnest and piercing look of her eyes, the simplicity of her attitude, is perfect nature.

The gamester replies to his honest friend : ' You are too busy, sir.' Mrs. Beverley rejoins : ' No, not too busy; mis- taken, perhaps—that had been milder.' The note on this runs :

Comes up to Beverley with a hasty anxiety and hurried voice, alarm and kind reproach in her look and manner.

The notes on the ' Gamester ' end here.

We are nowadays happily delivered from the false sentiment which required the ideal woman to love the more, the more she was ill-treated. We are rather in danger of shutting our eyes to the real beauty of patient Grisyld, the original of many copies, mostly, like Mrs. Beverley, caricatures. Chaucer's Grisyld fawns unpleasantly, but in the story of Griselda as

Boccaccio tells it, we find a very noble woman who thought herself of so small account in this great world, that she claimed nothing, while she held herself bound in all things to do her best. Her goodness is above all strong, whereas Mrs. Beverley is above all weak; her husband ruins, cheats, insults her, and she simply dotes on him all the time with slavish animal affection. No play can, however, be successful which has not some merit, and it is easy to recognise that in the conduct of the plot Moore shows skill, in so far that each scene reveals a deeper and deeper misery.

In Queen Katharine, Shakespeare has shown to what extent a woman of heroic mould might continue to love a husband who had mortally wronged her, and how fully the same woman could be just to a fallen enemy. Katharine, unlike Mrs. Beverley, is both good and strong.

Professor Bell wrote as follows on the fly-leaf of ' King Henry the Eighth : '

Mrs. Siddons's Queen Katharine is a perfect picture of a great, dignified, somewhat impatient spirit, conscious of rectitude, and adorned with every generous and every domestic virtue.

Her dignified contempt of Wolsey when comparing her own royal descent, her place and title as queen, her spotless honour, with the mean arts and machinations by which this man was driving her into the toils and breaking in upon her happiness ; her high spirit and impatient temper ; the energies of a strong and virtuous mind guarding the King at all hazards from popular discontent and defending her own fame with eloquence and dignity ; her energy subdued, but her queen-like dignity unimpaired by sickness ; and the candour and goodness of her heart in her dying conversation concerning her great enemy—all this, beautifully painted by Mrs. Siddons, making this one of the finest female characters in the English drama.

Our notes begin with the entrance of the Queen. The text, as before, is that of Mrs. Inchbald. The words on which the emphasis fell are underlined in the notes and are here printed in italics. An acute accent marks a word on which the voice was raised in pitch ; a grave accent marks a word on which the voice fell.

Act I. Scene 2.

Enter the Queen, *ushered by* Guildford, *who places a cushion on which she kneels. The* King *rises, takes her up, and places her by him.*

    King. Rise.

    Queen. Nay, we must longer kneel ; I am a suitor.

    King. Arise, and take place by us :—half your suit
Never name to us ; you have half our power ;
The other moiety, ere you ask, is given ;
Repeat your will, and take it.[2]

    Queen.   *Thank your Majesty.*
That you would love yourself, and, in that love,
Not unconsider'd leave your honour, nor
The dignity of your office, is the point
Of my petition.

    King.   Lady mine, proceed.

    Queen. I am solicited, not by a few,
And those of true condition, that your subjects
Are in great grievance : there have been commissions
Sent down among them, which have flaw'd the heart
Of all their loyalties :—wherein, although,
My good lord cardinal, they vent reproaches
Most bitterly ` on you, ´ as putter-on
Of these exactions, yet the king our master
(*Whose honour heaven shield from soil !* [3]) even he escapes not
Language unmannerly, yea, such which breaks
The sides of loyalty, and almost appears
In loud rebellion.

    Nor.   Not almost appears—
It doth appear : for, upon these taxations,
The clothiers all, not able to maintain
The many to them 'longing, have put off
The spinsters, carders, fullers, weavers, who,
Unfit for other life, compell'd by hunger,
And lack of other means, in desperate manner
Daring the event to the teeth, are all in uproar,
And danger serves among them.

    King.   Taxation !

    [2] Rises and sits by him. Then, in a composed and dignified tone, addresses him, very articulate and very earnest.

    [3] Tenderly and religiously.

Wherein ? and what taxation ?—My lord cardinal,
You that are blam'd for it alike with us,
Know you of this taxation ?
  Wol.   Please you, sir,
I know but of a single part in aught
Pertains to the state ; and front but in that file
Where others tell steps with me.
  Queen.   No, my lord,
You *know* no more than others :[4] but you frame
Things that are known alike, which are not wholesome
To those which would not know them, and yet must
Perforce be their acquaintance. These exactions
Whereof my sovereign would have note, they are
Most pestilent to the hearing ; and to bear them
The back is sacrifice to the load. They say
They are devis'd by you ; or else you suffer
Too hard an exclamation.
  King.   Still exaction !
The nature of it ? In what kind, let's know,
Is this exaction ?
  Queen. [5] I am much too venturous
In tempting of your patience ; but am bolden'd
Under your promis'd pardon.[5] [6] The subjects' grief
Comes through commissions, which compel from each
The sixth part of his substance, to be levied
Without delay ; and the pretence for this
Is nam'd, your wars in France :[6] this makes bold mouths :
Tongues spit their duties out, and cold hearts freeze
Allegiance in them ; their curses now
Live where their prayers did ; and it's come to pass,
This tractable obedience is a slave
To each incensed will. [7] I would your highness
Would give it quick consideration.
  King.   By my life,
This is against our pleasure.

The notes cease until the surveyor of the Duke of Buckingham enters, to whom Wolsey speaks :

  Wol. Stand forth ; and with bold spirit relate what you,

---

[4] Mildly, but very decidedly, accusing him.
[5] Gracious apology.   [6] Very articulate and clear.
[7] Very earnest.

Most like a careful subject, have collected
Out of the Duke of Buckingham.
　　King.　　　　　　　　　Speak freely.
　　Surv.　First, it was usual with him—every day
It would infect his speech—that if the king
Should without issue die, he'd carry it so
To make the sceptre his : these very words
I have heard him utter to his son-in-law,
Lord Aberga'ny ; to whom by oath he menac'd
Revenge upon the cardinal.[8]
　　Wol.　　　　　　Please your highness, note
This dangerous conception in this point.
Not friended by his wish, to your high person
His will is most malignant ; and it stretches
Beyond you, to your friends.
　　Queen.　　　　　[9] My learn'd lord cardinal,
Deliver all with *charity*.[9]
　　King.　　　　　　Speak on :
How grounded he his title to the crown,
Upon our fail ? to this point hast thou heard him
At any time speak aught ?

The Surveyor continues to give his evidence, stating that a
Chartreux friar had prophésied to the Duke that he should
govern England.　Then the Queen intervenes :

　　Queen.[10]　　　　　If I know you well,
You were the duke's surveyor, and lost your office
On the complaint o' the tenants ; [11] *take good heed
You charge not in your spleen a noble person,
And spoil your nobler soul : I say, take heed.*[11]
　　King.　Go forward.

The Surveyor continues his evidence, and states that
Buckingham had said that if he had been committed to the
Tower he would have put a knife into the King; on which the
King exclaims :

　　[8] She hears all this with a dignified, judge-like aspect, often darting a keen look of inquiry at the witness and the Cardinal.
　　[9] A grand sustained voice.　The emphasis on ' *charity* ' strong.
　　[10] A very penetrating look.　Looks very steadfastly and seriously in his face for some time, then speaks.
　　[11] The second part of this speech very severe tone of remonstrance.　Grand swell on ' *and spoil your nobler soul.*' ' I say,' &c., very emphatic.

 King. A giant traitor !

 Wol. Now, madam, may his highness live in freedom,

And this man out of prison ?

  Queen.   God mend *all* ! [12]

The scene shortly ends. Mrs. Siddons in this scene evidently brought into strong relief the intellect and power of the
Queen as well as her rectitude. In the fourth scene of the
second act the Queen enters, called into the court at Blackfriars.
The clerk of the court says, 'Katharine, queen of England, come
into the court.' Again Guildford precedes the Queen with a
cushion, and again she kneels.

ACT II. SCENE 4.

 Queen. [13] Sir, I desire you do me right and justice,

And to bestow your pity on me ; for

I am a most poor woman, and a stranger,

Born out of your dominions ; having here

No judge indifferent, nor no more assurance

Of equal friendship and proceeding. [13] [*She rises.*] [14] Alas, sir,

In what have I offended you ? what cause

Hath my behaviour given to your displeasure,

That thus you should proceed to put me off,

And take your good grace from me ? [14] [15] Heaven witness,

I have been to you a *true and humble* wife,

At all times to your will conformable. [15]

        [16] Sir, call to mind

That I have been your wife, in this obedience,

Upward of twenty years, and have been blest

With many children by you : if, in the course

And process of this time, you can report,

And *prove it too*, against mine honour aught,

My bond to wedlock, or my love and duty,

Against your sacred person,⁄ in God's name,

Turn me away ;⁀ and let the foul'st contempt

Shut door upon me, and so give *me* up

---

[12] A long emphasis, intimating that the Cardinal and his designs were
known to her.

 [13] A most sweet and gracious prelude, yet no departure from her dignity.

 [14] Remonstrance, dignified, without any bitterness.

 [15] Earnest protestation.

 [16] Dignified confidence in her own innocence.

To the sharpest kind of justice.[16]   [17] Please you, sir,
The king, your father, was reputed for
A prince´ most prudent,` of an excellent
And *unmatch'd* wit and judgment` : Ferdinand,
My father, king of Spain, was reckon'd one
The wisest prince, that *there* had reign'd by many
A year before : it is not to be question'd
That they had gather'd a wise council to them
Of *every* realm, that did debate this business,
Who *deem'd our marriage lawful* : wherefore I humbly
Beseech you, sir, to spare me, till I may
Be by my friends in Spain advis'd ; whose counsel
I will implore ; if not, i' the name of God,
Your pleasure be fulfill'd !
    Wol.   [18] You have here, lady
(And of your choice), these reverend fathers ; men
Of singular integrity and learning,
Yea, the elect of the land, who are assembled
To plead your cause : it shall be therefore bootless,
That longer you desire the court ; as well
For your own quiet, as to rectify
What is unsettled in the king.
    Cam.   [19] His grace
Hath spoken well and justly : therefore, madam,
It's fit this royal session do proceed ;
And that, without delay, their arguments
Be now produc'd and heard. [[20] CAMPEIUS *rises.*]
    Queen.   Lord Cardinal,
To you I speak.
    Wol.   Your pleasure, madam ?
    Queen.   Sir,
I am about to weep ; but, thinking that
We are a queen [21] (*or long have dream'd so*),[21] certain

[17] Pause. A new division of the discourse. The argument beautifully spoken, very distinct.

[18] This response taken by her with great impatience, very indignant at his interference.

[19] Surprise and grief when the legate speaks thus.

[20] When Campeius comes to her she turns from him impatiently; then makes a sweet bow of apology, but dignified. Then to Wolsey, turned and looking *from* him, with her hand pointing back to him, in a voice of thunder, 'to *you* I speak.' This too loud perhaps ; you must recollect her insulted dignity and impatience of spirit before fully sympathising with it.

[21] Great contempt in this parenthesis.

The *daughter of a king*,[22] my drops of tears
I'll turn to sparks of fire.
     Wol.   Be patient yet.
     Queen.   [23] *I will, when you are humble* ; nay, *before,*
Or God will punish me.[23]   [24] I do believe,
Induc'd by potent circumstances, that
You are mine enemy ; and make my challenge `.
You shall not be my judge : for it is you
Have blown this coal betwixt my lord and me,
Which God's dew quench ! [24]—[25] *Therefore,* I say again,
I utterly abhor, yea, from my soul
Refuse you for my judge : [26] whom, yet once more,
[26] I hold my most malicious foe, and think *not*
At all a friend to truth.[26]
     Wol.   [27] I do profess
You speak not like yourself ; who ever yet
Have stood to charity, and display'd the effects
Of disposition gentle, and of wisdom
O'ertopping woman's power. Madam, you do me wrong :
I have no spleen against you ; nor injustice
For you or any : how far I have proceeded,
Or how far further shall, is warranted
By a commission from the consistory,
Yea, the whole consistory of Rome. You charge me
That I have blown this coal : I do deny it :
The king is present : if it be known to him
That I gainsay my deed, how may he wound,
And worthily, my falsehood ! yea, as much
As you have done my truth. If he know
That I am free of your report, he knows
I am not of your wrong. Therefore in him
It lies to cure me : and the cure is, to
Remove these thoughts from you ; the which before
His highness shall speak in, I do beseech
You, gracious madam, to unthink your speaking,
And to say so no more.

[22] Very dignified.
[23] Great contempt. Her voice swelled, but monotonous.
[24] Very distinct articulate charge against him.
[25] Great swell.
[26] ' I hold,' &c., very pointed. ' Not *at all*,' &c., syllabic and most impressive.
[27] Great impatience and contempt during this speech of Wolsey.

Queen.[28]   My lord, my lord,
I am a single woman, much too weak
To oppose your cunning.   [29] You're *meek* and humble-*mouth'd* ;
You sign your place and calling, in full seeming
With meekness and humility ; but your *heart*
Is cramm'd with arrogancy, spleen, and pride.[29]
You have, by fortune and his highness' favours,
Gone slightly o'er low steps, and now are mounted
Where powers are your retainers ; and your words,
Domestics to you, serve your will, as 't please
Yourself pronounce their office. I must tell you,
You tender more your person's honour than
Your high profession spiritual : that again
I do refuse you for my judge ; and here,
Before you all, appeal unto the pope,
To bring my whole cause 'fore his holiness,
And to be judg'd by him.
                    [*She curtsies to the* KING, *and offers to depart.*
    Cam.   The queen is obstinate,
Stubborn to justice, apt to accuse it, and
Disdainful to be tried by 't ; 'tis not well
She's going away.
    King.   Call her again.
    Crier.   Katharine, queen of England, come into the court.
    Grief.   Madam, you are called back.
    Queen.   [30] What need you note it ? pray you, keep your way :
When you are call'd, return.[30]—[31] Now the Lord help,
They vex me past my patience ! [31]—Pray you, pass on :
[32] I will not tarry : [32] no, nor ever more,
Upon this business, my appearance make
In any of their courts.   [*Exeunt* GUILDFORD *and the* QUEEN.

Professor Bell was as good a hearer as actor or actress need
hope for.

The scene in the fourth act where Katharine is discovered
sick unto death is prefaced with these remarks :

Mrs. Siddons in this scene admirable in simplicity and pathos.
No affectation, not a more complete deception in dramatic art than

---

[28] Breaking impatiently through his speech.
[29] Contempt. Contrast strong between ' mouthed ' and ' heart.'
[30] Very impatient, angry, and loud.
[31] Peevish expression.   [32] Strong determination.

this of the sickness of Katharine. The voice subdued to softness, humility, and sweet calmness. The soul too much exhausted to endure or risk great emotion. The flash of indignation of her former spirit very fine at Guildford's interruption.

Unfortunately there is only one more remark; it is appended to Katharine's verdict on Wolsey, which in Mrs. Inchbald's edition runs as follows:

> *Queen.* So may he rest ; his faults lie gently on him !
> Yet thus far, Cromwell, give me leave to speak him,
> And yet with charity.—He was a man,
> Of an unbounded stomach, ever ranking
> Himself with princes.
> His promises were, as he then was, mighty ;
> But his performance, as he is now, nothing.
> Of his own body he was ill, and gave
> The clergy ill example.

Professor Bell says of this :

Beautifully spoken, with some mixture of energy ; but the subdued voice throughout.

Probably the writer was too much affected by this scene to be able to make minute critical observations.

Of Mrs. Siddons' readings, Professor Bell says :

Mrs. Siddons in her readings was like the tragic muse. She sat on a chair raised on a small platform, and the look and posture which always presents itself to me is that with which she contemplates the figure of Hamlet's ghost. Her eyes elevated, her head a little drawn back and inclined upwards, her fine countenance filled with reverential awe and horror, and the chilling whisper scarcely audible but horrific. Sir Joshua Reynolds's picture of Mrs. Siddons as the tragic muse gives a perfect conception of the general effect of her look and figure in these readings.

In her readings the under parts, which in acting are given offensively by some vile player, were read with a beauty and grace of utterance which was like the effect of very fine musical recitation, while the higher parts were the grand and moving airs. It was like a fine composition in painting : the general groundwork simple, the parts for effect raised and touched by a master's hand.

In the higher parts it was like the finest acting. The looks, the tones, the rapid hurry of the tumultuous emotions, the chilling whisper of horror, the scream of high-wrought passion, were given less strongly, but as affecting as on the stage.

The comic touches were light and pretty, but she has no comic power.

The graceful and sweet parts were quite enchanting. The mellow subdued voice of sorrow, to give variety, she kept much in whisper—very audible notwithstanding. *Her* whisper is more audible and intelligible than the loudest ranting of an ordinary player.

She read 'Hamlet' and the 'Merchant of Venice.' 'Lear,' I think, should be read by her, not acted.

There is special mention of her manner when reading Hamlet's speech beginning

    Angels and ministers of grace defend us.

Mrs. Siddons in reading gave, by her look of reverential awe and chilling whisper of horror, more fully the idea of a ghost's presence than any spectral illusion on the stage.

This was a whispered speech throughout, growing in energy and confidence as other ideas took the place of the first startle of horror and dismay. Kean speaks too loudly and boldly, not enough as in the withering presence of a supernatural being. The first line should be a whisper of horror, with a long pause before venturing to address the phantom.

It is believed that Sir Charles Bell made notes similar in character to those now published; but if so the books have been mislaid. There is a curious passage in a letter from him to his brother, dated the 10th of June, 1809, in which he says :

Jeffrey saw my 'Shakespeare' and liked it much, and talked to Mrs. Siddons about it. I said I intended some time to take a good play and make it so *in fancy*. He said he should like to do so too. He saw your pencillings in the margin : not knowing whether you would like it, and not knowing what they were, I told him they were all mine ; so perhaps his liking this kind of thing was owing to you. Do not forget to pursue it.

This appropriation by one man of another's work, reads oddly, though it is an indication of the absolute confidence of one brother in the other. We may all feel glad that Professor

G. J. Bell did pursue the plan, and wish he had pursued it further.

In reading of Mrs. Siddons one cannot but regret that her genius should have been employed in representing a Mrs. Beverley or even a Lady Randolph. It is a standing reproach to our literature that outside the roll of Shakespeare's characters a great actor can hardly find a great part. When we reflect that West and Haydon have been followed at no distant time by Millais, Leighton, Burne-Jones, and Watts, we cannot but hope that in a sister art a similar revival may occur. The time seems ripe, for the novel is in decadence, and coming writers must win distinction in a new field. A man who has sufficient talent to make a good novel would probably succeed in writing a good play if he went to work in the right way; but the art of the playwright has not been studied by our leading authors for many generations. This art is that of selecting proper subjects for stage representation and giving them such a form as will enable the actors to move their audience. The success of a play in stirring an audience depends less than is usually supposed on style, on the delineation of character, or even on the invention of an ingenious and probable plot. Plays succeed which are glaringly defective in all these respects; for instance, the 'Lady of Lyons.' The one necessary condition for success is that the scene represented shall move the audience; the emotion may be sad or merry, noble or ignoble, but emotion there must be. If this element be wanting, no depth of thought, no beauty of language, no variety of incidents will save the play. The skilled playwright knows what scenes will stir the hearers, and how best to frame each scene and the whole play with this purpose. If with this knowledge he possesses originality of conception and beauty of style, his plays become part of the literature of his country; without these higher qualities he remains a mere playwright, but we go to see his plays, built up as they are of old worn-out materials. The playwright is familiar with the materials used in his art; he knows the stage well on both sides of the footlights; he mixes with actors, managers, stage-managers, scene-painters, and stage-carpenters. From Æschylus downwards, all great drama-tists have had this practical knowledge of the instruments at their command. A drama should be written for the stage, as a song should be written to be sung. The author must sub-consciously—if such a word may be used—have the stage always in mind : the exits, the entrances, the time required to cross the stage, the positions of the actors, their very attitudes and dress. No author provides more admirably for all these stage exigencies than Shakespeare, as any one may see who will consider his inimitable contrivances for removing dead bodies from the stage. There is no doubt a danger that those who become familiar with stage-machinery may content themselves with remodelling the old puppets, rearranging stock incidents, and repatching old rags to produce good guaranteed old stage effects ; but a man of real talent would not be misled by the Mr. Worldly Wiseman of the stage.

We may learn much from French practice as to the frame-work of a drama. A great part of the success which is certainly achieved by modern French plays depends on the art shown in their construction. M. F. Legouvé, who is a skilful playwright, tells us frankly how a Frenchman proceeds. First, he chooses or conceives the situation which is to be the crisis of the play : from this he works backwards, considering how that situation is to be brought about, and what characters will be necessary for the purpose. His first act is devoted wholly to informing the audience of the relations between the characters at the begin-ning of the piece; his second act develops the plot; in his third act the plot thickens; his fourth act contains the crisis for which the play is written, and his fifth act gives the solution of the knot which has been tied in the fourth act.

These rules seem rather barren, but we shall see their signi-ficance if we consider what other courses may be followed.

A writer may begin by inventing an ingenious or interest-ing plot, or by choosing some historical period which he will dramatise, or by conceiving some marked characters whose feel-ings and thoughts he will expound. M. Legouvé tells us that none of these is the French method; that for the French author the motive of the play is essentially one situation ; that his characters are chosen so as to make this situation tell, and that his plot is a matter for after-consideration, devised so as to reveal the characters of the persons and lead up to the crisis.

Shakespeare did not work in this way, but in this one matter of construction it may be worth while to listen to maxims derived from the study of plays which in all other respects are greatly inferior to his. Moreover, these maxims are ultimately derived from the practice of Sophocles, no mean master.

The French, following the Greeks in this, look on a play as a representation of feelings rather than of actions. The incidents which occasion the feelings, and the actions they lead to, are alike kept in the background in French as in Greek plays. Rapid action in a play does not, in France, mean a rapid succession of events, but a rapid development of feeling in the persons of the drama. A scene in which the emotion represented is monotonous will be dull even if crammed with incidents.

The author who is penetrated with the belief that the aim of the drama is to produce emotion will be indifferent to beauty of language or of metaphor, to profound philosophy and to brilliant sayings, except when these help to move the audience. He will know that obscurity of language or of thought is fatal to his purpose. The knot, crisis, or motive of his play will be chosen by him to exhibit, not a striking event, but strong feelings. He will so contrive the story leading to the crisis as to exhibit a gradually culminating series of emotions, produced by incidents arranged so as powerfully to affect the personages of the drama, and through them the audience. The direct action of incidents on the audience is of importance only in that low form of art which aims at stirring the vulgar feeling of curiosity and the vulgar love of gaping.

The most telling play is that in which the feelings naturally exhibited by the persons of the drama are strongest. The greatest play is that which shows the feelings of the noblest men and women. This, in the opinion of Plato and Aristotle, is the object of the drama in its higher form.

Plato, in ' The Laws,' after saying that no freeborn man or woman should learn comic songs, grotesque dances, or burlesques, but that it might be well to have these things presented by slaves and hired strangers, in order better to understand by contrast that which is truly beautiful, speaks thus, referring to his ideal city : ' If any serious poets, such as write tragedies, should ask us, " Shall we, O strangers, come to your city and bring our poetry and act it? How stand your laws in this respect?" what answer ought we to give to these divine men? For myself I should reply thus: " Oh, most excellent of strangers, we are ourselves, to the utmost of our power, poets of a tragedy the most beautiful and the best; for the whole of our polity consists in an imitation of a life the most beautiful and best, which we may say is in reality the truest tragedy."' We here see that Plato thought the object of tragedy was to represent the noblest kind of life, and only rejected the imitation as unnecessary where this life itself was to be seen.

Aristotle defined what he meant by a tragedy with greater fulness. He points out that a certain magnitude is necessary in the event represented; that the spectator as he follows the action feels pity and a kind of awe which may be termed fear or terror, and that he comes away from the spectacle chastened and purified. The first part of his definition requires that the action shall be heroic, or such as represents the thoughts, deeds, and feelings of great men. By the last part of his definition he, like Plato, required that the action should have moral beauty. This does not imply that a play should be didactic, or deal only with the actions of well-behaved persons. The teaching of the dramatist is as the teaching of nature. See these heroes in their strength and their weakness, live with them, and you will learn from them. The function of the tragic poet, from Æschylus to Shakespeare, has been to show us the intense life of heroic men and women at the moment of their trial.

But not all heroic or beautiful actions can be made the subjects of a tragedy. Aristotle points out that the action must be such as will stir certain moral emotions—pity and fear he calls them; but the English words very imperfectly describe the feelings roused by a great tragedy; those feelings give keen pleasure, whereas pity and fear are painful. Sympathy may be a better word; the pleasure is to live a little while greatly with the great ones of the world, to feel their feelings, to experience their passions, to dare, to love, to hate with them, so that for a little while we too are great; but words fail to describe emotions to those who have not felt them. If it be suggested that the sensation experienced while watching a tragedy is rather a feeling *with* the persons of the drama than a feeling *for* them;

that when Othello cries out, ' O the pity of it!' we feel as he feels and what he feels, and are very far indeed from entertaining a pleasant and comfortable pity *for* him; that the strange pleasure depends on our recognition in ourselves of the power to feel as Othello feels, to suffer as he suffers, even to sin as he sins; this suggestion may awaken a memory of what the emotion was in those who have known it but can explain nothing to others.

The higher and lower forms of the drama differ simply in respect of the character of the feelings awakened. The highest may be our highest moral emotions; the lowest, the lowest animal passions. Either moral or immoral the stage must be, and always has been, for its very existence depends on its action upon this part of our nature.

The morality of a play depends on no exclusion of crime, no enumeration of maxims, no system of rewards or punishments; it flows from the heart of the author and is tested by its action on the audience.

It is in moral grandeur that Shakespeare, Æschylus, and Sophocles stand absolutely pre-eminent. It is to this that Racine and Corneille owe their hold on men. It is by this that the ' Misanthrope ' claims high rank. It is in this that the modern French stage chiefly fails.

The French dramatic authors of the Second Empire have succeeded in producing living plays, because, besides being skilled playwrights, they do in their works appeal to real and strong feelings. A certain moral poverty alone prevents the school from taking a very high rank. The authors have usually meant well; and if the verdict must be that their moral ideal is always poor and often false, this conclusion is forced upon us by the words and actions rather of their good than of their bad people. Even Victor Hugo's verse cannot make us believe that Ruy Blas is not a poor creature.

Our own writers show no similar moral ineptitude, and since they have created scores of types which in freshness, truth, power and interest surpass the men and women of French authors, we are driven to the conclusion that if the English do not write great plays it is rather because they do not know how, than that they lack power. Our best authors, when they attempt the drama, seem to be misled by a desire to appeal

rather to the intellect or to the æsthetic sense of their hearers
than to their moral emotions. If they were to mix with actors
on familiar terms they would soon learn the playwright's art;
for the actor knows what will succeed upon the stage. An actor
calls a part well written when the words and situations are such
as enable him powerfully to express strong feelings. He will,
if permitted, cut out every line which does not help him in this,
his art, and for stage purposes he is right. Charm of style,
beauty of metre, wisdom of thought, novelty of character, ingenuity of plot, poetry of conception, all these things may be
added to a play with much advantage; but they will not insure
success either singly or all together. A play which does not
move an audience, as neither intellectual nor artistic pleasure
ever can move them, must fail upon the stage.

Professor Bell's notes show what he felt when a Siddons
acted a Katharine. He was a man of hard intellect, whose dry
legal labours still guide shrewd lawyers. He was a man of
learning and taste; but when seeing a great actress in a great
play, no ingenious theories, no verbal emendations, no philosophical reflections, no analytical remarks occur to him. He
records his emotion, and, as far as he can, how that emotion was
produced. He may be taken as representing an ideal audience
—that which does not comment, but responds to author and to
actor.

TALMA ON THE ACTOR'S ART.[1]

MR. IRVING, in his preface to this remarkable essay, calls it ' a
kind of vade-mecum of the actor's calling, written by one of
themselves, and by an artist universally recognised as a competent
expositor ; ' ' a permanent embodiment of the principles of our
art.' We may then start with every confidence that we have
here a true explanation of the manner in which a great actor
works. Let us listen to his words. ' Every actor,' says Talma,
' ought to be his own tutor. If he has not in himself the
necessary faculties for expressing the passions and painting
characters, all the lessons in the world cannot give them to him.
The faculty of creating is born with us ; but if the actor possesses
it the counsel of persons of taste may then guide him ; and as
there is in the art of reciting verse a part in some degree
mechanical, the lessons of an actor profoundly versed in his art
may save him much study and time.' Here, we take it, is the
truth, the whole truth, and nothing but the truth as to dramatic
teaching. A man may be taught to speak and move well and
suitably ; then, if he has genius, he may in twenty years teach
himself to act, and during the process he may be much helped
by the counsel of persons of taste. And how is he to know
whether he has the necessary genius ? Talma answers, ' sensibility ' and ' intelligence ' are the two faculties pre-eminently
required, but under the general heading of sensibility he includes
much. He puts almost contemptuously on one side ' the faculty
which an actor possesses of being moved himself and of affecting
his being so far as to imprint on his features, and especially on
his voice, that expression and those accents of sorrow which
awake sympathy and extort tears.' No doubt the actor must

[1] Review of *Talma on the Actor's Art* : with preface by Henry Irving.
From the *Saturday Review*.

have this kind of sensibility; but to this extent sensibility is not
rare.   It may sometimes be recognised in amateurs acting for
the first time; and we take it that no moderately successful
actor, even on a second-rate provincial stage, ever wanted sensibility to this extent.   Let us call it, for the purposes of future
reference, sensibility in the first degree, and then pass to what
Talma further requires and still calls sensibility—namely, 'that
imagination which enables the actor to look on at the lives of
historical personages, or the impassioned figures created by genius,
which reveals to him as though by magic their physiognomy,
their heroic stature, their language, their habits, all the shades of
their character, all the movements of their soul, and even their
singularities.'   We begin to feel that sensibility in the second
degree is more difficult of attainment, and here it is well to
remark that Talma does not place this faculty under the heading
of ' intelligence.'   He does not tell the actor that he must
understand his author.   This insight which he so justly acquires
is to be a matter of feeling.   The revelation comes by magic, not
logic.   Fanny Kemble says, in perfect accord with Talma, perception rather than reflection reaches the aim proposed.   It is
the absence of this sensibility to the second degree that makes
many ordinary fairly good actors so insufferably bad in great
parts.   Probably they understand the words they speak, and
have a vague notion of what the person they represent may be
supposed to feel, but they have no insight into heroic thought
or feeling ; and, says Talma, ' if the actor is not endowed with
a sensibility at least equal to that of any of his audience, he can
move them but very little.'   Too often our actors have less of
this sensibility than many of those who hear them.   Why then,
it may be asked, do not audience and actors change places ?
Because the sensitive hearers lack sensibility in the third degree
—for Talma has not done with this word yet.   He includes in
this term ' the faculty of exaltation which agitates an actor,
takes possession of his senses, shakes even his very soul, and
enables him to enter into the most tragic situations and the
most terrible of the passions as if they were his own.'

Now not one of the audience which condemn the second-rate
actor in a great part because they have more sensibility than he has
will be found capable of the kind of exaltation here described.   We

think that here Talma has confused or blended two very different faculties under one name. To feel and to express were one to the great actor, but the vast majority of mankind is, we think, denied the gift of expressing emotion. And here it seems to us that Talma misses the very point which distinguishes the actor from other artists. All artists must have this sensibility he demands, but the form which each naturally employs to express his emotion determines whether he shall be author, painter, musician, or actor. Under the influence of this ' exaltation' the actor finds the tone, the look, the gesture required to express the feeling with which he is inspired, and this gift is, to some extent, possessed by all actors who can earn their bread. This is the faculty which is trained by stage practice. And here we may again refer for support to ' Notes on some of Shakespeare's Plays,' by F. A. Kemble. Speaking with the authority of tradition in a great family, she says, ' There is a specific comprehension of effect and the means of producing it, which in some persons is a distinct capacity, and this forms what actors call the study of their profession.' And although Talma mixed up expression and feeling when endeavouring in a brief way to write an analytical account of his own art, he takes precisely this view of study. Here is his method. ' The actor who possesses this double gift' (sensibility and intelligence) ' adopts a course of study peculiar to himself. In the first place, by repeated exercises he enters deeply into the emotions, and his speech acquires the accent proper to the situation of the personage he has to represent. This done, he goes to the theatre not only to give theatrical effect to his studies, but also to yield himself to the spontaneous flashes of his sensibility and all the emotions which it voluntarily produces in him. What does he then do? In order that his inspirations may not be lost, his memory, in the silence of repose, recalls the accent of his voice, the expression of his features, his action—in a word, the spontaneous workings of his mind which he had supposed to have free course, and, in effect, everything which in the moments of his exaltation contributed to the effect he had produced. His intelligence thus passes all these means in review, connecting them and fixing them in his memory, to re-employ them at pleasure in succeeding representations.' This passage expresses better than anything we have ever read what the actor's study really should be. After a certain amount of preparation, he yields in a state of exaltation to impulse; suggestions crowd upon him; tones, cries, gestures, expressions, actions, are created. The exaltation is extreme, and these moments when he is alone, and the god works in him, may be those of keenest pleasure. But this state is succeeded by a calm and critical mood, in which the true artist chooses, rejects, and groups the partial effects obtained so as to produce one great and consistent whole. In this work, he will be greatly aided if he has a sympathetic friend of sound judgment—Talma's 'person of taste'—whose counsel he may take. Those who know what this study means are driven almost to distraction when they hear an actor—perhaps a great actor—complimented on being able to remember the words of his part. But, on the other hand, it must be almost as galling when a great actor is told that he really understands his author's meaning. One great charm in this essay by Talma lies in the total absence of this contemptible worship of the human understanding—a very good thing in its way, though one of but small importance in mere art. To Talma intelligence meant a sound critical faculty, not logical, but perceptive, enabling its possessor to keep what was good in art and reject that which was less good. We find in this essay a clear solution of the question continually asked, whether the actor really feels what he is acting. Talma, as we understand him, only felt the emotion once in its full intensity—that is to say, at the moment of creation during the solitary rehearsal. Subsequently the effect was produced by the aid of memory; but the body is so constituted that if by the aid of memory we perfectly reproduce a tone or cry, that tone or cry brings back simultaneously a close reproduction of the feeling by which it was first created. Thus to act a great part a man must be capable of real greatness. As Talma says: ' He will never rise to excellence as an actor whose soul is not susceptible of the extremes of passion.' And yet the representation night after night of these great feelings may come to be almost mechanical, or, rather, the feelings of the actor can be almost mechanically reawakened by the excellence of his own art. Thus in describing Le Kain at his best period, when his art was ripe, he says:

Accent, inflections, action, attitudes, looks, all were reproduced at every representation with the same exactness, the same vigour ; and if there was any difference between one representation and another, it was always in favour of the last.

Spontaneity is an admirable gift, but you cannot be spontaneous a second time. Spontaneous movements are right and necessary at the moment of creation, but are wholly out of place before an audience.

Talma liked good scenery and correct dresses, but one feels that if he were alive now, he might say, ' Faut de la vertu, pas trop n'en faut.' His remarks on truth and nature are true and natural. He points out, taught by the scenes he had witnessed during the Reign of Terror, that ' the man of the world and the man of the people, so opposite in their language, frequently express the great agitations of the mind in the same way,' and that ' the great movements of the soul elevate man to an ideal nature in whatever rank fate may have placed him.' While, however, he recommends the observation of passion in others, it is clear that he never condescended to mimicry. Some talent for mimicry is very common among actors, and is indeed a useful accomplishment, especially in the lower walks of the profession; but no man can ever hope to play Coriolanus by mimicking some statesman.

Talma's chief observations were made upon himself. He attended to his own tones, his own face, when in real grief; he is half ashamed and half proud of having done so. We imagine that all artists are alike on this point, and that in this fact lies a certain compensation for the extra keenness of their feelings. They suffer more than other men, and get more good from suffering. Talma observed that an emotion truly expressed moved an audience which did not understand the words. Most people would attribute this to gesture ; but he, rightly as we think, considered the effect as due to the voice, and as an instance he speaks of Miss O'Neil moving Frenchmen who did not understand her to tears. The point is a curious one, for we have observed that a foreigner can judge artistic truth in acting with fair success when he is wholly incapable of appreciating any little niceties of accent or elocution. Thus too we allow foreigners to act on our stage who cannot speak one word so as

to be acceptable in our ears as English.   Yet their tones will
bring tears almost as readily as if they spoke with English
tongues.   We believe that this admits of explanation; but the
theory would demand too much space to be developed here.   Let
all who are interested in acting read Talma's essay; and then,
if they wish for a little amusement, they may turn to the 'Actor's
Art,' by Mr. Gustave Garcia.   Talma tells his readers what a
great actor must learn, Mr. Garcia explains what small actors
can be taught and do learn.

93

ARTIST AND CRITIC.[1]

ARTISTS—whether they be painters, actors, writers, musicians, or what not—are usually dissatisfied with their critics; and we will not deny that they have reasons for their discontent. Good criticism is at least as rare as good art of any other kind. But the artist when grumbling at his critics often asserts what we hold to be untrue—namely, that criticism to be valuable must proceed from a man who is himself skilled in the art of which he speaks. He says that the judgment of the ignorant should be disregarded, and he counts all men ignorant who cannot execute a masterpiece at least equal to his own best work. The artist, in fact, claims to be judged by his peers. The claim seems so reasonable that, if we dare to challenge it, we must do so with many reservations. In the mechanical arts, such as the forging of horseshoes or in arts of mere skill, such as cricket, the judgment of the laity, as we may call the unskilled public, is really worthless. To be a judge in these matters a man must have forged iron or played cricket, and in respect of the finer arts also we take it that as regards mere skill of workmanship, deftness of execution, the artist has such great advantages over the layman, that his verdict in respect of skill must be received without appeal. We here grant the artist a clear supremacy, and admit that a very bad sculptor will be a better judge of the skill shown in carving marble than the most learned connoisseur. He will recognise distinctions of touch, style, and method which are invisible to the multitude, and seen but dimly by unskilled lovers of the art. In painting, in music, in writing, in all fine arts, the same answer holds good. The skilful are the sole judges of skill; but in the fine arts skill is not all in all. The

[1] From the *Saturday Review*, October 11, 1884.

layman may say to the artist, your knowledge of technique is a snare. In judging a work of art you examine the workmanship, and forget to look at the work produced. Are you a writer? You allow the merit or demerit of word arrangements to distract your attention from the idea they should suggest. Are you a painter? You see paint, and not a picture. Are you a musician? You hear combinations, not music. The layman who says such things is guilty of gross exaggeration, and yet his words indicate a real danger to which an artist is liable when he plays the critic.

In mechanical arts the craftsman uses his skill to produce something useful, but (except in the rare case when he is at liberty to choose what he shall produce) his sole merit lies in skill. In the fine arts the student uses skill to produce something beautiful; he is free to choose what that something shall be, and the layman claims that he may and must judge the artist chiefly by the value in beauty of the thing done. Artistic skill contributes to beauty, or it would not be skill; but beauty is the result of many elements, and the nobler the art the lower is the rank which skill takes among them. The intense enjoyment which the artist takes in the exercise of his own skill, and from the generous and sympathetic perception of skill in others, tempts him to overvalue this element of beauty, whereas most laymen are apt to undervalue the skill which they only half perceive. Passing to higher walks of criticism it is often said, and with some truth, that our judgment of artistic work should not depend on our mere personal perception, but should be based on some acknowledged principles or canons of art. Now the word principle is here used by the critic and artist in a very different sense from that attached to it by his scientific brethren; and a very pretty, but uninstructive, wrangle ensues when the man of science challenges the artist to put his principles into clear language. Nevertheless, it may be granted that the so-called principles of art current in each school at a given moment, though vaguely defined, imperfectly apprehended, illogical, and variable with the development of the art, do nevertheless exercise a healthy influence, especially on the great mass of ordinary men who get their living by that art; nay, it must be allowed that many statements about art made by Aristotle and Plato still seem true, but the sources from which a knowledge of these eternal canons can be gained are open to the layman as to the artist. Many men who have no practical skill in any art take great pleasure in studying the laws of art, and of these some are far more competent than most artists to collect experience, to analyse emotions, and to arrange in logical sequence the facts observed. The artist is seldom a man of clear thought, though instances to the contrary may be found. Nevertheless, here we think that the seeming advantages of the layman tend really to his disadvantage. In the mind of the artist the vaguely apprehended principle may really live and guide his work; as religious faith may guide a man who could neither defend a dogma by logic nor even apprehend its meaning with accuracy. The layman, precisely because he formulates his principles more clearly, is in greater danger of using them like formulæ. He then obtains his judgment by a sort of calculation and can no longer trust his senses. In other words, he becomes a pedant. No form of criticism, not even the dogmatic, is so barren as that which endeavours to test the merit of a composition by a series of comparisons with a series of assumed standards. The elements of beauty cannot be weighed in a balance like chemical ingredients. We are not aware that any dramatic critic (with the possible exception of Mr. William Winter) has spoken of a play as containing fifteen measures of variety, ten of repose, six of style, and so forth, though we suspect that professorial examiners in literature have before now endeavoured to estimate the comparative excellence of essays by some such simple numerical process; but even when numbers are eschewed, the critic who systematically considers a work of art under a series of heads, and endeavours to appraise its value by ascertaining how far it squares with each successive rule, is trying to measure beauty with a tape line. A man's judgment of a work of art, be he artist or layman, should come to him as a direct perception, although when he desires to understand his sensation or to explain it to others, he may have recourse to analysis and comparison. The artist is in less danger than the layman of inverting the proper order.

Yet this is rather a dangerous argument to use in favour of the artist—that he is less likely to go wrong because he trusts less to his intellect.   If we leave the mere principles of art and consider other kinds of knowledge, such as the history of art, the literature of art, the lives of artists, and so forth, the average artist certainly stands at no advantage over the connoisseur.

Besides skill and knowledge there is a third qualification which a man must have who would judge soundly of any art.   He must be capable of deriving intense pleasure from what the art produces.   He must really enjoy that particular kind of beauty —the delight a man feels in looking at a picture, in hearing music, in seeing acting, is a proof that he possesses at least the rudiments of that sense which, when duly cultivated, may enable him to be a good critic of the given art.   Here, again, the artist seems at first to stand at great advantage.   The mere choice of his profession proves that, far from being indifferent to the contemplation of the works in question, he so loves them that he is willing to spend his life in trying to acquire the power to produce these good things.   Yet other men can show that they love beauty.   The professed critic may say that he, no less than the artist, gives his life to the contemplation of art.   Many buyers of pictures buy them because they enjoy them.   And the price they give is evidence, so far as it goes, that the enjoyment must be considerable.   And, indeed, there are fortunately thousands of men and women, neither artists nor critics, who find a considerable portion of their happiness in the enjoyment of some form of art.   The first impulse which made the artist choose his profession was probably due to an enjoyment keener than that of any among this crowd ; but, as life wears on, the use of this fine art-sense of his to make bread numbs the nerves.   Praise, blame, hope, fear, rivalry, habit, the influence of his set—all these things warp his art-sense, and many of them lessen the pleasure he derives from art.   Whereas the layman keeps his art-sense fresh.   The pettiness of life cannot taint the pleasure he takes in beauty.   It is, indeed, true that the regular and wise use of any faculty tends to its improvement.   This is that cultivation by which taste becomes good taste.   But here, again, the layman has some advantage.   The artist almost invariably becomes a specialist.   And his taste, no less than his execution, is thereby specialised.   He attaches overwhelming importance to some few qualities of which he is an unrivalled judge.   To

other beauties he is blind, deaf, and callous. If you hear a man
who is clearly interested in pictures, plays, or music openly
proclaim that Raffaelle could not paint, that Shakespeare was a
poor dramatist, or that Beethoven was grossly ignorant of music,
you may guess that man to be an artist. His thesis is all the
more amusing because usually he is right so far as he goes—
that is to say, the qualities he admires are in all probability
good qualities enough, and are more or less absent from the
work which he rejects. He is only wrong in attributing overwhelming value to minor merits.

The question we are discussing is of far more practical
importance to artists than to laymen. The many will have their
way whatever artists may think of the value of their judgment.
The public holds the purse-strings, and so commands a hearing;
we commonly hear artists lament this. The painter, for instance,
speaks contemptuously of pot-boilers, and grieves over the sad
necessity which binds him to give, not that which he knows to
be best, but that which the ignorant public will buy. To these
men we would say there is an ignorant public for whom you
may at your choice write or paint, and so make large sums of
money; but there is also a cultivated public whom your very
best work cannot satisfy. This cultivated public is now so large
that no really good work ever fails to produce a livelihood for
the artist. If, then, you produce pot-boilers knowing you can
do better, this is not the fault of the public, but your own fault.
This, however, is a mere side issue to the general question
raised, whether the artist should work mainly with a view to
satisfy other artists, or to satisfy that portion of the public
which is interested in his art and has a cultivated taste.

We have no doubt which solution to prefer. The great
artist must, indeed, satisfy the priesthood of his art, but he must
work for mankind, not merely for his fellow-craftsmen. He
must not be content with a neat garden-plot, he must reign
over a great world. Yet we know how difficult it is for him to
believe in the existence of an artist's world outside his little
ring of friends. The painter hears the world speak of his
pictures, dismissing the work of months with some curt and
pert remark. The great actor sees the world neglect him for
some pretty girl—some noisy fool; musicians see the world

enjoying barrel-organs and brass bands. What wonder if the
artist is tempted to despise the world! But among his friends
the artist finds knowledge, taste, and courtesy combined, with
some appreciation of the effect required to produce even a pass-
able result. He and his friends loathe the poor vulgar work
which deceives the great mass of men, and they despise a public
which is cheated so easily. The artist has tangible proofs that
real criticism exists in his own circle, and it is hard indeed for
him to believe that in that vague, unknown, incoherent mass of
creatures outside live his real masters, his true judges. Yet
this is so. Every priesthood is similarly tempted to believe that
the priest is not to be judged by the people; but all forms
of religion, including that of art, are in the end judged by the
religious laity, and surely this is well. Why should religion be,
if not for the people? The relation of the artist to the public
is that of the priest to mankind.

But here some artist might say, ' I agree that I shall work
for mankind. I desire strongly that a large part of mankind
shall be so cultivated as to be able to judge and appreciate my
work. I admit that among the laity there are men whose judgment I value very highly, and who are therefore in one sense
my critics; but these men should not seek to impose their
opinions on others. They have a right to their private judgment, which is often excellent; but they have no right to preach.
I court private criticism; but these laics must not hold forth in
books and journals professing to act as accredited guides. For
this purpose, at least, it is wise to have a priesthood, so that
the really ignorant may not be led away by strange doctrines.'

It is, indeed, highly convenient that there should exist some
recognised critical body in each art, and this body should, we
take it, consist wholly of artists. In the Royal Academy, with
its power of selection and rejection, we have one body of this
kind. The French Academy is another instance; the Comédie
Française is a third; but, except in the case in which the artists
are writers, it would seem desirable that the expression of
opinion given by the artist judges should be mute. If painters
or musicians tried to formulate a corporate judgment they would
meet with much difficulty. Even the writers cannot accomplish
this. They have to delegate the duty of writing an authoritative

criticism to each member of their body in turn, and this plan could hardly be adopted in other arts. In fact, the right to print one's judgment depends in great measure on the power of expressing it, and the power of doing this belongs to the writer caste alone. Hence we find that of necessity the professed critics are literary men. Painters, musicians, actors, are all judged by journalists; not because journalists know more of these things by nature than other people, but because they can write what they feel in such a way that people will read it, and whether this be right or wrong it is inevitable. There are indeed some writers who treat criticism itself as a fine art, or rather as one branch of literary art. Taking a book, a picture, or a song as text, they write a graceful essay. Whether the opinions of such a critic as this are right or wrong matters little —the essay is ingenious, learned, suggestive. It contains something for the artist, something for the connoisseur, something for the public. It has a beginning, a middle, and an end. Yet the painter or musician who reads the essay must often writhe with indignation at the judgments given. They may find comfort in the thought of the larger public who cannot write neat essays, but who answer to the magic call of art whenever the piping is true. This public reads the neat essays because in their way they, too, are artistic; but when a great play, a great book, or a great picture shines out through the daily fog of life, not all the little literary essays of all the little literary men on earth can hide the new sun.

Beyond and above the journalist, the essayist, the prophet, and the artist there is a wiser judge—not any one man living now or in the past or future, but that section of mankind which, by an extension of the analogy we have used in contrasting artist and layman, we may call the Art Church. All true art believers of all times are members of this church. Each honest art lover in his day brings some little atom of his soul to nourish this great judge, and each little element building the spiritual whole coheres in virtue of its fitness.

The Art Church has its schisms, offshoots, heresies, reformations. Even in her bosom there is no abiding rest; but poor fallible mankind can provide no better champion for beauty.

GRISELDA.

A STAGE PLAY IN THREE ACTS.[1]

PERSONAGES OF THE DRAMA.

WALTER, *styled Marquis of Saluce, an absolute prince.*
TANCRED, *son to the Marquis of Saluce, styled in the third Act Count Roland of Saluce.*
GIAN, *a peasant, father to Griselda.*
SERGEANT, *a trusted soldier of the Marquis of Saluce.*
JOHN, *the Majordomo of the Palace.*
BARNABY ⎱ *Servants.*
RICHARD ⎰
GRISELDA, *wife to the Marquis of Saluce, separated from her husband.*
FILOMENE, *daughter to the Marquis ; styled in the last Act the Countess Dora of Saluce.*

Lords, Ladies, Courtiers, Soldiers, and Servants.

ACT I.—A Cottage in the Abruzzi.

ACT II.—*Scene I.*—A Chamber in the Palace of the Marquis of Saluce.
Scene II.—A Bedchamber in the same Palace.

ACT III.—Banqueting Hall in the Palace of the Marquis of Saluce.
Time.—That of Boccaccio's Story on the same subject.

ACT I.

SCENE.—*Interior of* GIAN'S *Cottage.*

GIAN *and* GRISELDA *discovered.*

Gian. You work too hard, Griselda. I have worked hard all my life, and I can work hard still, because I have always taken proper rest, food, and wine. Not too much, but enough. Rest is what you want most ; but a cup of wine would do you good.

[1] Privately performed in Professor Jenkin's house at Edinburgh, January 1882.

Gris. What good would it do me, father?

Gian. It would cheer you, child. A sad heart makes a sour face.

Gris. Is mine sour, father?

Gian. It might well be sour. Certainly no woman ever had a better right to be miserable; and so, I think, a cup of wine would do you good. No? (*Drinks wine.*) I admit this is poor tipple after such liquor as you drank at your husband's court. He must have a princely cellar! What growths does the Marquis of Saluce favour most?

Gris. He prefers Tuscan wine.

Gian. A marquis should know good wine. I think this good, but he would not swallow a drop of it. He would call it poor country stuff. My talk, too, is poor stuff for you; you have lived with ladies, and conferred about State affairs with great lords. I like to hear their names. Tell me the styles and titles of the ladies who have waited on you.

Gris. Father, you know the roll by heart. Forgive me— words which tell of that old time come unwillingly, even when called to do you pleasure.

Gian. True! true! A grief once stilled should not be stirred. I am dull and blundering—quite unworthy to be father to a queen. For my part, I never understood why a reigning prince should have married you, but I understand still less why he unmarried you years after. Men say, too, you were a good queen. Well, great folk do strange things; but small folk, such as you and I, do just what we must.

Gris. True, father; and now our cows must have their fodder, and I must give it them. Where is the stable key?

Gian. You have an angel's patience. I never could have borne half what you have borne.

Gris. The key.

Gian. Nay; I shall keep the key. A stable is no place for a gentlewoman.

Gris. Gentlewomen, father, have good care of all who serve them. Pray give me the key.

Gian. Gentle or simple, women must have their way. I warrant that in obstinacy you could match the best of them, Grisyld.

Gris. Ah! you have soon forgot my angelic patience.

Gian. No whit, Grisyld. You are obstinately patient. (*A knock is heard at the door.*)   Come in.

TANCRED *comes in hastily,* C, *looks round, and sees* GRISELDA.

Tanc. Lady, if your name be Griselda, as the fairness of your presence teaches me to believe, honour me, I pray you, with your quick attention, and pardon the absence of that full courtesy which your worth deserves, but which danger, both rude and imminent, forbids.

Gris. In this house, sir, courtesy is chiefly due to my father. He will, no doubt, give you such aid as his power allows and your need may justify.

Tanc. Old sir, I crave your pardon for that which seemed discourtesy, but which was truly haste to do you service. I ask no help, but bring your daughter a warning and the means of safety.   Have I your leave to speak with her?

Gian. A humble house can give a hearty welcome. Pray be seated, sir. I see that you are hot, and, as I think, tired. Before telling us your story, you might relish a cup of our thin country wine.

Tanc. Sir, present delay is dangerous, even that which lets me thank you for your kind welcome. I beseech you, give me leave to tell your daughter my errand.

Gian. Nay—speak—your speech is fair, and I will stay no man's speech so long as he speaks fair. Your court language will tickle my daughter's ear. She has been used to it. I like it too.   You are noble, sir?

Tanc. Truly I know not. At daybreak, lady, I took the best horse owned by the Marquis of Saluce——

Gris. You come from him?

Tanc. Although I bring no message from your husband, noble lady——

Gris. I have no husband, sir.   I am a peasant woman.

Tanc. For me you are still a queen. The decree by which the marquis put you from him weighs no more with me than so much paper.

Gris. Such paper can do much.   Who are you, sir?

Tanc. One who brings you news and help.

Gris. I need no help. You bring me pain. Father, may I go?

Gian. By no means, Grisyld.

Tanc. Let your anger visit me hereafter for those faults which my youthful inexperience has put upon me; but in the present moment grant me that faith which a loyal intention merits. Believe me, lady, I come with true reverence to serve you.

Gris. Young sir, I believe you, but I need no service. Until you came all was well with me.

Tanc. Yet hear me. I left Saluce before dawn, and I have spurred hard all day, for behind me I heard the rattle of horses' hoofs. My purpose is to save you. If you will not listen, the rude messengers who follow will bring death.

Gris. Well, then, speak.

Tanc. The Marquis of Saluce bribed the Pope to sanction the divorce by which you were discrowned; this you know. Now he goes farther. He will marry again. The bride he has chosen is a certain Lady Filomene, and already she is at his court.

Gris. Is this all your news?

Tanc. You say no more?

Gris. What should I say?

Gian. Say?—say with me that such a marriage is impossible and infamous. The Marquis was and is my daughter's husband, and I say——

Gris. Father, words are useless.

Tanc. I should not of my free will have played the messenger merely to bring bad tidings. I have more to tell.

Gris. More?

Tanc. A troop of horse sent by the Marquis follows me, and will soon be here. Their sergeant brings a letter which summons you to court.

Gris. A letter from my lord! How should you know what he will write to me? You are too busy. From his letter I shall learn what my lord bids me do, and that I will do. You have spent one welcome, sir; your news may buy another from some gossip. We will not traffic with you.

Gian. Grisyld, Grisyld! To my ear the news rings true. I would fain have more of it.

Tanc. If I leave you now I shall throw away men's lives—lives risked for you.

Gris. Men's lives risked for me ?　Father, if it please you, I will hear this gentleman alone.

Gian. I will go.　Men's lives !　I have neither skill nor will to meddle with such gear; I am an ignorant old man—your servant, sir.　Be prudent, Grisyld.　　　　　　　[*Exit* GIAN, L.

Gris. Now, sir, I will hear you; but first tell me who you are; I never saw you at the court.

Tanc. Why speak of me ?　I am nothing.

Gris. You wish me to believe you, and to act on my belief. Tell me, then, who and what are you ?

Tanc. I know neither who nor what I am.　I have inherited no father's name, and as yet I have won no honour by which men should know me; but henceforth they shall say, ' He saved Griselda.'　Lady, I cannot buy your faith, but I entreat you give it me as a free gift; trust me, and I will soon prove that your trust has been well bestowed.

Gris. Young sir, these are wild words.　I am a very simple person, and I wish for a simple answer.　Who and what are you ?

Tanc. I am called Tancred, and I have no other name.　My guardians would not tell my parentage, but they gave me kindly nurture in Bologna.　There we, the student youth, loved no theme so well as argument concerning noble women.　One we judged to be the worthiest of all who live.　She was a peasant girl, whom the wild Marquis of Saluce, in a mad freak, made his wife.　Men mocked at first, but by gentle dignity she taught them both to love and honour her.　Then, as though in spite that she whom he had raised should stand upon the solid ground of her own worth, the Marquis killed the two children she had borne him, calling them base-born brats, because she was their mother.　She said he could not kill her love, whether for them or him.　Thereupon his mad mood changed, and he gave her power, which she so used that for many years the weak had peace, while the strong loved the law which bound them.　Then the moody Marquis took back all that he had given, and robbed her even of the name of wife.　Discrowned, divorced, stripped to her smock, and cast out from the palace like a beggar, she went barefoot to her old father in the forest. On that day, the Marquis would have been toppled down into the mire, if she had said the word; but she was silent, for she loved him. As we in Bologna reasoned on her story, each man's judgment mirrored the man himself. By the crafty her straight simplicity was wrested into crooked cunning, but the noble gave her back a noble recognition of nobility. As for me, I held that she was wrong in having patience with such wickedness, and yet I worshipped her; and, as I now learn, my faith was such that she wrought miracles on me. Nay, the facts will fit the words. When heat has prompted me to wrath, I have often heard a voice that soothed me—my mother's, as I then believed; and once, when folly would have had me enter a dark gate which led to sin, I saw a vision on the open blackness barring me out with light—the Virgin Mary, as I thought. But now I know the truth; the saint's face was yours; the saving light shone from your eyes. I called your voice my mother's. Fancy could not trick me, for until this hour Griselda's face and voice were both unknown to me. Yet what I knew led me, with many others, to lay plans by which we hoped to draw you to us in Bologna, when suddenly my guardians, giving no reason, sent me to the Marquis of Saluce. I made no long stay with him, and this day come from his court.

Gris. Is he well?

Tanc. In body he is well. He received me kindly, saying that my father had been his dear friend. He spoke much with me, and showed so keen and sound a wit, that if I had not borne in mind your wrongs I should have loved him.

Gris. I understand——. Did he —— ?

Tanc. At times he spoke of you, and I was so foolish as to hope that I, a boy, might turn his heart back to you.

Gris. Hope may not always be wise; yet we love hope, and those who hope—forgive me;—you see I am not very patient. I spoke cruelly to you.

Tanc. I told my errand with such want of skill as deserved reproof. But now you believe me?

Gris. I do believe you. I think your nature noble; your heart kind—— And when you spoke of me to the Marquis—— ?

Tanc. He would smile, and often he would give no answer; but I thought I understood him better than his courtiers, and that his harshness was a mask hiding a kind heart. I thought he really loved you.

Gris. He did so once.

Tanc. Is there any pain like losing faith in one whom we have loved?

Gris. I have never felt that pain.

Tanc. Seven days since Lady Filomene arrived. The Marquis changed. I thought he had put on a mask; but I soon learnt that his seeming goodness had been the mask. What I now saw was his evil face.

Gris. Remember—— I was his wife.

Tanc. And you believe—— ?

Gris. Believe with me. The Marquis is a noble gentleman, how strange soever his actions seem.

Tanc. Madam, what should I say? He is your husband in the sight of God; yet he will marry this Lady Filomene.

Gris. Marry her! Who is she?

Tanc. No name is given her but Filomene. No one could tell her lineage, nor say whence she came; nor dared any man inquire. The courtiers laughed and said the Marquis held curiosity to be a trespass on his royal chase.

Gris. Is she very beautiful—and young?

Tanc. Young, beautiful, learned, and very proud.

Gris. Is this marriage—to her mind?

Tanc. Truly, no. The Marquis has wooed her in no loving or even courteous fashion, and he is old.

Gris. Ah!—yes—old. You had more to tell me? You said my lord would summon me to court, and that there was danger to men's lives. I could not understand.

Tanc. In a few hours a messenger will come, sent by the Marquis, to bring you to Saluce by force. There he will make you servant to the Lady Filomene.

Gris. No force is needed. I will go.

Tanc. You need not go.

Gris. I need not?

Tanc. I came here to secure your freedom. Thirty students, friends of mine, have at my summons crossed the frontier secretly and wait for us hard by, at the little wood by the cross-roads. Come with us, and in six hours you will be on Tuscan soil, and free.

Gris. Free! I cannot understand—let me think. Silence — let me speak. You speak too quickly. My lord is about to marry. He will send for me to his court to serve the lady. You, a stranger, come to me and ask me to follow you and disobey the noble Marquis. Why?

Tanc. That you may escape from him; that you may reach safety and happiness.

Gris. My boy, you know not what you say. I can reach no happiness by flight or disobedience. I will go to the Marquis.

Tanc. To the Marquis? Say to persecution, insult, death. A moment since you asked why you should follow me. I ask, now, Why go to the Marquis?

Gris. (*a pause*) I think it right.

Tanc. Right! I know well that as wide a gulf sunders wrong from right as hell from heaven. Only, let us not give the name of right to every common custom. In women obedience is a custom, and this custom would make you obey your Marquis, serve his concubine, and kiss his hand—the hand which struck you. This custom would have you kneel and stretch your throat to meet his knife, rusty with your children's blood. Can you call this custom right?

Gris. You hurt me. You say you come to serve me. Be gentle, then.

Tanc. That is the voice. When I hear that voice I must be calm; but the voice even of an angel cannot make truth false. Go to Saluce, and you do wrong.

Gris. You say that going to Saluce I shall do wrong and suffer much, and you would bribe me from this wrong-doing and misery with happiness. What happiness? Daily food? Shelter from heat and cold? More? A servant—many? Learned Tuscans, too, to talk with—friends, you would call them; wealth, perhaps, and leisure—much leisure to think upon the past. It may be wrong, young sir—I am a poor casuist—but I will go to the Marquis.

Tanc. Your words sting, but they are not just. In our

Tuscan town your life would have a larger range than in this village.

Gris. Here I have obeyed my lord.

Tanc. There you shall rule many better men.

Gris. How should that be?

Tanc. I speak in the name of hundreds such as I, and nobler; we offer you a brighter crown than that which you have lost. By our faith in you, and by your faith in God, I summon you to be our leader, our prophet, our martyr, if needs be.

Gris. What rhapsody is this?

Tanc. You think that I speak wildly, but I tell you simple facts. Would men who are not in earnest defy the Marquis in arms? We few, who are here to-day, were chosen out of many hundreds—bound secretly by oath to good deeds and to each other. I cannot yet make all clear to you, but we are strong. Our aim is that of knights who in old days would right every wrong or die. We cannot die; each man of us may fall, but the beauty of his death fires twenty more to take his place; so that we thrive by loss. Truth is our guide, justice is our aim, and our glaive slays like the wrath of heaven—none knows where or how. We will make crime rare in Italy.

Gris. You will add fresh crime; you are raw youths—secret murderers. You are no knights; you stab like cowards, safe in the dark. Look to yourselves if you would right wrongs. You wrong the kingdom's peace; that is the greatest wrong of all.

Tanc. Only the wicked fear us.

Gris. You in turn will fear those you hurt; soon you will hate those whom you fear, and hate will make you cruel. Then begins a monstrous game. The open and the secret tyrant brag against each other, whose torture shall be most feared, while poor honest men are ground between two stones, pitiless law, and pitiless rebellion. Would you have me take part in this?

Tanc. Your words sound true. But to persuade me is little; come and teach us all. We have gone astray; come and guide us. Lead us.

Gris. How should a woman lead?

Tanc. As a saint—by beauty in her life. We hold you to be a saint, and we will follow you. We have worked in the dark—you shall bring daylight to us; we have been blind—you shall give us eyes; we have known what to hate you shall teach us what to love; and we will worship the divinity of womanhood in you. In us you shall have many sons, and our one strife shall be to make the whole world call you blessed.

Gris. You have thought this of me, planned this for me! I have known a strange happiness. I am childless, yet I see in you the ever-budding youth of the world, and hear from your lips the praise a noble son might render a good mother. In a sad hour, sir, you have brought me a great joy. But I cannot follow you.

Tanc. Why not?

Gris. I would not grant your prayer even if I could. You think I have lived nobly; if that be so, the faith which I have kindled in you is that help you claim from me; that other help you ask is not mine to give. I am no oracle ready to solve every tangled riddle set by men's lives, nor would a single man among you take an answer from my lips. You made a goddess of me at a distance; but now I am before you, you see only the woman, and it is you who would command me, not I you. Oh, it is so. Even now, I tell you that to me it seems right that I should obey my lord, go to Saluce, and serve his bride; but you will not believe me.

Tanc. None can be judge in their own cause. I fear you are in love with suffering, and would choose whatever lot gave you most pain. Surely it is better to lead good men seeking guidance, than as a willing victim to let a bad man make sport of you.

Gris. I think your goodness pains me more than that badness you speak of. I cannot argue. Let me be, I pray you.

Tanc. If you are right, why do I not recognise your present will as good?

Gris. Because it thwarts your plans, and men are ever masterful, even the best.

Tanc. In a good cause I would not spare my life; no, nor my friend's—not yours! But this obedience serves no purpose. To obey a villain wilfully is wrong, unholy.

Gris. You speak as a man. Man can strive. Man should believe no evil to be without a remedy; this he must win at the sword's point if no less will serve. But a woman cannot battle with hard hearts and hard hands. If submission, if endurance be unholy, there is no martyr's palm for women.   And he is no villain.

Tanc. Until now you have shown strength by your sub-mission, but to submit farther, and for no good end, is weakness.

Gris. I will submit.

Tanc. Nay, then, you shall submit to me.   My will is your good.   His purpose is your degradation. (*Goes out, back.*) We are ready.   Bid them come.   [*Re-enters.*

Gris. What have you done?

Tanc. I have called my brethren; you shall come with us whether you will or not.

Gris. I will not come.   [*A confused noise is heard without.*

Tanc. You shall not go to your death—for he needs your death.   This new marriage, if it be contracted while you live, may be set aside any day.   Death is your only stop game, said the Marquis.   Marriage and divorce may play see-saw, quoth he; and then he laughed.   I hoped then that he might take you back.   He shall not have you back.

Gris. He shall have his will.

Tanc. Right!   All I said was folly.   How could I dream that you would flinch from martyrdom while you were free! but you shall not be free; you shall be saved.   Force shall save you. And I thank God I am a man, and can use force.

Gris. You will not save me.   Have some mercy.   Let me at least die near him—by his hand.   Do not kill me here; for I will die sooner than go with you. (*He seizes her.*)   Oh, bind my hands; you cannot bind life in me.   I will neither eat nor drink.   That's too slow.   I will not breathe: I will die—here— now! (*A trumpet call is heard.*)   No, I shall live; for he has come!   That trumpet sounded his call—I hear his voice. Walter!   I am here.   [*Bursts from* TANCRED *to the door.*

> *Door opens,* SERGEANT *appears,* C, *and salutes.*

Tanc. How is this?   Tancred!—friends!—a Tancred.

Gris. Soldier, speak.

> GIAN *enters,* L.

Gian. What is this noise?   Who may you be?

Serg. Sir, I am a sergeant in the bodyguard of the most

noble Marquis of Saluce, and I come to bring this letter to the Lady Grisyld.

Gris. From my lord ?

Serg. Ay, madam.

Gris. Give it me. (*Takes letter—reads :*) Madam, follow the bearer : he will bring you to our court. Walter, Marquis of Saluce. (*Kisses letter.*) Soldier, I come gladly. Sir Tancred, I thank you for playing herald to good news.

Tanc. I have still to earn your thanks. Sergeant, Lady Griselda shall not go with you to Saluce.

Gian. Sir, she must go if the Marquis bids her. This is good news indeed.

Tanc. I say she shall not go. [*Whistles again.*

Serg. Your friends have got beyond that whistle, Sir Tancred, judging by the pace at which they started.

Tanc. Cowards !

Serg. Ten to one is eight or nine too many. The Marquis knew the very corner of the wood where we should find them.

Tanc. The Marquis knew ?

Serg. The Marquis always does know. He gave me orders to take you prisoner, but to let the others go—or rather make them go, and they are gone.

Gian. Grisyld, child, are my wits wandering ? Who are gone ?

Gris. To take him prisoner ! Soldier, are you sure that this is so ? He has harmed no one. I come with you freely. Were you in any case to bring Sir Tancred with you ?

Serg. My orders are clear, lady, to bring Sir Tancred back —alive.

Tanc. You may find that difficult.

Gian. I cannot understand.

Gris. Father, it is better you should not be concerned with what has happened ; this remains ; I am sent for by the Marquis, and, as I believe, I am to be the servant of his new bride.

Gian. Well ! it is good for you to go to court again, but I shall miss you. Still, you may be of use to me there. Probably the Marquis, though all is over between you and him, might grant such a trifling favour as I want—for old remembrance' sake.

Gris. For old remembrance' sake!

Gian. What I wish would benefit you too. If I were made a bailie, your position would be better. I think it hardly right that, having been a queen, you should be nobody. Now, as the bailie's daughter, you would have a certain rank.

Tanc. Sir, there is no question here of rank. The question now——

Gris. Silence, sir; my father asks but a small favour, which, if I can, I will obtain.

Serg. If it please you, we had better start.

Tanc. What think you is to be my fate?

Serg. A plain man would say you had earned a hanging, but no one knows what the Marquis will or will not do. Just now, you must come with me.

Tanc. Must may be too soon spoken. Madam, farewell. For men at least, you did not say submission was a virtue. (*Draws his sword.*) Room there for man or sword!

Serg. (*steps aside*) Room enough for both.

     [T*ancred*, *while rushing out*, C, *is tripped up and pinioned by two soldiers behind the door.*

Tanc. Treachery!   [*He is borne away.*

Gris. He is not hurt?

Serg. No, madam. Will it please you, come.

Gris. Father, farewell. You will need help when I am gone. Bridget will serve you best. The cows have waited long to-day; remember them. (*To the* S*ergeant*) Take me to my lord.   [*Exeunt* G*ris.* *and* S*erg.* C *as Curtain falls.*

ACT II.

S*cene* I.—*Well-furnished room in the palace at Saluce. Table with fruit, wine, and silver cups. The* M*arquis* *is discovered sitting. To him enters the* S*ergeant*, R.

Marq. How fares your prisoner?

Serg. Meagrely, my lord. Sir Tancred eats little and drinks nothing.

Marq. Nothing!

Serg. Nothing but water, my lord, which a soldier may call nought.

Marq. So you drank the wine we sent him ?

Serg. I will never let good wine spoil, my lord, for lack of a dry throat.

Marq. Go back to Sir Tancred. Say, we regret that our good cheer should pall on him for lack of better company than drunken soldiers, and that we therefore bid him to our own table. Go, and bring him here. (*Exit* SERG. R.) The lad is eighteen. If I cannot win his confidence before he knows I am his father, constancy in hating me will seem to him a point of honour. I like the boy. He meant to set Grisyld on a throne, maugre me and all my men. That was folly—but the folly of a prince. Filomene seems venomous. She thinks me a devil, and despises Grisyld. Her head was turned by young Count Malatesta's flattery. She has seen the boy four times, and believes that she could never love another man. Well, he shall marry her. I wish him joy. (*Enter* SERG. R.) Where is Sir Tancred ?

Serg. My lord, Sir Tancred bade me say that, if it please you, he prefers silence to your speech and hunger to your food.

Marq. It does not please me. Bring him with you. (*Exit* SERG. R.) His indignation has the true antique Roman flavour. Virtue is his goddess, and he saw that ancient lady smile on him when he sent that message. Good old Virtue ! She is so incompetent, so weak, that I feel bound to help the poor dame. (*Enter* TANCRED *and* SERG.) He has no weapon ?

Serg. No, my lord.

Marq. Leave us. (*Exit* SERG. R.) Your case, sir, is not so gloomy as to warrant those black looks. You shall not be hanged to-day, no, nor to-morrow.

Tanc. What matters what day ?

Marq. What matters but the day, man ? I should take hanging to be a speedy, peaceable kind of death—good enough for any man if it came at the right time, which for you is not yet. Let this wine prove to you that life may be enjoyed, no matter how near Death may lurk.

Tanc. My life is forfeited, but I will not use wine to drown thought ; nor will I pledge my gaoler.

Marq. Now, in like case I should pledge my gaoler roundly. He should swear 'twere pity if such a jolly fellow might not live—at least until the cask was out.

Tanc. I have no wish to resemble you.

Marq. Ought I rather to make you my model? You broke my laws, you violated my hospitality, and you betrayed my counsels, though I made you my friend. Can you defend yourself?

Tanc. I will not justify myself to you. You would not even understand what my words meant.

Marq. Youth is sometimes unjust to age.

Tanc. Age is never just to youth.

Marq. Age, for instance, plays the gaoler when youth would carry off his wife.

Tanc. That speech is, indeed, a foul instance of injustice.

Marq. Solemn Dame Justice must work hard if she weighs that infinity of trifles, all we say and all we think.

Tanc. Trifles! Well, yes. Life and Death are trifles to you. Why have you sent for me?

Marq. I sent for you to win your friendship. I defy you to believe me; therefore I can speak frankly. I intend that you should respect me. I mean you to approve of my life.

Tanc. You ask much; but I understand you better than you think. You have as your prisoner a boy who, in a blind way, tried to do a good action, and failed ridiculously. If you can make him talk, his crotchets will give you sport, and good sport should end with the quarry's death.

Marq. A pathetic picture, but not true to nature. Let me play the painter in my turn. My canvas shows a lusty lad, whose friend, desiring some pleasant talk, offers him a cup of wine; but the boy's brow is black with twisted knots, dark folds, wrinkled plaits, for no clear reason. You mislike your host? abuse me then—hail arguments upon me. Let each pellet sting me to the quick; let each phrase prove me a villain. I'll face the storm.

Tanc. A sword's point is the one argument I should care to use. You were wise to ask if I had no weapon. If a knife lay on that table I would kill you like a beast of prey.

Marq. Is it wise to tell a man you wish to kill him? But to my purpose. I know that you will give your friendship to no man whom you cannot honour; therefore I must praise myself. I govern well: you have seen the proofs. Women smile in Saluce, men walk upright in the streets, and children laugh in our gutters, but at court my great lords are humble. I grant, they tell you in a whisper I am mad. They believe it, partly because my follies are not theirs, partly because from time to time I have devised certain quaint punishments for rogues and villains, whereby, let me say, the breed has been notably discouraged. But these freaks would not scare you from my wine. You would not refuse the hand I proffer, but for one sin which you think stains it: you believe I wronged Griselda.

Tanc. You are more hateful when you play the hypocrite than when you are openly a fiend.

Marq. You crow loud for a caged fowl. Some grain of the fiend may lurk in me, but I am no hypocrite. To prove that I am worth your friendship, I ask for no better evidence than I can draw from Grisyld's story.

Tanc. Will you slander her? I have seen her eyes and yours. She is true. You are false.

Marq. She is as true as the steel of my sword-blade, and my flaws lie not where you find a blemish. You say you know by a man's eyes when he speaks the truth. Look in mine, young sir. What my reason has called right, I do, shall do, and have done all my life. I own that a complete record of that life would furnish matter for a quarto tome of acts, called sins by priests and women; but I have never done the deed nor said the word I will not justify before a man. Show me base stuff in me, and I'll not flinch before the knife. I will cut the rotten matter out, though I bring life's root with it; but cant me no cant; prate of no old customs such as cumber the world's drudges like old fusty clothes. When have I been base?

Tanc. In your youth—first.

Marq. You mean that I drank hard and loved women. Neither my palate nor my stomach were the worse; and as for the women, they liked it.

Tanc. You spoilt a finer sense than taste.

Marq. That finer sense was not so blunted as to let me buy a court puppet for a wife. I loathed great ladies. Their gracious manners were a mere aping of real grace, as their fine clothes mocked real beauty; but in the forest one spring day I found Griselda, a woman to my liking. Was my ideal base?

Tanc. No, my lord.

Marq. My friends were of a different opinion—then. Now —you are wiser. I set Griselda on a throne. You will hardly call that act base.

Tanc. She was unlike the women you knew, and your appetite was tickled by the novelty.

Marq. Tickled!—Appetite! Commend me to the purity of this precisian. I loved her, man, as I think you will never love. Your nature cannot even apprehend my passion. Can a fair body and sweet soul only tickle men like you? Is appetite the name you give to what men feel for a Griselda? I call that feeling love; and I glory in it. You have not found the base spot yet.

Tanc. Your feeling was not love. A man's love is an honour to her he loves—an honour to all women; but your words would make good women blush.

Marq. Would they? Let me hear, then, what you mean by love?

Tanc. Love is man's worship of a woman's excellence. That worship cleanses him from all base thoughts. How should a lover not be pure when his whole desire is set on excellence? That desire wakes his strength, and at a touch from Hope his courage rises till he will outdare demigods. When his prize is won, her soul and his twine into a tree of life, stretching from earth to heaven, from time into eternity. Love is a miracle, and I have faith that God will work this miracle for me.

Marq. Well said; you have faith, but I have experienced the miracle. We expect more than miracles. We long to burst from the solitude where we are one only self; and for an hour Nature tricks us with a dream in which we reach heaven's shore, and walk there with our mate, hand in hand, soul in soul; but we wake to find we are still battling in life's sea— alone.

Tanc. You think love cannot last.

Marq. Love lasts while we hope; and when the woman is Griselda, hope lasts long. You spoke of souls; we have bodies too, and strange ones. Love is less simple than you think. That sphinx, Nature, sets each woman as a riddle. Her lover must solve the problem or he dies. I thought of no such matters when I found Griselda. Her peasant bodice clipt the body of a queen. That body held a noble woman's soul. Body and soul matched mine; and I claimed my mate.

Tanc. My lord, you seem even now to speak of her with triumph.

Marq. Seem! You think much of Grisyld. I made her. You owe her to me.

Tanc. Do we owe the martyr to the hangman?

Marq. That tongue-thrust was well put in; but your weapon galls me not, for your point lacks truth. You owe Grisyld to me, for I am the man who saw the girl had stuff in her to make a queen. As queen she was perfect, but as woman she was incomplete. I felt alone when I was with her—she was not open with me. She never spoke of herself, never asked me of myself; never wondered why one day I was moody and another in good heart. She would as soon have questioned why the breeze blew yesterday and not to-day. She seemed not a woman, but a plant. I was not a man for her. I was the sun, and when I shone she lifted up her head.

Tanc. Put shortly, goodness wearied you.

Marq. Far from it; her calm excited me. I made my wife my idol: she became a statue. I would have strained her warm and living to my breast, and I found her marble. My arms slipped from the smooth cold stone—chilled, powerless. Anger gave me heat and strength again; I cast down—I broke my idol, and I was left without a god.

Tanc. You killed your children to see if their mother had a heart.

Marq. I killed no children. A palace is no place for the young, and I therefore sent my boy and girl to Tuscany, where they might grow in the open air. When I took the children from her, Grisyld was still marble. My anger said that her patience might be mere plebeian dulness, and I told her I had killed them;—I set a strange problem: What was her duty? To abhor me? or still to love and honour me?

Tanc. Her duty was to loathe you.

Marq. Why then should we both admire her for not loathing me? If I had foreseen what she would suffer, I could have spared her and myself much pain; but without trial we know nothing. If Grisyld had no wants, small thanks to her for patience. I worked—wept—stormed—fought for what I wanted. She was silent; yet now I know that my most passionate desire was a boy's greed, measured by the mother's yearning for her lost little ones.

Tanc. My lord, I must believe that you once loved Griselda, and I think that, even now, you do not love evil. Be again the man you were when you first won her——

Marq. I should like that much.

Tanc. Act now as you would have acted then, and you will feel now as then. Give Griselda back her throne, her children, and her husband; her husband not in name only, but the man she once loved—the man who must once have been worth her love.

Marq. I may take your advice; you think those children are alive?

Tanc. I am fooled again; yet I would rather be the fool and killed than you who kill.

Marq. To convince Grisyld that they lived I must have brought them back. She would not have taken my mere word, and yet she trusted me. What I did was still for her well done.

Tanc. By heaven, it was base in her!

Marq. By solid earth, I now first feel anger! Pray, sir, how would you, so ready to teach me, have taught Griselda? Are you silent? Grisyld was right, I say; she nursed no grief to give herself the pleasure of self-pity. She abjured sorrow, and lived usefully. Grant me bad: her hate would have damned me. Would you have had her in a nunnery to mumble prayers for me in hell? I have good cause to know the part she chose was better, for by her trust in me she has redeemed my soul. Diana's body! you have heated me——

[*Unbuckles sword, throws it on table, rises and walks to end of room, with his back to* TANCRED, *then turns.*

Tanc. Why have you tempted me? You cast the sword there purposely.

Marq. True. And you have not murdered me.

Tanc. You play easily with your life.

Marq. I thought the object worth the risk. I wished you to trust me, so I trusted you.

Tanc. You trusted to my weakness.

Marq. No; I honoured you so far as to believe that when you heard the truth you would know it to be true, and I was not deceived.

Tanc. My lord, when you praise the Lady Grisyld, you are transfigured, so that I must believe you—even when, as now, your deeds cry out loud that every word is a lie. I surely know that you discrowned this wise queen, and martyred this pure saint. Did you not cast her out with ignominy? How can I believe you when you tell me her trust saved your soul?

Marq. Many said that she loved me little, but my crown much; that her patience was mere cunning, and her cheerfulness the proof of cold self-satisfaction. Do you think this?

Tanc. No, by day's light!

Marq. Nor I—nor any one now. Tancred, I pawned my life to purchase your faith; therefore let me have it. I own that in my house of life there are some crooked passages and unknown chambers, but to-morrow the sun's light shall blaze through its darkest vaults; you shall come in and judge whether each stone has been well shaped. As for the master of the house, some trace of cat or monkey may disfigure his fair lineaments; yet he is a man worthy of Grisyld's trust, and much more of yours.

Tanc. Marquis, I wish I could believe you.

Marq. Let the will give the power. When the sun sets to-morrow you shall judge me. Until then give me your trust as a favour if not as a right.

Tanc. I will rather be deceived ten times than once fail to return faith for faith. You have my word.

Marq. I thank you. Now pledge your gaoler (*hands him the cup of wine*), and then come with me where you shall see the Lady Grisyld well employed.

Tanc. (*Takes cup and is about to drink—stops.*) Is this poison?

Marq. I asked you for your faith ; you gave it.

Tanc. Your true health, my lord.

Marq. We may be friends yet

[*Exeunt,* L.

SCENE II.—*Bedchamber in the palace of the Marquis of Saluce.*
LADY FILOMENE *discovered sitting. To her enter* GRISELDA
and SERGEANT, C.

Serg. My Lady Filomene, the Marquis sends you greeting.
By his orders I have brought Lady Grisyld to your chamber.
He bade me say that she would serve as your tirewoman against
the marriage, and obey you in all things.

Filo. Obey me but in one thing, madam ; leave me.

Gris. Am I free to go ?

Serg. I grieve to say, madam, you are not. My instructions
are precise. To-night, neither you nor Lady Filomene may
leave this room.

Filo. Will you take a message to the Marquis ?

Serg. Madam, I may not.

Filo. The world's turned miser, and grudges me one hour's
peace.

Gris. Teach me how to make my presence least unwelcome,
lady, and so far as my power goes I will obey.

Filo. Obey! Ah, that's the spaniel trick in which you're
perfect. In my lessons I was taught how well you could obey.
I learnt to despise obedience, and to hate you.

Serg. We soldiers, madam, are proud to obey, and we were
never prouder than when we could obey Lady Grisyld.

Filo. Were you ordered to stay here and prate ?

Serg. No, madam.

Filo. Then go. (*Exit* SERG. C. GRISYLD *arranges some disorder in the room.*) I see you are a busy body.

Gris. Madam, if you will look upon me simply as a servant
—and I am nothing more—I shall trouble you little.

Filo. Your hypocrisy apes truth to the life—or are you no
hypocrite, but a true slave—one who will fawn on any hand
that holds a whip ?

Gris. I am made your servant, madam, by the prince who

governs here. I have no power to disobey, even if I had the will, which I own I have not.

Filo. I believe you tell the truth, and that you play the spy right cheerfully for your kind master. Can you obey his vilest orders with no sense of degradation?

Gris. He would not give, nor would I obey, orders that were vile. To serve you is no degradation. Women have served me. They felt no shame.

Filo. Slaves may feel pride in serving; but you have been free—more than free—men have called you queen. A queen should not turn servant.

Gris. What should a woman do, neither queen nor slave?

Filo. Resist.

Gris. How can she?

Filo. As I resist. I will not yield. Men may bind, starve, torture me, but I will not yield.

Gris. To me—— But I have no right to speak.

Filo. I give you leave. Show me how low you can fall.

Gris. I think women are too weak to resist. Useless resistance is not honest.

Filo. The weaklings can die; is that not honest? But snared birds flutter, fish writhe in the net. Will you laugh at them because they are so weak?

Gris. They do not know their weakness.

Filo. You mean, a woman knows; and when a woman is in the toils and shows impatience with the strings that strangle her, patient Grisyld sneers.

Gris. If I could, I would cut the strings.

Filo. You would help me?

Gris. Most willingly.

Filo. You lie. You are tempting me. You seek for some way to betray me.

Gris. Why should I wish to hurt you?

Filo. Why? Because I am your rival. Am I not a fair, young, fortunate rival? Is not every word I speak an insult? But words are not whips, and a leathern thong is barely hard enough to gall your tough peasant's skin.

Gris. Suppose I am in pain, which indeed is true; suppose, also, that you had caused the pain, which is not true;—why should I wish to give pain ?

Filo. Your wounds are skin-deep, or you would not stand there smiling.

Gris. I think if I were a martyr, I could be quiet.

Filo. If you were a martyr ! You are a martyr.

Gris. I cannot think so. I have suffered ; but not more than many women, and my joys have been great.

Filo. You speak as if past joy were a possession. Dante says there is no greater grief than past happiness remembered when we are in misery.

Gris. The saying does not please me. When a gift perishes, our gratitude should not end.

Filo. Ah! you are one of those who would thank God for daily bread, and never curse Him when He lets you starve.

Gris. I am.

Filo. Then you are a fool.

Gris. I cannot judge God, and to me His yoke is easy; for I rejoice in His law.

Filo. I hate the law by which I suffer, and I defy the maker of such laws—if there be any maker.

Gris. Your defiance will not change His laws.

Filo. Now you laugh at me.

Gris. God forbid ! for, as I think, you are in deadly pain.

Filo. I am, I am : be you fiend or fool, you speak the truth in that. Oh, pardon me! I am not wicked, but the pain makes me mad. Men here say that you are good, and I half think you are; but I almost forget the meaning of the word.

Gris. Poor child ! poor child !

Filo. You would really help me ?

Gris. Indeed I would.

Filo. Oh, you are a woman! I have seen no woman for many days. Men's eyes are horrible; they look at me as if I were a beast which they could buy and kill for food. Your eyes are different (*looks away*). What am I saying ? You will spurn me ?

Gris. My child!

Filo. Now, you pity me. I will not be pitied. You have cast a spell upon me. I was strong before you came. Men had not tamed me. I had outstared death. Now I am weak, crushed, despised, pitied. I hate you. Oh, no, no! I am too miserable. (*Falls on* GRISYLD'S *neck.*) I meant to be silent, to go and say no word, make no sign to man or God; but I was alone, and I felt very cold. Now—I have spoken. Will you come with me?

Gris. Where, my child?

Filo. Into the grave.

Gris. No, no! not yet—nor so. There is a time for death; but not yet. I will hold you; you will not go from me while I hold you.

Filo. I have not slept for days—I could sleep now. I could sleep if I were certain not to wake. But to-morrow! I must not wake to-morrow: you and I are poor weak women: we cannot resist—you said so—we can only die; let me die while I am happy. To-morrow, instead of you that man would come —I should feel his hands, breathe his breath. No, no! now is my time to die. Why is he so wicked?

Gris. He is not wicked, Lady Filomene.

Filo. (*Disengages herself.*) Not wicked!

Gris. Twenty years have I known the noble Marquis of Saluce, whom you first saw ten days since: I have loved him as I could love none wicked; I love him still.

Filo. Have you come here to praise him? to play the go-between?

Gris. When I heard you call him wicked I could not choose but speak.

Filo. If you loved him, you would rather see me dead than his wife.

Gris. That you should be his wife is not my grief; my pain —oh, my pain is old and manifold!

Filo. Then you have suffered. Men said you could not feel.

Gris. Yes, I have suffered.

Filo. Have you any hope?

Gris. None.

Filo. Then come with me.

Gris. Why should you die?

Filo. You must not hinder me; you can bear much because

life promised you so little. I was promised all that earth can
give—rank, learning, beauty, wealth, power, love, even virtue—
all were to be mine. I believed they were mine already—
would be mine eternally. Was that a sin? Surely not, only a
child's folly. I wished for what was good and beautiful. I
thought poets knew the truth; they told me of Love. I dreamt
that one day he would build me a charmed palace where nothing
foul could enter. I saw the first stones laid. I cannot explain
all. For a little while I was in fairy-land—now I am back
on earth, where I will rather be food for worms than that man's
wife.

Gris. You are sorely wounded; but will you fly in your first
battle?

Filo. Death is no coward's choice. Why should I bear
misery? If I had any hope—yes, I would bear some pain; but
without hope—how can I? And there is none.

Gris. Experience would give you faith instead of hope.

Filo. Faith!—faith in what?

Gris. In God—in man—even in yourself.

Filo. Is that the lesson your life has taught you?

Gris. It is.

Filo. Then I have not been told the truth. I heard that
you were made a wife in mockery; that your husband first
murdered your children, and then, in sheer wantonness, divorced
you and cast you out a beggar. Is this not true?

Gris. That would seem the truth to many.

Filo. How could experience of such a life give you faith?

Gris. Your words make no true picture of the life I see when
I look back.

Filo. Show me what you see. I cannot wait for more experience. Let me learn from yours. There is not much time,
but if you will tell me your real life I will wait till the day
breaks. No one ever told me what they really felt. I think
you will. If I knew what you meant by faith, perhaps I might
believe.

Gris. Then you would live.

Filo. No, no! Let me rest: I am tired; I long for rest.
Tell me your life, and I shall forget my miserable self a little
while. I will believe that I am dead already, and that you are

an angel comforting me, teaching me to forget all my own false
life. Oh, speak! speak! If you leave me time to think about
myself, I must be gone at once.

Gris. I cannot readily find words—my suffering grew from
that; yet, if you will be calm and gentle, you shall hear the
story of my life, for I love you, child.

Filo. Even your touch soothes me.

Gris. I never knew my mother; my father was wealthy for
a peasant, and I grew up strong and well favoured. One day I
was gathering fodder, when a bear ran from the wood into the
little clearing where I was. Seeing that I could not run so fast
as he, I sat and waited. The great beast stopped short, gazed
in my face, sniffed round me like a dog, and then went slowly
from me. He had not gone ten paces when a young forester
sprang to my side, and the bear turned upon us. The young
man took two steps forward, waited till the bear rose, and with
one spear-stroke slew the wild thing, who died in pain. I cried,
' You should have spared him as he spared me;' but the forester
laughed merrily, and said, 'You did well, and you speak well,
as a woman; but men and wild beasts do not spare each other,
and I cannot think they should.' We talked long. I have
never heard or seen a man who looked or spoke like that young
forester.

Filo. Was it the Marquis?

Gris. Truly, but I thought him a forester. We met often,
and one evening in the greenwood he asked me if I loved him.
Then I gave him all my heart, and we sat for a while happy;
but I could not make him feel my love. He thought me cold.
He longed for more than I could give or promise. Men's love
is not like ours. He went in anger, saying I should never see
him more. Then first I knew grief.

Filo. I would that he had kept his word.

Gris. Many days went by in silence; then in full noon,
amid the summer hay, he stood beside me, habited like himself,
a prince. 'Grisyld,' he said, 'we have been sad, but we shall
be so no more. I am the Marquis of Saluce, master of the
land and of its people. You alone of women have my love; you
are worthy of it, and you love me alone of men. I know it
well. I think myself no fool, and yet we have been sad; but it

is over now, and you will come with me.' I trembled, and he, impatient, flashed his eyes, and said, 'Come, and come now, for you shall be my queen, though not my wife; and never will I have wife, you living. I love you, girl, peasant as you are, and I so glory in my love, that I will set you on a throne of gold, and honour you before all men. You shall help me to rule the land.' 'Yet not your wife, my lord?' I said; and he then spoke with scorn of priests, and called me poor of heart, and said I did not love him, and was silent. 'My lord,' I said, 'had you been the forester you seemed, should I have done well to live with you, not being your wife? I will not give the prince that which I could not give the man; my joy has come and gone.' Then he was very wroth, and made as he would kill me, but turned and ran.

Filo. You would have suffered less if he had killed you then.

Gris. After long weeks, in harvest, as I worked among the stacks at home, I heard the blare of trumpets and the tramp of horses. Then a wild voice I knew well hallooed, 'Grisyld! come, Griselda!' And I stepped forward and met the prince with his train, cavaliers and dames—a noble sight. He dis- mounted, took me by the hand, and spoke these words: 'Ladies and nobles of my realm, you have much urged marriage on me, which I deferred until I should have found a wife who pleased me—not alone my eye, but my ripe judgment; and here the woman stands. Nay; start not. Whosoever smiles but in thought shall die, and rot unburied;' and as he looked round faces paled with fear. 'Your purblind eyes,' he said, 'see nothing they are not taught to see. This girl is true queen by nature, and I am wise in that I know it. Griselda, will you be my married queen?' Now, when I heard the trumpet tones in which he praised me, my heart said, 'I will be worthy of this great love, the greatest ever shown by man to woman;' and I answered boldly, 'Yea, my lord.' 'It is well,' he said. 'Mark me then, O queen! Had you been my mistress, I would have served you in the foolish fashion of my court with adulation, sighs, and meek obedience. Being your husband, I remain your master. Will you obey?' Looking on him, I thought there was but one man on earth, and I said, 'I will obey.' 'Grisyld,' he said, 'put off those peasant clothes, there where you stand, and don these robes. The ladies of your court will aid you.' As they came to me, one laughed; the Marquis struck her with the back of his hand so that her mouth bled, and she wept, but no man stirred. When my clothes had fallen from me, the Marquis bade them pause, and asked me if I was ashamed. I answered, 'No; I will never be ashamed to do what you command. I trust in you.' And as I spoke a shout went up from all the men, and my lord smiled. As the ladies robed me, he took a necklace, and speaking to her who still wept, said, 'Lady Florizel, this will last when your bruise is past, and as you wear it remember to chafe no man when he is roused.' She took the necklace with thanks, and then I better understood why he had chosen me.

Filo. He was well named the wild Marquis.

Gris. I went with him to his court, and there with much state was married; and my lord heard willingly from me all the sufferings of the poor; and as he learnt the truth he was not slack in doing justice. He was well pleased with me. No happiness on earth could be greater than mine was then, and I owed all to him.

Filo. It did not last.

Gris. I had years of happiness, and each year my blessings grew, for I bore him children—one boy—one girl. They were not with me long.

Filo. Blessings soon leave us, and the blessings we have lost are curses.

Gris. I willingly pay all the pain my life can hold for the joy my children gave me. They are my chief blessing now.

Filo. And your Marquis killed them. He said, 'No peasant's child shall sit on my throne.'

Gris. He never said so in my hearing.

Filo. You know he killed them.

Gris. The world says so.

Filo. And their father?

Gris. He sent them far away; afterwards he told me they were dead.

Filo. Not that he had killed them? What were his very words?

Gris. 'Griselda,' he said, looking strangely on me, 'I have learnt to-day your children are both dead.' I was dumb. I could not think about my children for his eyes. They were the eyes I saw when he had slain the bear, and again when he would have killed me. I hardly heard his words. I only saw his eyes.

Filo. Said he no more?

Gris. 'Did you not love your children, woman?' he said, frowning; and I found no answer. Then he cried out loud, 'I think you are a stone, insensible; do you love me?' I said, 'Most truly;' and his anger waxed, as I think, because my voice was cold. He went aside a little way, and suddenly turning, said, 'I killed your children. Can you love me now?' I said, 'Yes, my lord.' 'You lie!' he cried.

Filo. Was that no lie?

Gris. How could my heart change in a moment? His words were mere air. I heard of death and children, and felt mazed; but he was there, before me, miserable. So when he asked me if I loved him, my heart said Yes. Was I wicked?

Filo. Some minutes since I should have called you wicked. Then he—— ?

Gris. He drew his sword, and when I saw the blade, and in his hand, I cried, 'Ah! yes;' for I thought I saw the way to my children; and I rose and threw myself upon him and kissed him, and then trembled; for the steel had not pierced my heart, but he was kissing me, and his eyes were raining tears. He put me softly down, and ran out into the darkness weeping.

Filo. Strange!

Gris. That night in my chamber the devil tempted me. Lady, you would die because you lack hope. Children are hope made flesh, and they were both gone. Little Roland: he could barely lisp, but his young soul shone with love of honour; and my baby girl—born princess, with her father's eyes—Dora——

Filo. My second name.

Gris. They were gone. Their father said that he had killed them. The devil lurked all night in my heart, whispering, 'There is no God who lets such things be done; there is no God who lets such things be felt; there are no men but only fiends, since your man of men is but a fiend. Earth, heaven, and hell are all a mockery. Go back, Griselda, to the dust from which you came.' But when the dawn brought light Satan fled; for God spoke to me, saying, ' Have I not these many years nourished your body and your soul ? Have faith, faint not.' Then strength came, and I could remember all my husband's goodness. At sunrise he summoned me.

Filo. What was his mood then ?

Gris. He was gentle. ' Grisyld,' he said,' you are the best or worst of women. If, as you say, you love me, conquer sorrow. Let my people be your children. If you can find means to make them happy, you shall not lack power.' I obeyed him.

Filo. Yet he divorced you. Why ?

Gris. His first doubt was his last. He thought I could not love him. I fear that I grew wearisome; but I know now I have kept faith; I love him still.

Filo. He was cruel.

Gris. He said there was no middle way for him and me. He must have all my love or none. I should have all his power, or no crust of bread from him. I think him right.

Filo. And your children ? You learnt nothing more of them ?

Gris. Nothing.

Filo. Have you ever spoken of them ?

Gris. First to you.

Filo. Do you think them really dead ?

Gris. Why do you ask me if I think them dead ?

Filo. His words were strange, and not like the truth.

Gris. Then I am not a fool. You—you, the first who hear the facts—the very first—you whose judgment is not warped like mine, but clear, dispassionate—you say they are not dead. Now I am certain. Now I know they live. What can be more sure ? He could not murder them. He, so good, so kind; the tender little ones! How could he ? He loved them, and he loved me too, their mother, once. It is impossible that he could murder them; and I know they are not dead. They would have come for me, and I have never seen them once—not in a dream. Say it again; say they live—say it! Will you not speak ?

Filo. I cannot believe they live.

Gris. What can you believe ? You can believe any folly.

You can believe he is a murderer. Why cannot you believe they live? Speak!

Filo. I cannot.

Gris. Think again. I have told you all quite truly. I have hidden nothing, and your first thought was my children lived; but now you will not say it. He is no murderer. Men tried to scare me from Saluce. They threatened me with shame and death; yet I had no fear. I should have feared a murderer. The trumpets of his messengers sounded the very call which summoned me to meet him in my youth. That music heralds hope, not fear; and as we came the whole long road smiled with children—their eyes shone hope on me; and while they ran and laughed and leapt, again and again the clarion notes rang out and sang to make hope bold, and always at the close I heard his voice call, ' Grisyld, come, Griselda; ' yet since I came I have not heard even his footfall; and when I tell you all my life, you cannot believe my children live, and you call my love, my soul, their murderer. Why do you stare at me? I am not mad. You are mad. You credit infamy. I know that my children live.

Filo. This world is too strange for me. I can stay no longer. (*Sucks ring.*)

Gris. What have I done? What have I said?

Filo. This ring held poison. Your story did not tempt me to stay. There is a little left for you.

Gris. I have killed her. Help!—sergeant—soldiers—help! Lady Filomene lies dying.

Filo. No, no!—Let me be alone with you. Be as you were at first. Peace—peace—not all this tumult. While I die you might be gentle with me.

Gris. You may be saved. When he knows the truth he will set you free.

Filo. I will not be saved. They will torture, they will degrade me. Oh, I am afraid—I am afraid. Why can I not die? The rogue lied; he said instant death. Will you not kill me—quickly—out of mercy? [*Swoons.*

Enter MARQUIS, TANCRED, *and train.*

Gris. Quick, quick, most noble lord! Lady Filomene lies

poisoned. This ring held the poison. You may learn the nature of the drug from the strong smell. You have skill in antidotes; save her, my lord!

Marq. Griselda, I greatly fear that you have murdered her.

Gris. Oh, waste no moment now, my lord; the poison works—speed, speed!

Tanc. My lord——

Marq. Silence. (*Aside*) Sunrise is not sunset. (*Aloud*) I sent her to serve Lady Filomene. Is it not most like that out of jealousy Griselda poisoned her? Nay, as I think, some one heard her use the words, 'I have killed her.'

A Courtier. I heard that, my lord.

Gris. What matters how she got the poison? My lord, I never knew this slackness in you. Is she dead?

Marq. (*To courtiers*) You are passing wise. Yet not so wise as Grisyld there, whom you have known for years to be a true and faithful woman, the noblest in the land and the most prudent; but you let one black doubt in one minute blot the whole fair record. Cheerly, Grisyld! The girl is not dead, nor like to die.

Gris. God's blessing on you! but be quick.

Marq. I will not touch her, or these wiseacres will say I cured her with some antidote—to screen you, perhaps.

Gris. Oh, let me be, my lord.

Marq. Silence, Grisyld. In good sooth, to-day patient Grisyld is impatient. See! the girl revives—stand where you are; 'twas a mere swoon.

Filo. (*Flies to* GRISYLD.) Save me, save me from him!

[GRISYLD *clasps her.*

Marq. Now, mark me. This foolish girl wore a ring in which she stored poison, as she thought. The jeweller in Florence who sold the ring, as in duty bound, told me, her guardian. I took good heed the ring should hold mere scented water. Young lady, you are richer by a strange experience. I envy you. Sirs, are you certain that Griselda has not poisoned Lady Filomene?

Courtiers. We are indeed, my lord.

Marq. Day dawns. Our marriage feast begins shortly.

Wait a few hours, Filomene, in this poor world, which you may then find worth endurance. If not, at nightfall you shall have your choice of poisons. Grisyld, be of good cheer. I heard your story. If you could have spoken so long since, the world would have been the poorer by a tale. Obey me once more; serve at our festival. Let your dress be rich to suit the day. Look cheerfully. Order the house; set the tables; usher in the guests. This service done, you shall be free—if a woman can be free. Will you do this?

Gris. My lord, right cheerfully.

ACT III.

Scene.—*Banqueting-room in the palace of the Marquis of Saluce. Daïs and table at the back of the stage. Two thrones behind this table. Door centre, with curtains. Side tables on a lower level right and left. Banquet partly set.* Griselda *and two servants discovered; two soldiers on guard* R. *and* L.

Gris. Barnaby, these seats must not stand here. Place them behind the tables.

Bar. There will be no room for us to pass.

Gris. Nay, my friend, I knew this house of old, before you came here. There will be room.

Bar. Old ways are not new fashions. We were told to put them where they stand by John, our majordomo; and when John bids, John's men had best do his bidding.

Gris. I will answer him for you. These chairs must be moved.

Richard. We had best do it; I heard the Marquis say she was to order this feast, and the Marquis is a bigger man than John, though John thinks not.

Bar. I fear John his little finger more than the Marquis his whole hand. The Marquis never heeds poor knaves like us; but I'll do her will, if thou'lt bear me out. Look you, mistress, you must answer it to John. (*Removes chairs to back of tables.*)

Gris. You shall take no harm.

Bar. Here John comes; now see to it.

Enter MAJORDOMO, C., *with two large bouquets, which he places in front of the two thrones on the table.*

Gris. Those flowers, sir, make a fair show——

Major. They will please——

Gris. But let them stand more apart, so that the Marquis and his bride may see their guests. (MAJORDOMO *walks away.*) You hear me, sir?

Major. Hearing is not heeding. If you will serve in my household, Madam Grisyld, learn to spare me your advice when I ask none.

Gris. My lord will be ill pleased if he finds the flowers there. I know his liking, and he commanded me——

Major. No more words, I pray. I command here, and none but I. You scurvy loons, why have you displaced those chairs?

Bar. and Rich. The lady gave us orders, John.

Major. The lady! Orders! Why, look you, what a pitiful thing it is to be a fool. Now, I will set the whole hall in a roar telling of you three; how the cast-off country wench gave orders, and how the two big foolish knaves did as they were bid and called her lady!

Gris. Sir, you mistake; my authority, though small, is real; given by the Marquis. I but do his pleasure.

Major. The Marquis takes small pleasure in you. His pleasure is to take a young wife, and use the old one for a serving-wench. He's a merry fellow the Marquis, and he sends you here to make us merry. That's his pleasure. Ha! ha! ha!

[*The servants gather round and laugh. The* MARQUIS *enters unseen at the back,* C., *with* TANCRED, FILOMENE, *and suite; they pause.*

Gris. We lose time. The tables are not furnished, and guests will be here immediately. Let us do the work.

Major. To it, wench, to it. Do thou work, we'll look on and admire; but let those flowers be, else thine ears shall tingle for it.

Serg. Shame!

Major. What, old tosspot! wilt thou meddle? Hast thou still the humour to play the bully for thy lass?

Serg. When my watch is out, my staff shall play the bully on thy back, an' I be cashiered for't.

Major. Now am I so sweet a man, that for thy ill threat I'll give thee good counsel. Make a match on't; marry thy master's leavings—'tis a known way for old cullies to buy promotion.

[*The* MARQUIS *advances.*]

Marq. How is this, fellows? You are not ready.

Major. My lord, this madam hath hindered us.

Marq. And what has been done is not well done.

Major. She is answerable, my lord.

Marq. Why, you have set a bush here; a very thicket to hide my guests from me. Griselda, you know how I mislike all tawdry ornaments hindering good fellowship.

Major. I bade her shift those very flowers, my lord, but she would take no orders from me—not she.

Marq. You said she was answerable. Was it not rather your duty then to do her bidding?

Major. Your lordship will be merry.

Marq. Would they not obey you, Grisyld?

Gris. My lord, they thought you did not really mean that I should be obeyed.

Marq. Ah! (*smiling*) you thought I sent the lady here to make you sport.

Major. We're not so dull but we can take a joke, my lord.

Marq. But it seems you have a spirit yet, Griselda, and would not move the flowers at this fellow's bidding.

Gris My lord, I was about to move them, not indeed at his bidding, but knowing the custom of the house. Shall I move them now?

Marq. Too late, Griselda. Those who speak in my name must be obeyed. You, fellow, say that she refused to shift those flowers; name her punishment yourself.

Major. Some merry gibe, my lord, would do well for a wedding day. Let me see. Marry, I have it. My knaves might paint her face, my lord, and smirch her fine clothes, and turn her out among the common sort of people. They would laugh, my lord.

Serg. Hound and liar! Pardon my speech, my lord; the lady never disobeyed.

Major. You must not credit him, my lord. He and she will

swear anything for one another, being something love-sick, an'
it please you.

Marq. Silence, old soldier; I saw and heard him. Sirrah, your
sentence falls on yourself. You shall not lose your merry gibe.
We will brand it on your face in good lasting colours as a token
that you are the author of the jest—a jest so good that it shall
make the common sort of people laugh all your life. Our hangman shall be the artist; his style is rough, but tells its story.
His graving tools make bold work; he uses them redhot, with
a firm touch. Soldiers, take this false witness to our question
chamber. Bid our executioner burn a mouth upon him fit
to utter foul lies. Show him me a month hence grinning
merrily, like a half-rotten skull. If the brand you stamp on
him fail to make me laugh, he shall try his hand on you.

Major. (*kneeling*) Mercy! Oh, my lord, have mercy! We
mistook, and I was only one of many.

Marq. Will you still accuse? Would you have companions
of the brand? No; they shall grin at you, but not so merrily
as you will grin at them. Laugh, hounds. Are my jests worse
than his? (*Servants try to laugh.*)

Tanc. My lord, I beseech you, pause at least for one hour.
You are not master of yourself. Your sentence is more
monstrous than his villainy. The more so, since you are the
real culprit, in that you sent the noble lady to consort with
lackeys.

Marq. Sir, I once knew a knight who let a noble lady run
some risk by his own fault; but he killed the churl who wronged
her. Was he not right, my Lady Filomene? By your face,
vengeance and you should be close cronies. Do I not well to
punish this hound?

Filo. I cannot tell. Punishment should not be merciless.

Marq. Merciless punishment may be true mercy to the
innocent. Is that not true, Griselda?

Gris. Sire, you once taught me that punishment is not
vengeance. I remember that you said those punishments were
just which roused just so much fear as would deter men from the
crime. Will not this sentence make my fellow-servants fear too
much?

Marq. The good need have no fear.

Gris. The best will fear most. These men are your lackeys. Their rough mockery is venial. They may be good men

Marq. You cannot call this man good.

Gris. As yet he is no criminal; but if you carry out your threat, he will be jeered and flouted into crime. This punishment will make him wicked.

Marq. Grisyld, Grisyld, you are barely honest. You plead this man's cause not for its justice, but because you cannot bear to think of even vile things in pain. I know that weakness of old in you. Yet your argument was good, and you gained time. I am cooler now. Go, fellow. I do not like your face. If you show it me again I will carve it till it please me better.

Major. May Heaven bless you, lady, and make me a better man.

[MAJOR *hurries off*, R.; *the* MARQUIS *turns to his guests, who take their places behind the side tables. The* MARQUIS *leads* FILOMENE *to one throne, and remains standing. Servants and* GRISELDA *in centre.*

Marq. Gentles all, your pardon. Take your places. I would say our feast had begun badly, but that the lady here spoke well. Think you not so?

Tanc. Would that my lord might ever have so good a counsellor.

Marq. Good wishes are ever welcome, and they are most welcome on a marriage day. Would the Lady Filomene have us keep Griselda as our counsellor?

Filo. My lord, if you would take her counsel in the greater matter of our marriage, you would be spared a far greater sin.

Marq. What sin? But on this day I must do your will. Griselda, what think you of this lady? Is she not fair and noble, and fit in all ways to be my bride?

Gris. My lord, no man can desire a fairer wife. I know her nature to be noble. She is worthy to be a queen. Only give me leave to pray that you will not crown her with such sharp thorns as your former wife bore patiently. She was a peasant, but this lady is a princess.

Marq. True! I was a prince and you a peasant; but was that the reason you obeyed me?

Gris. If you too had been a peasant, I should willingly have obeyed you as your wife. Yet to us who toil and bear hardship from our childhood, patience and obedience come more readily than to women who are nurtured delicately. My prayer is that you will remember this.

Filo. That prayer will not save me.

Marq. And you, my lords, I trust that you all think Lady Filomene a bride worthy of your prince.

Courtier. Her worth is such, my lord, that had you not been secret in this matter, each princely bachelor in Italy would have sent heralds to forbid the banns, and claim the lady for himself.

Marq. Yet Grisyld here is wise, and fair too.

Courtier. True, my lord, but Lady Filomene is twenty years the better bride.

Marq. Excellent valuer! My lords and ladies, pardon me if I delay our feast yet a little while; I have that to say which may give it great relish. Nay, keep your seats—for once my privilege shall be to stand. You must not steal away, Grisyld: you are wanted here. I have but half a will to speak. Now, I know what no man else knows, and one minute more will transform my mystery into vulgar news. Yet my news will bring me much profit. I shall gain a loving wife—oh! 'tis true, girl; the words that I shall use are magical. As for you, good youth, when I have spoken you will crave a blessing from me; though now I see your hands quiver like leashed hounds when they scent the deer : hold them from my throat one minute longer ; then you may slip them if you list. You, my friends, who at last know with certainty that I am mad—stark mad—you will crack the air with shouts proclaiming that I am the wisest prince who ever lived, and your applause will be sincere. You, Griselda, shall be happy. Not one of you believes me. Common courtesy almost forbids that I should show you how much mistaken you all are. When I have told my story, you will think me good.

Courtier. My lord!

Marq. Your surprise is natural. You imagine me to be a murderer, but in my courts we have one wise rule. We never hang a man unless to make the pair with a dead body there already.　Griselda, I heard you say that you thought your children were not dead.

Gris. Oh, sire, have mercy on me!

Marq. Other children might be born to me, and if the first two be alive, they have rights to the succession in our marquisate.　What proof is there that the children died?　Which among you ever met a man who had even heard another say that he had seen those children lying dead?　But, if he be not dead, most certainly Count Roland is our heir.

Gris. If he lives he can resign his claim.　You said I should be happy.　I believed you, though they all thought you mocked me.

Marq. The blindness of this fair assembly far surpasses all my expectation.　Would a son like Tancred here content you?

Gris. Is he my son?

Marq. Come here, young man.　(TANCRED *joins* MARQUIS *and* GRISELDA *in centre.*)　If you had your choice of mothers would you take this lady?

Tanc. Griselda my mother!

Gris. Sir, my strength fails.　I cannot hear.　What was that you said?　Will you not speak clearly?　Pray speak loud.

Marq. Count Roland of Saluce, kneel down before your mother.

Gris. Ah! and Dora?　She is Lady Filomene.

Marq. Even so.

[*The courtiers have all risen.*　FILOMENE *runs and kneels
　　　before* GRISELDA, *who stares at her children.*

Gris. True, you might be Roland; you are what he would have been.　I could not wish more—wish—I wish I could see my little boy again.　You must not be angry.　I will be happy. I cannot yet; but I see happiness shooting towards me from heaven's gate.　God's angel brings me a new heart.　Gentlemen, there is no doubt that this is true; you all know it.　Why should he deceive you in a matter of such great moment? That is Dora.　I know her eyes well.　Are you glad I am your mother, Dora?　Will you kiss me?

[FILOMENE *rises to receive the kiss.*　GRISELDA *takes her head
　　in her hands, kisses her slowly, plays with the head,
　　looking earnestly in her eyes, then, as the hair is dis-
　　arranged, cries suddenly.*

Gris. Ah! I see the scar; child, I have kissed that wound a thousand times. Walter, there, look where your sword— Ah! (*Falls on* FILOMENE *and devours her with kisses.*)

Marq. Yes, gentlemen, I tried one day whether the child, having no experience of wounds, would yet fear steel. That scar tells the answer. Our little Dora flung herself against the blade. I had forgotten it.

Tanc. Mother, I remember you; I knew your voice.

Gris. My son, my son, my Roland! No more doubt, Roland. Gone, gone—doubt is gone. Doubt was a heavy load; it grew and grew, and weighed me down. The thing lived: it had a voice—a cuckoo voice—always in my ear— doubt, doubt, doubt, doubt, unceasingly. Dead now. I am free. You are no murderer, my lord. I knew that always; you heard me say I knew that you were not a murderer; but no one gave me any comfort; no one said that what I knew to be the truth was true. I might have been mad, you see; but God made my faith firm, and now I can thank God—Ah me!—I am tired. Forgive me, children; I will love you. You are strong, Roland; help me to bear happiness, my son.

Marq. Rest, Grisyld; lean on Roland; I have more news yet—none bad. Friends and counsellors, that divorce I showed you was a forgery—good parchment spoiled. This lady has been queen all the while. I trust none of you remember words spoken to her which a queen might not hear. Griselda, have you followed me?

Gris. Not clearly, my lord; but it is well with me.

Marq. Gentlemen, in time past certain of our subjects caballed against our queen, and denounced her son as base born. For his safety I concealed him, even from his mother. That danger is now past. He stands there, a worthy prince, while his mother's fame is such that before long heralds will forge pedigrees to prove kings her kinsmen. Once I crowned her; now I hold that she crowns me, for her glory outshines mine. Say I well, my lords?

Courtiers. Most nobly. [*Shouts.*

Marq. Griselda, take my hand; let me lead you to your throne. I felt less joy on the first day I brought you here, a younger bride. Last night you spoke words I had almost despaired of hearing ; words for whose sake I could make you welcome to a thousand thrones.

Gris. What words, my lord?

Marq. Words that are far too precious, Grisyld, to be wasted on the common ear. When we are alone you will smile to hear how well I remember all you said. When you spoke of love last night I heard passion in your voice, my wife. My doubt is at an end.

Gris. Wife! my lord. I am not your wife. I saw the papal bull, where it was plainly written that from that time forward I should no longer be your wife.

Marq. The bull was forged, Grisyld. Board, bed, and throne, all are yours.

Gris. Will nothing hold? My lord, you promised that I should be free. Will you keep your word? If I am your wife, I am not free.

Marq. You shall be free. Henceforth I claim no obedience. Have no fear. I will make no further trial of your love; but queen and wife you are, and queen and wife you must be.

Gris. If I must, my lord——

Marq. What is this? Friends, we have been somewhat sudden. I pray you leave us for a little. The Queen needs rest. Ladies, make you merry—you shall be summoned to the banquet presently. (*Exeunt courtiers, servants, and soldiers talking,* C., R., *and* L.) Bring a chair and cushions, Roland ; your mother faints. Filomene, wine !

Gris. I am not faint, my lord.

Marq. Call me Walter, Grisyld.

Gris. I must not.

Marq. Nay, wife ; for you I was Walter long before you knew me as a prince, and I will be Walter now.

Gris. You are not he, my lord ; nor am I Grisyld. These are not my baby children. All is changed, my good lord.

Marq. We are not changed since last night, Grisyld.

Gris. Are we not? I do not feel as if we were the same. Sir, I cannot be your wife.

Tanc. Mother, mother !

Filo. She is right, Tancred ; he has sinned past forgiveness.

Marq. Your words are wild, Grisyld. Last night you still

loved me, though you feared I was a murderer. To-day you know that I am almost innocent.

Gris. The relief that knowledge brought me is past words ; you must not accuse him, Dora. I have been much to blame. He meant nobly. I have suffered, but suffering once past is ended. If I am free I will go back to the forest; there I shall be happy. There, my lord, I will bless·you every hour of each day; you, children, will come to me sometimes; but not you, my lord. I will love you in the forest, but I will not see you.

Tanc. My father's doubts were folly; his trials of your love were insults—sins; but your present action, mother, is not of a piece with your past life. You called for death sooner than freedom. when I would have set you free, although my father then seemed to threaten ; while now, when he owns his fault— gives you back your children and strives to make amends—you will not stay with him ! Why not ?

Gris. I am not what I was.

Tanc. Surely you are less noble.

Gris. That may be ; but I am what I am.

Filo. All your life you have chosen what was best, most right. Is this your choice now ?

Gris. I never chose. I did the one thing I could do. Would it be what you call best, most right, to say I love the Marquis when I do not love him ?

Tanc. I know the full depth of his sin ; but you, mother, can you not forgive ?

Gris. I have nothing to forgive. You all say well. I am in the wrong ; I beseech you all to forgive me.

Marq. Forgiveness is a word. Nor you nor I can be as if what has been had not been. I could regret much that I have done, but I ask for no forgiveness. When I did what I could most regret, my love's sea was at the flood. That sea was wild and cruel—cruel both to you and me ; yet the passion was love —a wild deep sea of love, whose waves still buffet me ; and my strength is failing. I fear that I shall sink.

Gris. I know that you have loved me ; I think I loved you, long ago. Last night I thought that I still loved you, but to-day I cannot. I must tell the truth. Oh, let me go. You torture me.

Marq. Griselda, you are full of pity. I have known you pity thieves and murderers. Pity me. Stay with me. I will not ask for love; but all misery, even mine, claims pity. When you turn from me, my life's light is quenched. Last night the sun still shone on the green meadows in the wood where we first met—where we were young—where we loved. All's dark now. My past is dead. I had not killed the past for you; it lived in your voice last night.

Gris. I wish that I had died last night.

Marq. Although to-day brought back your children! What is my new crime?

Gris. None. The crime is mine. That is why I wish that I had died.

Filo. There is no crime. We honour you. Why should you wish to die when we love you so?

Gris. Why would you have killed yourself last night?

Tanc. Marquis, your good deeds are but new born, and they are far too weak to cope with your old sins. Mother, I grant that you must go. I will not say that you are wrong, nor can I say that you are right. I will go with· you, protect you, honour you, obey you.

Gris. And condemn me. I will go alone, if I may go, my lord.

Marq. Look up, Grisyld. This some time past I have not seen your true eyes. Grisyld, you are in the right. We have all been wrong, as men always are when they doubt you. Last night, children, your mother spoke of me as I once was, and you heard that then she loved me. She even said she loved me now; but she was thinking of me as I was long since. Now, I am strange to her, and this stranger comes, gives her a strange man and woman for her children, and says: 'Quick! love me. This old rogue is that young forester you loved long ago.' You are right. I am not he, Grisyld, though I remember him, and I am glad to think you once loved him.

Gris. That is true, my lord; but must a woman's love fade as she withers?

Marq. Not so. As we ripen, love grows; but growth means change. 'Twas sheer folly when I begged for love as boys beg from girls. Your forest story set me dreaming. Now I am awake. I say that you are right.

Gris. I cannot think so; but my faith in you is justified.

Marq. Let this be clear. You owe me no debt. Love can never be a debt. Love is a free gift, or is not love. So, then you are free. I set no limit to your freedom; choose and act. The forest shall be yours in fee if you choose its silence. Or in Tuscany Roland can find you work that will reward your pains. But if you can freely choose the lot, remain with me. Rule in Saluce—my queen and my wife.

Gris. Sir, your great trust gives me courage. I will try to tell you what I feel. My lord, for many long years you left me in pain, when by one word you could have healed my bitterest wound. All through those years I hoped—more—I believed that your ways might be justified, though they were dark. Now I know the truth. But that's not it. The sin of those we love gives us pain, but the pain cannot kill our love, and my love is not dead; but you and I use that word with a difference. Long ago you told me in the forest that such love as I could give was of no worth to you; yet I thought I loved you well, and when I had become your wife with such joy as I must not think of, you still feared, no less than before, that I could not love you. Even when I gave our babe into your hands in my first hour of motherhood, your greeting was to tell me in a whisper how you hoped that, with our child, at last, love had come. You took my children from me—murdered them, you said—and with the same breath you asked, ' Can you love me now ? ' I answered, ' Yes,' and my answer was the truth. But after that I saw men's eyes question me. ' Can she love ? ' they said, ' or does she lie ? ' The smiles of women, as they passed me in the street, told of the same doubt. Perhaps I read my own thought in their looks, for not a dog could fawn on his master in my sight but I would ask myself whether my love were as true as his. Even dead things seemed to mock me. In the dusk, love's goddess laughed from the hangings of our room and said, ' You cannot love; we can love—we who are merry. You have a broken heart.' Her wanton boy would chatter to me of dead children, not of love. Fear mastered me, so that I believed I was a liar; but when you sent me to the forest I found my own true self there. There I loved you. I loved you last night. Only to-day, when I heard you call me wife—when I heard your voice speak that word love—a chill struck through me such as falls upon a wretch entering his prison cell. What I feared was your love. I fear that love now. Wolves, met in winter, have scared me less. My lord, you know the truth.

Marq. I had said that I would not ask for love. I have. learned to rate the passion I so cherished no higher than an old antic fool—past service. The discarded knave should not have frighted you as you sat upon your throne. A crust to mumble in some unfrequented nook about your palace would have made him almost happy.

Gris. You are angry. A woman cannot help her fear.

Marq. No, lady, no! I am not angry with the sun if he should blister me. I will not quarrel with the rain although it wets me. Sun and rain are good. I think you good.

Gris. I thank you for your gentleness, my lord.

Marq. You need have had no fear, Grisyld. I have no quarrel with the truth.

Gris. How base a thing fear is! Day by day I have set this hour before my eyes, night by night I have dreamt that I had told all, and my wicked fear always showed me cruel phantoms wearing your shape. I saw you kill me. I saw you die by your own hand. I heard you curse me with such words as blacken memory, and now the day has come, and in the light of day I see no more phantoms. I see you. You stand before me kind and gentle. I have been a fool.

Marq. Neither fool nor coward, Grisyld. You have been a woman, and we men hurt women much more than we know ; but I will do justice, now I know the truth. Your place is upon the throne, but mine is not beside you. Saluce is somewhat stale to me. I will go see the world.

Gris. I rejoiced too soon.

Marq. Teach our son how to govern. He will be an apter pupil than your husband. One maxim take from me, Roland : the law must be feared—your mother is too tender-hearted to rule men wisely. Speech is useless, Grisyld : you cannot change my purpose, nor follow out what I see to be your own. You must not leave Saluce. Your son needs you ; he has no experience. Be comforted. This my act shows no despair, but plain wisdom, simple justice. I deserve neither wife nor children—not even my people's love. If I were to stay, I should be alone and a tyrant. Let this be my praise, that I could be just, even to myself. Roland, call our friends back. They will say that my last news betters all I told before.

Gris. My lord, my lord, I will not stay an hour in this palace if you leave. I have sinned. You drive home the knowledge as if it were a sword.

Marq. You have not sinned, Grisyld. We have been like two ships drifting in the dark. We clashed and are sorely shattered; but the light has come, and we will hurt each other no more. Leaving you upon the throne, I can face my remnant of this life gladly.

Gris. Sir, can you still love me, now that you know the truth?

Marq. Ay, Grisyld, we will love each other from far off. I feel older since an hour—wiser too perhaps. You said well, we loved with a difference; your love shamed mine, and I am your convert.

Gris. I fear that now you do not love me. Long since I used sometimes to ask that question. You answered differently.

Marq. Grisyld, you are cruel!

Gris. Am I? Husband!—Walter!—I will be your wife.

Marq. I feared that offer. Would you have me find my joy in your pain, Grisyld? Be just to me.

Gris. You are very blind. Sir, in other matters you are shrewd, clear-witted, and experienced; but in love I think you are a greater novice than myself. You need have no fear. Your joy never shall be my pain. I wonder if I ever shall feel pain. Oh, I will be true! I will not unspeak one word I spoke. I do not wish one word unspoken; but when we next stand before the picture in our room, you shall hear me say, 'O Lady Venus, I have told all the truth, and your spell is broken; you and your boy, you are shadows of false love: it was you who lied. I have living children, and I can love; I love my husband.'

Marq. Grisyld, Grisyld, is this true?

Gris. Walter, you were not wont to be so cold when I made you welcome. (*Holds out both hands.*)

Marq. My wife! My wife! (*They embrace.*) .Be in this mood once a twelvemonth, Grisyld, and I will think the whole year happy.

Gris. This is no mood. This is myself. I have come. back to life. I rejoice to live. All sights show me beauty. All sounds are music chiming with the song my soul sings within me.

Marq. Your voice has the old ring. I believe that you are happy, though I have somewhat lost my faith in my power to confer a blessing.

Gris. Blessing! I am a mother and a wife.

Marq. Queen too, lady. Small blessing for the nonce, since for one long hour at least I fear the queen must listen to the foolish babble of her guests. Summon them, Roland.

[*Exit* ROLAND.

Gris. I will meet our friends gladly. Among them I saw many who were dear to me.

Courtiers enter, R., C., *and* L. *The* MARQUIS *and* GRISELDA *welcome them.* TANCRED *comes down to* FILOMENE.

Marq. Ah! womanlike, they have brought the queen's robes for you, Grisyld.

Gris. I love my peasant clothes, Walter. You remember them?

Marq. They shall be set in gold like the relics of a saint. Right, Dora : it is your turn to play the tirewoman while I crown this linen coif.

Gris. You will never again lose hope, Dora.

Filo. Mother, I know now that pain may be a blessing. Mine has taught me how to love you.

Tanc. Sister, shall we be friends? (*Leads her up.*)

Filo. More than friends, Roland. I am glad to have a brother. I am glad that you are he.

Marq. I've no more surprises for you, lords and ladies. You know all. My tale is told. It will be retold often, and for the most part wrongly. Men will praise you, Grisyld (*leads her to throne*), and some women will jeer at you. No woman will praise me, nor will any man take me for his model, though I am he who found and crowned you. But let the world jog as it lists. Our hearts are glad. With full cups, my lords, pledge your queen, and make the farthest echo in the palace ring out, ' Welcome to Griselda!'

Courtiers. Welcome to Griselda! Long life to our queen!

Gris. My lords and counsellors, I prize your welcome very dearly, for it is given by the voices of old friends—friends whom I ever found rich in love and serviceable deeds. I see younger faces, too, no less kindly in their look, although as yet unknown to me. These new friends I shall in some sort link with my children in my heart, as gained to-day. Old friends and new friends, I thank you all.

Marq. Have you no thanks for me, Grisyld? I shouted louder than the youngest of them.

Gris. For so much, my lord, I thank you, but your other gifts pass all thanks. [*Applause.*

Marq. Our friends endorse your verdict, but young Tancred Roland here is silent.

Tanc. Father, you are a noble gentleman, but you wronged my mother bitterly.

Marq. Yet you will take my hand, and when Dora learns that she is promised to Count Guido Malatesta, Filomene will kiss me. (TANCRED *takes his hand, and* DORA *kisses him.*) What noise is this?

Courtier. A fellow here, my lord, has brought a letter for the queen.

Marq. Read it, Grisyld. They have begun to beg of you.

Gris. The man, my lord, who was your majordomo writes craving that on this most auspicious day he may not alone be outcast. He begs my intercession in favour of his reappointment. In a postscript he says his profits from the place were too small.

Marq. Will you have the fellow back?

Gris. In good sooth, no, my lord.

Marq. Right. One of you let him know I am hard at work devising tortures. There is one boon which you must ask, Griselda, or break faith.

Gris. What boon, my lord?

L 2

Marq. Are you so forgetful? Nay, then, 'tis too late—the place is given. Madam, your father has this day been made Mayor of his Bailiewick.

Gris. The thought and deed are kindly, and deserve requital. My lord, your queen drinks to the health of the most noble Walter, Marquis of Saluce.

[*Shouts. The courtiers pledge the* MARQUIS.

CURTAIN.

ON RHYTHM IN ENGLISH VERSE.[1]

I.

THE reprint of Dr. Guest's well-known book will be welcome to all students of English verse. The present editor truly says that its numerous and well-arranged quotations give the work a great and permanent value; we will add that this value is much enhanced by the copious index which Mr. Skeat has compiled, and by the complete references which he gives to the source from which each quotation is drawn. The title is somewhat misleading, although many historical facts of interest are to be found in the book, more especially with reference to Anglo-Saxon and Early English metres. We should rather describe the work as treating of the analysis and classification of English metres according to a new system based on Anglo-Saxon practice. Mr. Skeat does not insist upon the merit of this system, which, to our thinking, is in itself of small value, although it led Dr. Guest to write a valuable and interesting work. It must, however, be conceded that no two persons ever yet agreed concerning the theory of English verse.

The older writers assumed that each line was composed of feet analogous to those employed in classical metres, and their theory is not wholly abandoned even in the present day. Accented and unaccented syllables are, however, now usually accepted as the elements of the English metrical foot in place of the longs and shorts of our gradus, but the word accent is somewhat loosely used to denote any prominence given to any syllable. It is clear that feet consisting of elements which differ merely by their strength and weakness are not metrically

---

[1] Abridged from three articles in the *Saturday Review* for February and March 1883; the first a review of *A History of English Rhythms*. By Edwin Guest, LL.D. D.C.L. F.R.S. New Edition, edited by the Rev Walter W. Skeat, M.A.

equivalent to feet composed of long and short syllables. The ancient foot measured an interval of time, whereas in English verse we allow, and indeed demand, that successive feet called by the same name shall occupy dissimilar and irregular periods. Notwithstanding this broad distinction, it is found that our English iambs, trochees, and anapæsts arranged in accordance with classical laws produce lines possessing many of the qualities which ancient grammarians attribute to the analogous classical metres ; but lines in which an attempt is made to combine spondees and dactyls in classical fashion are not very successful.

So long as scanning was looked upon as a formal matter, having very little connection with the sound of well-spoken sentences, the heroic line commonly used in blank verse was with no hesitation treated as a simple iambic of five feet. With the aid of a little licence, all difficulty found in scanning lines in this or any other metre was easily explained away. Indeed, our language is so wonderfully flexible that no theorist has any difficulty in bending the vast majority of examples under his own special yoke ; and when he comes upon some more than usually stubborn verse he says the line is bad, though Milton, Pope, or Shakespeare may have written it. Mr. Goold Brown, in his 'Grammar of English Grammars,' gives examples of all sorts of metres classically scanned, and quotes the following lines from 'Paradise Lost' to illustrate the catalectic iambic pentameter. A vertical line is used both by Dr. Guest and Mr. Goold Brown to denote that an accent falls on the *preceding* syllable :

> No soon|-er had| th' Almight|-y ceas'd|—but all|
> The mul|-titude| of an|-gels with| a shout
> Loud as| from num|-bers with|-out num|ber, sweet
> As from| blest voi·ces ut|-tering joy| heav'n rung, &c.

Far be it from us to decide which of these so-called feet the grammarian considered to be iambs, which trochees, and which perhaps spondees. By an effort of the will we may conceive that 'titude' and 'gels with' are in some way like iambs ; though, if we are to call the second syllables of these feet accented, the word accent must receive a definition of much-embracing amplitude.

Dr. Guest pays no regard to scansion such as this. In dealing with modern verse he never mentions feet, but substitutes for the old-fashioned scansion a wholly different method of analysis based on the final and middle pause. According to him, the rhythmical element in all English verse is a *section* or group of syllables bounded by a pause at either end. The shortest section must, he says, contain at least two, and the longest section at most three, accented syllables. It is, for Dr. Guest, a self-obvious axiomatic law for all forms of English verse that two consecutive syllables cannot both be accented. Each accented syllable must be separated by one or by two unaccented syllables, and the section may begin or close in three ways—with an accented syllable, with an unaccented syllable, or with two unaccented syllables. These laws admit of thirty-six forms of the section, no one of which is called better or worse than its neighbour. Each line of English verse is said to be built up of two out of these thirty-six sections; and very numerous examples are given to show that all metres can in this way be analysed and classified. At first one is inclined to think the method at least novel; but the novelty is less obvious on further consideration. Stripped of disguise, Dr. Guest's analysis, applied to the heroic line, amounts to this.

Each line of blank verse has five accented syllables, and usually five unaccented syllables separating the others; but now and then an extra unaccented syllable is added, and at other times one is left out. Moreover, a pause occurs between the second and third accents, or between the third and fourth. We think these statements neither new nor wholly true. Three hundred and twenty-four distinct verses of five accents can be built by combining two of the thirty-six permitted sections; but even the author of the theory does not venture to say that all these will form good heroic lines, or good lines of any sort, nor can he give us any clue as to which will or will not be successful; he merely notes which of these combinations have been used by poets with success, and in these we easily discover our old friends the iambs and the trochees. Dr. Guest has observed no new fact; he simply offers us a new and complex notation by which we may name and classify many varieties of verse. Critical examination is not aided by this new notation,

for indeed our author makes no attempt to show why one combination should sound better than another, nor does the new
method remove any of the difficulties which confessedly arise in
scanning.   Like Mr. Brown, Dr. Guest must find five accents
in Milton's line :

> The multitude of angels with a shout ;

whereas a plain man without a theory would surely say there
were but three.   But Dr. Guest plays lightly with accents,
tossing them with much ease from syllable to syllable.   Ben
Jonson wrote :

> A third thought wise and learned, a fourth rich,

and we think this good stout line well able to withstand all the
assaults of its enemies ; but unfortunately there are two consecutive accents on the last two syllables in plain defiance of
the Doctor's rule.   He condemns the line as in duty bound, but
not for this reason.   He selects it as an example of the vicious
practice of putting an accent on the article ' a,' not a little to
the bewilderment of the straightforward reader who, at this
page of the book, has not yet learnt the law of the composition
of sections, according to which no two accented syllables ever
can come together ; since the line is undoubtedly verse, Dr.
Guest logically concluded that ' a ' was accented and ' fourth '
was not.   Jonson might allege that he had no intention of
putting an accent on the article ; but then he knew nothing of
the new rules of verse.   Again, Milton wrote :

> Thus at their shady lodge arrived, both stood,
> Both turned.

One line ends with three strong syllables, and the next begins
with two—we have here apparently five consecutive accents—
but Dr. Guest escapes from this difficulty with the greatest
readiness.   In defiance of Dr. Johnson, who thought differently, he will not allow that ' both ' receives an accent in
either line.

He assumes as incontrovertible that there are always five
accents in each heroic line, neither more nor less.   We have
already quoted a line from Milton with only three accents, and

to our ear there are no less than seven accented syllables in the following example from Pope, as in many others :

Go, measure earth, weigh air, and state the tides.

At least we are certain that seven of these syllables should be pronounced with emphasis, and that any person beating time while he speaks the verse with dramatic effect will strike seven blows. The old practice of scanning fits all these lines better than the new law of five accents adopted by Mr. Goold Brown, Dr. Guest, and many others.

The hard and fast rule for the position of the middle pause leads to difficulties quite as great as those raised by the laws for the formation of each section. Thus, our author writes, ' There are many instances, and some of high authority, in which the middle pause falls in the midst of a word ; these, however, should not be imitated.' As an example he gives Milton's line :—

My ang|er un| : appea|sable| still ra|ges.

Again, the unprepared reader might imagine that Milton never intended a pause to come where the colon is placed, but in that case where would the law of sections be ? Indeed, even with the pause placed to suit our theorist, the second section with an accent on ' ble ' and none on ' still ' must have seemed to him a very tough morsel. Yet the line is in Milton's noblest style.

A supporter of Dr. Guest, while admitting that the condemned lines are good, might perhaps urge that the new theory has been found to fit a vast number of examples, and need not be rejected because here and there an exceptional verse falls outside the rules. To this we answer that it would indeed be strange if some one of the three hundred and twenty-four modes of analysing a simple line of five feet should not be found applicable to most cases ; but, as the theory is absolutely general, it must, if true, fit all examples ; and in this it fails. We shall readily grant that some of the lines which defy the new analysis are not of the normal heroic type ; but to satisfy Dr. Guest we must go much further and admit that, unless the line is mispronounced so as to let the pause and accents fall after his fashion, the lines are not verse of any kind, but prose, which is absurd.

Our author is quite fearless in applying his theory even to those examples which have been most obviously written to scan. Thus he prints a verse from Gray as follows :

> When the British : warrior Queen,
>   Bleeding from the : Roman rods,
> Sought with an : indignant mien
> Counsel of her : country's gods.

The two dots indicate that Anglo-Saxon pause which he always finds even when, as in the last three lines, any such pause in the delivery would make the verse ridiculous.

In fine, the new theory requires that we should often pause where no pause is possible, call syllables accented on which no stress falls, and others unaccented on which the plain meaning of the words demands emphasis. It offers no criterion of excellence nor any clue by which we might recover the almost lost art of elocution. Under a new name we meet with the old false law, classifying verse by the mere number of accents ; and in place of scansion we are offered new and far more complex rules which, notwithstanding their great laxity, are yet inapplicable to much good verse. We conclude that the new theory is of small value. And yet we hold that Dr. Guest was guided by historical research to the very threshold of the door, which, had he opened it, would have disclosed all the secrets of English rhythm and metre.

In discussing the arrangement of his subject, he promises to treat of a ' metre which resulted from modifying the longer Anglo-Saxon rhythms by the accentual rhythm of the Latin chants,' and again of other metres ' which appear to be the natural growth of the Latin rhythm modified by the native rhythm of our language.' Here, as we think, is the root of the whole matter. Two independent verse-systems have endowed English poetry with power and beauty. Two series of rhythmical elements, one classical and one native to the soil, co-exist in each verse ; but this idea did not occur to Dr. Guest, for, as we find in later chapters, he simply meant that certain modes of old-fashioned verse were possibly suggested by Latin and others by English rhythms. He makes no attempt to show how Anglo-Saxon rhythm and classic metre became blended. He simply abandons all scansion, and in the place of any law of living rhythm which our ear can recognise, he offers meaningless rules, based, so far as they have any base, upon the look of lines as written, not upon their sound as heard.

One object in passing this strong condemnation on the proposed theory is to set the reader free fully to enjoy the charming book in which this fallacy is set forth. If he will pass carelessly over all references to the new dogma, he may wander with untired mind in a delightful maze of history, poetry, and criticism. When he reaches certain translations from the Anglo-Saxon, their rugged sections may not improbably inspire him with awe, and yet in the rifts even of their middle pauses he may find matter for pleasant cogitation. Elsewhere an almost endless series of lines quoted from our best poets will lure him onward by a charm comparable with that which we experience as we dreamily peruse an early edition of Johnson's Dictionary, and that charm is great.

II.

Nowhere are the difficulties of analysing English rhythms more obvious and more perplexing than in the heroic measure. If, indeed, we attempt to scan this form of verse, we shall find that many lines may be divided into feet arranged as in an iambic pentameter. In other cases, in order to scan the line without greatly forcing the pronunciation, we must employ spondees and, perhaps, pyrrhics ; further, we must occasionally allow two unaccented syllables to count as one, and at other times we must let a pause do duty for a syllable. Still further, we must grant the poet leave to lengthen his lines by adding at their close one, or even two, unaccented syllables, when the normal five feet are already there. When all these privileges have been granted, our rules hardly aid us to distinguish blank verse thus licensed from prose. Nor can laws so loose aid us in judging of the excellence of verse. Hot controversy has raged as to whether long and short or accented and unaccented syllables

should be considered as the primary elements of the English
metrical foot. Victory appears to have declared for the school
which puts accent in the place of quantity ; but the controversy
was rather about terms than facts. No sane man ever pronounced English verse by the ancient rules of quantity ; and,
on the other hand, those who classify by accents are fain to call
many syllables accented whose sole claim to that honour is
given by their length. We doubt whether the nominal scanning
of a single line was altered by those who fought most violently
for accent against quantity. The difficulties of scansion are not
to be removed by a mere change of names.

Probably no new statement about verse will be found to be
true ; but some important truths have been imperfectly stated,
and others have met with neglect, so that no one complete
theory is now generally accepted. Instead of wearily picking
out small modicums of truth from this or that half-forgotten
author, let us search for the main laws of rhythm by listening
to the actual sound of prose and verse as spoken nowadays.
Both in prose and verse we habitually run words together, so as
to form a group of sounds as continuous as those in any single
word. The pause dividing these groups sometimes separates
the members of a sentence ; sometimes it is used for purposes
of emphasis ; but in many cases the pause we speak of is involuntary, being made while we rearrange the organs of speech, so
as to allow a fresh word to begin with a clear, well-cut sound.
We make the pause, in fact, to avoid what Mr. Melville Bell
calls a glide. This continuous group we will call a *section*. The
letter 'r' is a great cementer of words, and perhaps the meaning
of this term section may be best explained by examples of faulty
sections, such as ' Mariar Ann,' or ' idear of,' where words which
ought not to form a continuous group are nevertheless welded
by the vulgar into a section with a distinct rhythmical character.

We will first consider the function or properties of these
sections in prose, for prose has its rhythm as well as verse ; and
if we can find the simpler laws of rhythm in prose, we shall
then more easily ascertain the precise difference between prose
and verse. To our ear the following prose passage falls into
eight sections, separated by seven pauses, of which the second and
sixth are very short, and would in rapid speech be omitted :

¹I beseéch you : ²púnish me not . ³with your hárd thóughts :
whereiń : ⁵I conféss me múch gúilty : ⁶to dený . ⁷so fáir and éx-
cellent ládies : ⁸ánything.

The shorter pauses are here and hereafter indicated by a dot,
and the longer pauses by two dots, or a colon. The strong syllables are indicated by an accent. Other readers might divide
the passage otherwise, as by pausing after 'fair,' and not after
'deny;' but all Englishmen would break up the sentence in
some such way as is here indicated. Each group given above
has a distinct rhythmical character, which, avoiding all controversy as to length or accent, we may indicate in time-honoured
terms as follows :

¹Tititumti ²tumtititi ³tititumtum ⁴titum ⁵tititum-titumtumti
⁶tititum ⁷titum-titumtiti-tumti ⁸tumtiti.

The hyphens introduced in the longer sections may, perhaps,
enable the reader more easily to catch the rhythm indicated;
they are not intended to denote any peculiarity in the rhythmical
character of the groups. No one will deny that the short groups
are rhythmical, for they are closely analogous to classical feet.
The longer groups owe their rhythmical character to the regular
beat which falls on each strong syllable—each of these is, within
the section, separated from its neighbour by a constant time-
interval. If we were to employ musical notation to express the
time occupied in delivery, each strong syllable would begin a
fresh bar. The fact that the beats are regular is best observed
in the larger groups; as, 'I con|féss me| múch| gúilty,' or,
'So| fáir and| éxcellent|ládies.' The bar here is placed as a
bar is placed in musical notation. At the end of each section a
pause of uncertain length may be introduced; the time begins,
or may begin, *de novo*. No effort is required to secure rhythmical character for each section as we speak or write. This character belongs essentially to every sentence, depending wholly
on our native mode of delivery. A Frenchman hardly ever
masters the art; indeed, he seems to be without the organ
which recognises a difference between tumtiti and tititum.
His rhythm, in English at least, consists of a series of equal
beats on all syllables, followed by a long uncertain pause,
sometimes on a syllable, sometimes on the stops or chief

pauses. Even the Americans are apt to lose the true English rhythm, and to break our sentences into short sections which strike the English ear as quaint. Thus we may imitate one American mode of speech as follows : 'I confess me : much guilty : to deny : so fair : and excellent ladies : anything.' But, although no art is necessary to insure that each phrase we speak shall be rhythmical, yet the rhythm may be good or bad Our worst or weakest rhythms arise from monotonous reduplication, as titum titum, or tumti tumti. Sections of this character take all force out of any phrase in which they recur frequently. There is in prose no law for the collocation of successive sections; but unless they vary in length and character the sentence will ring poorly, while if they balance and answer one another too obviously, the effect is artificial and pompous. Observe how admirably Orlando's compliment stands all tests— no section poor, no repetition, no antithesis; some sections long enough to let us hear the rhythmical pulse; and these followed by a break in the time, due to a natural pause.

This last condition is essential in strong prose. We like occasionally to hear long sweeping sections; but if one continuous beat is maintained for several successive sections without a check, in place of flowing prose we hear weak verse. In contrast to Shakespeare's prose let us take a sentence such as any of us may write when we are so intent on saying what we mean as to be quite indifferent to rhythmical effect:

[1] In fáct : [2] we múst : [3] as will shórtly appeár : [4] meásure lábour : [5] by the amóunt of páin : [6] which attáches to it.

[1] Titum [2] titum [3] tititum-tititum [4] tumti-tumti [5] tititum-titum [6] tititumtiti.

The first two sections are identical; the third consists of a repetition of two identical parts; the fourth is one of the poorest groups possible; and the sixth is weak, in consequence of the symmetry of its form. Only the fifth can escape censure. Then, as to their collocation; no pause of sufficient importance occurs to warrant our breaking the time from first to last; and if the reader will beat time so as to bring one stroke on each accented syllable, he will speák like a professor painfully explaining his point.

In verse the section or group of undivided sounds is an element of no less importance than in prose. In the smooth line used by Pope, the rhythmical section very rarely extends beyond four syllables. Milton often uses sections of five syllables; but even in his sweeping lines we have met with none longer. Short sections are the first characteristic of verse as compared with prose. In the heroic line we habitually find four pauses, and in the normal arrangement the short and long pauses occur alternately. The two longer pauses are known as the final pause and middle pause or cæsure. By some writers the slighter pauses are called semipauses or demicæsures. Thus we find each line split into four sections, of which the shortest may contain one syllable, while the longest will contain four, or, by exception, five. If we limit ourselves to sections of four syllables, we may designate each variety of section by that letter of the alphabet which is represented by the same rhythmical group in the Morse code. In verse these sections are not joined at random. In beating time from beginning to end of each line one stroke will fall on each accented syllable, and even in passing from line to line the beat is very generally continued unbroken. (By accent we mean the primary or strong accent, not that stress which in a word of many syllables is sometimes called the secondary accent.) Not unfrequently a beat falls on the final pause, and occasionally a beat falls on some pause in the middle of a line, or at its close; but neither in verse nor prose does a beat ever fall on a weak unaccented syllable. In this continuity of pulse or beat we find a second characteristic of verse. In some loose forms the lines have no other characters, but in most verse a third rule prevails. A definite number of sections are so joined as to form lines which can be scanned according to laws borrowed from the rules of classic metre; we then have in each line two co-existing rhythms, one due to the grouping of sections, and one to the grouping of feet; the number of sections never coincides with that of the feet; so that one pause at least must divide a foot, as in the classic cæsure.

An example from Pope will assist to make our meaning clear. An accent is shown where we consider that a beat should fall. A colon marks the cæsure and a dot the semi-cæsure:

Gó · wóndrous créature : móunt · where scíence gúides :
Gó · méasure eárth : wéigh áir · and státe the tídes,
Instrúct · the plánets : in what órbs · to rún :
Corréct · old tíme : and régulate · the sún.

We will now to the best of our ability scan these lines, using
the ordinary classical symbols. In scanning we shall frequently
count syllables as long which receive no primary accent or beat;
but every syllable which does receive a beat will be counted as
long. For instance, we count the first syllable of ' regulate ' as
a long element in virtue of its accent, although it is very quickly
pronounced, and we are ready to give equal rank in scanning
to its last syllable in virtue of the time required for its pronunciation. It will be seen that the lines closely resemble iambics :

$$- \cdot - \mid \cup - \mid \cup : - \cdot \mid \cup - \mid \cup - \mid :$$
$$- \cdot - \mid \cup - \mid : - - \mid \cdot \cup - \mid \cup - \mid :$$
$$\cup - \cdot \mid \cup - \mid \cup : \cup \mid \cup - \mid \cdot \cup - \mid :$$
$$\cup - \cdot \mid \cup - \mid : \cup - \mid \cup - \mid \cdot \cup - \mid :$$

With very little persuasion we might be led to consider even
the spondees and pyrrhic as merely strong and weak varieties
of the iamb.

In the second mode of analysis by sections we will employ
long and short upright marks to denote that relation which in
prose we call tumti. The long marks correspond exactly with
the accented syllables :

$$ ı \cdot \mid ı \mid ı : \mid ı \cdot ı \mid ı \mid $$
$$ ı \cdot \mid ı ı \mid : \mid ı \mid ı \cdot ı \mid ı \mid $$
$$ ı \mid ı \cdot ı \mid ı : ı ı \mid ı \cdot ı \mid $$
$$ ı \mid ı \cdot ı \mid : ı \mid ı ı \cdot ı \mid $$

Those who know the Morse alphabet might readily write down
this scheme as

     tc : tá,   tk : mä,   ar : ua,   aa : la.

No one who merely saw the scheme of feet could form any clear
idea of the character of the lines, whereas a man to whom the
scheme of sections was given could almost recognise the style of
Pope.

In the lines quoted above there is no possible ambiguity in

either mode of analysis; but the reader may doubt whether the grouping by sections is always so well defined. The experiment is simply made. Let two persons select any passage of classical verse, and sitting in separate rooms, so as to avoid discussion, mark three places in each line where pauses, however slight, might conceivably be made. They must bear in mind that very few sections can exceed four syllables, and that no words are to be called one section which will not run fluently together with a continuous sound. We have repeatedly found the agreement between two such versions to be almost perfect.

If our observation of the facts has been accurate, we are now able to see clearly the essential characteristics which distinguish verse from prose. In prose we have long and short sections, grouped according to the taste of the writer; and our sense of prose rhythm is due to the individual sections more than to the groups which these form. In verse we are restricted to the use of comparatively few frequently recurring short sections; but these are grouped according to some law or laws so as to form the rhythmical unit which we call a line. The number of sections in a normal line is constant, and the line is delivered so that the interval between successive accented syllables is habitually constant; in other words, we have one beat per accent throughout the line, the only exception being that occasionally a beat falls on a pause. In certain looser forms of verse no other law holds good; but in the stricter forms the further law is added that the syllables in the section shall scan according to a more or less rigid scheme. When the scanning is strict this leads to a constant number of syllables in each line, and even when the scanning is lax the number of syllables does not greatly vary. In this complete form of verse we have time, number, and rhythm. The beat upon the accents marks the time, the feeling of number is given by the constant number of feet and the constant number of sections. The group of syllables within the section, and the group which these sections form within the line, give the primary sense of rhythm, and underlying this varying rhythm we have the secondary rhythm due to feet which by their approximately uniform arrangement assist in giving that sense of unity by which we recognise a series of lines as belonging to one species.

So long as this sense of unity is preserved, much licence is permitted to the poet and to those who speak his lines. If the verse is otherwise strongly marked, the speaker may break the beat by a random pause. If the scanning is strict, the number of sections may be allowed occasionally to vary. If the length and the number of sections are normal, the scanning may be lax. English scansion is often indefinite and lax, probably because its laws are not of indigenous growth. Moreover, our sense of number is weaker than that of time orr hythm. The schoolboy trying to write verse beats time naturally, and finds some rhythm, good or bad, by instinct, but he counts the syllables on his fingers. Further examples are needed to show how far these theories may assist in the critical examination of style, and what aid they bring to elocution.

III.

The normal English heroic line is an iambic of five feet, broken into four sections by one major and two minor pauses. The time of its delivery is measured by beats, each falling on an accented syllable or on a pause, and the number of these beats may vary from three to seven or even eight, although the number of long elements in the feet is only five.

The beauty of a line cannot be determined by noting how closely it approaches to the central type. We require indeed that no line shall be so abnormal as to disturb our sense of unity ; but there is no formula for a good line, and no criticism is more absurd than that which condemns famous verses because some stress or pause may fall in an unusual place. Nevertheless, if a theory of metre is to be of any value, it must help us to give praise where praise is due, and to point out the cause of failure when a cultivated ear rejects a line as bad. Our object at present is to show how far the thesis already stated will give us aid of this sort.

The character of a line varies much, according to the number of beats required for its delivery, or, in other words, according to the number of syllables which receive a strong or primary accent. Five is the number most frequently employed, but four beats per line are also common. For the purpose of analysis in this and other respects we have taken a hundred lines by Shakespeare, the same number by Milton, and the same number by Pope. On counting the number of beats when these lines were spoken with due emphasis we obtained the following results:

Lines of 5 beats	Pope 54	Milton 53	Shakespeare 45
Lines of 4 beats	Pope 38	Milton 34	Shakespeare 43
Lines of 6 beats	Pope 8	Milton 8	Shakespeare 7
Lines of 3 beats	Pope 0	Milton 3	Shakespeare 4
Lines of 7 beats	Pope 0	Milton 2	Shakespeare 1

There is nothing absolute in these numbers. Other readers would obtain other results. Even the same reader will not on different days obtain identical results; but all readers who make the experiment will obtain a somewhat similar series of figures. The deduction to be drawn is that the number of beats in a line is not employed to enhance the sense of unity, but is freely varied. The number will be good when it suits the intention of the author, contributing to the sense of aptness which, to quote Milton, gives ' musical delight.' Thus the seven beats required for the delivery of the following line add much to its solemnity:

Aš ōne' grēa't fŭr'năce · flā'med : yēt' · frŏm thō'se flā'mes.

How different is the sense of joyous motion produced by a line with three beats instead of seven:

Thĕ mūl'tĭtūde · ŏf Añ'gĕls : wīth ă shōu't.

In these examples, as in all which follow, an accent denotes a beat, a single dot a slight pause, and a colon a longer pause. The scanning is shown in the ordinary way; but many syllables are counted as long in the scansion which are weak in the sectional rhythm—as, for instance, in the case of the word ' with ' above.

If five successive beats come on the five long syllables of five successive iambs, we have a smooth but rather commonplace verse. Thirty-two of the hundred lines from Pope have this peculiarity, as:

Whĕre āll' · mŭst fāll' : ŏr nōt' cŏhē'rĕnt · bē' :
Ặnd āll' · thăt rī'sĕs : rī'se · ĭn dūe' dĕgrēe'.

Milton, in our sample lines, employs this form only five times,
and Shakespeare only thrice. In one of these three lines, the
uniform beat is admirably adapted to indicate the stealthy
creeping pace of Rumour:

Ặnd whō' · bŭt Rŭ'mŏur : whō' · bŭt ōn'lЎ Ī'.

In a second example it is used with a semi-humorous purpose:

Thĕy sēll' · thĕ pās'tŭre : nōw' · tŏ būy' thĕ hŏr'se.

The following line by Ambrose Phillips may show how unsuited
this form is to express strong emotion:

Erĕ yōn' · mĕrīd'iăn sūn' : dĕclī'nes · hĕ dī'es.

Yet this is the form which is sometimes held up as the perfect
type, and a line is praised or blamed as it more or less thoroughly
imitates this model of monotony.

The following from Pope is commonplace enough to prove
that four beats are met with almost as frequently as five:

Ị̆n doū'bt · hĭs mī'nd : ŏr bōd'Ў · to prĕfēr'.

Like Milton, Shakespeare, telling us of speed, uses three beats:

Ị̆n mō'tiŏn : ōf nŏ lēss' · cĕlēr'ĭtӮ :
Thăn thāt' · ŏf thōu'ght.

Monosyllables receive an accent only when they are to be pronounced with emphasis, and to these no rules apply, such as in
classical languages make certain syllables long and others short;
but in polysyllabic words one syllable is in English usually so
prominent as, in verse, to demand an accent or beat. Nevertheless, words of minor importance, such as ' being,' ' under,' or
' above,' although they often receive an accent, are also often
spoken rapidly with none. Thus we read the following line
with five beats:

Fō'ld · ăbōve fō'ld : ă sūr'gĭng mā'ze · hĭs hēad'.

The beat upon a pause may often be used with excellent effect.
Thus in two of the following lines we like to hear the middle
pause prolonged as follows:

Ăssū'me thĕ pō'rt ŏf Mār's —' : ānd ăt hĭs hēe'ls
Leā'shed īn' lĭke hōun'ds : shŏuld fām'inĕ · swō'rd · ănd fī're
Crōu'ch · fŏr ĕmplōy'mĕnt —' : bŭt pār'dŏn · gēn'tlĕs āll'.

These noble lines will serve aptly to introduce the difficulties
met with in English scansion, and to illustrate the mode now
suggested of meeting them. The practice according to which a
stress would be thrown on the conjunction 'and' in the first
line is to our ear singularly disagreeable; but, if a pause
followed by a weak syllable be allowed to count as a long
element (no new idea), this line and many which are similar to
it will be found to scan with perfect regularity. If the reader
will look back and forward he will find in this rule the explanation of the long marks which appear over several insignificant
words, such as 'to' and 'of' in examples already quoted.
Looking upon scansion as derived from a system of longs and
shorts, we shall see nothing forced in this claim, and we shall be
prepared to grant a further demand that when necessary a pause
may be allowed to do duty for a syllable, long or short. Thus,
in the third line above, scansion seems to us impossible unless
the pause be allowed to count as a long element. With this
licence we scan the line as an ordinary alexandrine.

This conception of a pause as sometimes equivalent to a long
syllable helps us to understand why we contentedly accept an
eleventh and even a twelfth short syllable after the long element
of the fifth iamb. These short superabundant syllables do not
break the iambic flow of the verse, for with the final pause they
make one iamb the more, and yet this foot is not sufficiently
prominent to disturb our sense of number requiring the feet to
be grouped in sets of five. To scan an English line we must
further have leave to count any syllable long which receives
a secondary accent, or is in any way slightly more prominent
than its neighbour. We must have leave to count two short
syllables as one—to treat elision as a reality and not a fiction,
which it certainly is; and, finally, to count spondees as equivalent to iambs, possibly on the plea that we may, if we please,
lay rather more stress on the second syllable than the first. It
is probably better frankly to admit the spondee, and scan
Milton's well-known line as follows:

Rŏck's cā'ves · lā'kes fēn's · bōg's dēn's ; ănd shā'des ŏf dēa'th.

If these demands be all granted, very little difficulty will be
found in showing that the mere arrangement of the heroic feet
in verse is extremely regular; the laxity is found in the great
freedom of choice as to what we may call a long and what a
short element, and yet, lax as we are in this respect, there is
seldom much difference as to the scanning of a line by different
grammarians; and good lines will be found to scan well.

A single trochee in place of an iamb is very common. It is
usually placed at the beginning of the line, but is often met
with immediately after the middle pause or cæsure. Thus we
may say in general that a trochee is used to begin a major
section. In the hundred sample lines we find a trochee substituted for the first iamb in 28 lines by Shakespeare, 20 by
Milton, and 22 by Pope. At the beginning of the second
major section we find a trochee 8 times in Shakespeare, 7
times in Milton, and once in Pope. Shakespeare twice has
two trochees in one line; one at the beginning of each major
section. Milton twice uses a weak trochee for the second foot.
The ear will not tolerate a strong trochee between two iambs
in a major section; it breaks the flow and the line must halt, as
in the following strange example from Pope :

Īs′ thĕ grēa′t · chāi′n : thăt drāw′s āll′ tŏ ăgrēe′.

Two successive trochees at the beginning of the second section
make a rough line, as in this example from ' Paradise Lost : '

Aňd dŭs′t · shălt ēa′t : āll′ thĕ dāy′s · ŏf thў lī′fe.

Notwithstanding all the licence which we are forced to claim
before we dare promise to scan the vast majority of heroic lines,
scansion is a reality; but the *beauty* of a line depends less on
scansion than on any other quality. There are certain rules to
be observed; but the simple observance of these rules will no
more result in beauty than the performance of steps in accurate
time will make dancing beautiful. Prosody is a kind of grammar.
We must learn the law and observe the law, but the law will
give no grace of style. At most we may learn by its aid why
some lines displease us. Above all, we must never force the
pronunciation so as to bring the arrangement of the feet into
prominence. The rules of scansion are like the rules of etiquette
—best kept when they are kept well out of sight.

That in English two short syllables often count as one need in no way surprise us, for scansion was originally based on a measurement of time, and in English two, or even three, short syllables are often so swiftly spoken as to occupy no more time than a single unaccented syllable in other parts of the same line. Elision is not wanted to make Milton's lines scan—the two syllables may count as one element in a foot, although both vowel sounds are perfectly heard; they really produce a diphthong. The English letter I, which counts as one syllable, is composed of two successive sounds quite as distinct as the final *e* and the initial A in the following :

Thĕ wāy′ · hĕ wēn′t : ănd ōn′ · thĕ Assȳr′iăn moūn′t.

Thassyrian is not English and is completely hostile to the music of the line. We believe that Milton used this extra syllable systematically to avoid the rhythmical sections tumti-tumti and titum-titum, which his ear loathed. With the elision the rhythmical scheme of the above line becomes very poor. The extra syllables just save it :

ı| ı| ı| ı ı|ı ı|

This may be observed again and again. Thus in the much finer line

Hūr′led hēad′lŏng · flā′mĭng : from thĕ ethē′riăl · skȳ′,

the third section with the elisions would again be tumti-tumti, instead of a fine rhythmical phrase contrasting by its vigorous rapidity with the strong solemn sections preceding it. These long sections in Milton often have a beauty like that of a trumpet call, as :

Thăt′ ĭnvīn′cĭblĕ · Sām′′sŏn : fār′ · rĕnōw′ned.

It is on the character and grouping of the sections that the rhythmical beauty of a line chiefly depends. The most commonplace sections are built of two iambs or two trochees ; the most commonplace lines are those in which the end of each section corresponds with the end of a foot. An example of this bad subdivision is to be found in Ambrose Phillips's verse already quoted.

The four sections are usually grouped in pairs separated by

the chief pause, and usually the pause between lines is longer
than the pause between sections; but these common characteristics are not laws, and any departure from the practice is in
itself neither good nor bad, but is good or bad as it serves to
produce or mar the musical and dramatic effect at which the
poet aimed :

> Dī're wăs · thĕ tōs'sĭng · dēe'p thĕ grōa'ns : Dĕs'pair
> Tēn'dĕd thĕ sĭck' : būs'iĕst · frŏm coū'ch tŏ coū'ch :
> Aṅd · ō'vĕr thĕm · trĭŭm'phănt dēa'th : hĭs dār't ·
> Shōok' : bŭt dĕlāy'ed · tŏ strī'ke · thŏ' ŏf't ĭnvō'ked.

In these lines the scanning is normal. The phrasing (as division
into sections may be named) divides each line into four normal
groups, but the relative length of the pauses is abnormal except
in the second line. The departure from ordinary custom is one
means employed to produce the effect, and we are sorry for the
critic who would condemn the *enjambement* at the end of the
third line.

On examining our sample hundred lines we find that in
Pope all but two have four sections—one of the exceptions has
three and the other five sections—and only three of Pope's sections have five syllables; Milton has four lines of three sections,
but two of these contain a compound word which might very
well be divided in pronunciation, as

> Bĕēl'zĕbūb : tŏ whŏm' · thĕ Aṝch' · ēn'ĕmy.

Milton uses sections of five syllables sixteen times, and the
rhythm of these is characteristic of his style. 'In adamantine.'
'The Omnipotent' (there are people who would say 'thomni-
potent'). Shakespeare allows himself great freedom in phrasing
as in everything else. In the hundred lines we find no fewer
than fifteen with only three sections, and the deviation from the
normal path is not accidental; the change is employed to vary
the character of the verse :

> Wĭthĭn' · thĭs wŏod'ĕn Ō' : thĕ vēr'ў̆ cās'ques :
> Thăt dĭd ăffrī'ght · thĕ āi'r : ăt Āg'ĭncōurt.

We have here a rapid, almost bounding, effect, very different
from the normal stately march resulting from the fourfold
division in such lines as these :

Ĭ' · frŏm thĕ ō'rĭent : tō thĕ drōop'ĭng · Wē'st
Mā'kĭng · thĕ wĭn'd : mȳ pōst'hŏrse · stĭll' ŭnfōl'd.

In seventeen cases Shakespeare used sections of five syllables;
but they seem to be accidentally employed, contributing little to
any special effect, as—

Ŭpōn mȳ tōn'gues : cŏntĭn'uăl · slān'dĕrs · rī'de
Thĕ whĭch' · ĭn ēv'erȳ lān'gŭage : Ī prŏnōun'ce.

The division of lines into sections, although well known long
since (*vide* Sheridan's 'Art of Reading'), has fallen much into
oblivion; so much so, that we may be accused of putting
arbitrary pauses where the ear detects none. We submitted,
however, farther back, an experiment by which the reality of the
sectional pause can be proved; and we will now give another
experiment, showing that it is of the very essence of verse. If
we wish to read blank verse as prose, to take all verse quality
out of it, we have only to cancel the greater number of the
pauses, running several short verse sections into long prose
sections. The iambic grouping of the syllables will not then be
detected by the ear. Thus, let us read the following passage in
a natural straightforward way, with no pauses except those
rigidly required to make the meaning clear : 'Síng Heavenly
Múse—that on the sécret tóp of O'reb—or of Sínai—didst
inspíre that shépherd.' Where is the verse? Or let us try
another plan, keeping time to a regular beat on each long
syllable of the iamb or trochee, but still omitting all unnecessary
pauses : 'Sing Héavenly Múse—that ón the sécret tóp of
Óreb,—ór of Sínai,—didst inspíre that shépherd.' All the
iambs are there much clearer than when the verse is properly
read; the continuous beat is painfully obtrusive, but the sound
is not even English, much less is it that of English verse. Now
let us try the effect of making the prescribed pauses, but with
no conscious attention to beat or scanning : 'Sing · Heavenly
Muse : that · on the secret top : of Oreb · or of Sinai : didst
inspire : that shepherd.' We think that, however badly the words
are read, some trace of blank verse will be discernible. Some
readers neglect the true phrasing, and introduce false pauses at
the end of each foot, as : 'Sing Héav—nly Múse—that ón—the
sécret tóp.' This style is sometimes heard upon the stage, and

is then generally enriched by the addition of a rise in the pitch of the voice on each accented syllable. In the ‘American Review’ for May 1848 this mode of division is frankly advocated. ‘In the line,’ says the reviewer, ‘ “ Full many a tale their music tells,” there are at least four accents or stresses of the voice with faint *pauses* after them just enough to separate the continuous stream of sounds into four parts, to be read thus: “ Fullman—yataleth—eirmus—ictells,” by which new combinations of sound are produced of a singularly musical character.’ We do not agree with this author, but he was a man of sufficient intelligence to observe and record his own practice and that of far less clever men.

There are many beauties in a fine line besides that of rhythm, and even rhythm cannot be accurately recorded by those coarse methods which simply distinguish between long and short, or weak and strong. One iamb may differ from another in character almost as much as one verse differs from its neighbour. Two sections with the same nominal rhythm represented by the same letter in the Morse alphabet may differ in character and beauty with all the difference expressible by the words good and bad. No analysis will enable anyone but a poet to write a single good verse, but a true theory would serve to protect us against certain pestilent diseases which at times afflict actors, readers, and critics. Even an imperfect theory may beget a better.

A FRAGMENT ON GEORGE ELIOT.[1]

IT is difficult to lay down the 'Life' of George Eliot without making some effort to record the impression produced upon the mind.

In one respect the book attains its object. No candid man or woman can lay it down without thinking, 'This was a good woman—a woman with warm feelings and strong just thoughts.' Unfortunately the book is in the truest sense uninteresting. George Eliot in her everyday correspondence shows hardly a trace of the great creator who gave us Adam Bede, Dinah, Romola and Dorothea.

There are wise and good sayings to be found here and there at long intervals, but these are not of such merit as to add materially to the fame of the writer. Of mere gossiping interest there is probably enough to carry many readers of the present day from the first page to the last, but this interest depends almost wholly on the great personality of the writer, not on the intrinsic merit of her doings or sayings. The position of the Queen gives an exceptional interest to the record of her likes or dislikes, and to all honest statement of what she does or thinks ; and so the great position of George Eliot gives her a like claim to the attention of us all : but intrinsically the merit of the great writer's autobiography is hardly greater than that of 'Leaves from our Highland Journal,' to which George Eliot's journals have a strange resemblance.

Well, it may be said, the 'Life' seems to have given this outsider a fairly clear conception of a person whom he admires and respects the more for what he reads there, and what more can a 'Life' do ? A biography may reveal to us a friend or at least a companion—someone whom we know as we know Dorothea

---

[1] MS.—The beginning of a review which was never completed of the *Life* of George Eliot.

or Romola. The present book falls wholly short of any such revelation. Possibly George Eliot never revealed herself except by her books, yet how we long to know what that narrow Low Church creed meant to the girl; how we long to understand how she felt when the faith to which she clung left her wholly within a few days. No picture of this mental and moral development could be presented without the highest artistic skill; but this woman had the skill—she experienced the change—she was one of thé great ones of womankind, and we are left without a vestige of a record of the stupendous change.

Passing on, we know that this woman of pure heart and right moral feeling lived as wife to a man who had no right to marry her. How did she feel? How did passion speak to this great soul? We see the possibility of a daily record of thoughts, hopes, fears, doubts, faith, love in fact, beside which the ' Confessions of Rousseau ' or the ' Diary of Pepys ' would seem very poor reading ; but, as it is, we know no more of all this than if taking Mr. Lewes for life had been no more than taking a house on lease. Conse- quently Pepys and Rousseau, though one was no very great man and the other was a mean fellow, will be of interest to men and women when George Eliot's ' Life ' interests historians only. No doubt what we have got is much more decorous, respectable, nice, in fact, than what we should have got if we had a true record of what this woman felt; but, somehow, one does now and then hanker for something else than niceness, and the appetite is not unwholesome ; it is merely the appetite for intimate communion with a great person.

We can go further. The ' Life ' shows a moral growth throughout her life which inspires admiration. The narrow- minded, somewhat pert methodistic school-girl becomes a clear- headed, businesslike freethinker, doing work diligently with dis- content. This rather shallow literary hack fell in love with George Lewes, and thereupon the possibility of true sacred passion became manifest to her, and remained an unshaken article of faith. But for this she would probably never have become an artist. Her books reveal not so much the hard-headed thinker of whom women are proud because she could think dispassionately like a man, but a warm, large-hearted woman, seeking for a higher life than any intellectual exercise can occupy. In reading her artistic work we feel the broad sympathy with all true human nature which she proclaims in her letters, but which those letters hardly show. Her artistic birth seems to have been a regeneration. The more she achieved the more humble she became. There is indeed no trace of self-abasement in the 'Life '; she knew that she meant to do good work and had great powers, but her mind was clearer, her sympathies larger, than most even of the ablest men she met. But yet, in relation to the possibilities of art and of life, the journal shows a fine simple humility. The reluctance to read criticism was not a fear of wounds to her self-love, but a dread lest to see her efforts misunderstood might render her incapable of further effort. She cherished a hearty faith in the great public; contact with petty criticism tended to destroy that faith. She did well to pass it by..

The necessity for religion as an artistic element impressed her strongly ; what she most cared to teach and to show was best set forth in the words of men and women with deep religious feelings. This seems to have shocked the little sect which had the honour of destroying her dogmatic faith, and the good hard raps she administers to those who would have kept her to the narrow way of orthodox atheism are among the few amusing things in the book. She endeavoured with much success to present a profoundly religious aspect of things and men while rejecting all dogmatic faith. In this she represents the generation, and represents it nobly. But while the 'Life ' confirms us in this view of her character it would do little to exhibit the nobility of her conceptions unless read by the light of those works—and the works acquire no new or deeper significance by the light shown in the 'Life.' On the contrary, a feeling is produced that she was hampered in her life by the set among whom she lived—men and women of narrow minds setting up science and logic as better than religion and fatal to it. There is nothing in the 'Life' to show that she had any real sympathy with scientific research, or even knew what it meant. She loved Lewes faithfully, and physiology for his sake, as she might have loved his dog because it was his. There is not a trace of living interest in the pursuit, either for the pleasure of the chase or for the hope of booty in the form of scientific spoils won from the dark country of the unknown.

The lady herself, to one who did not know her books, would be as uninteresting, when seen by the light of these writings only, as any famous author must be when met in the beaten way of social converse.

If we were not forewarned by the example of two or three wonderful written lives, we should be in much danger of believing that biography was worse than useless except in the case of those whose actions, rather than their lives, were interesting. But in truth we get no record of the lives of our thinkers or artists; we learn their actions and we read their opinions well or ill expressed, but the record of actions is no record of the true life of a man; and opinions of novelist or philosopher are of small importance except within that limited range where they have probably expressed their views far more perfectly in their well-considered writings. Even if we had a novelist's carefully considered verdict well expressed on architecture, law, medicine, oil and water-colour painting, the decalogue, spirit-rapping, and the home policy of statesmen, this would be no record of the life of that man or woman; it would possess, except as to style, no other interest than that felt for the record of the interviews published by clever journals——.

SPECULATIVE SCIENCE.

LUCRETIUS AND THE ATOMIC THEORY.[1]

I know not whether this inquiry I speak of concerning the first condition of seeds or atoms be not the most useful of all.—BACON.

THE popular conception of any philosophical doctrine is necessarily imperfect, and very generally unjust. Lucretius is often alluded to as an atheistical writer, who held the silly opinion that the universe was the result of a fortuitous concourse of atoms; readers are asked to consider how long letters must be shaken in a bag before a complete annotated edition of Shakespeare could result from the process; and after being reminded how much more complex the universe is than the works of Shakespeare, they are expected to hold Lucretius, with his teachers and his followers, in derision. A nickname which sticks has generally some truth in it, and so has the above view, but it would be unjust to form our judgment of a man from his nickname alone, and we may profitably consider what the real tenets of Lucretius were, especially now that men of science are beginning, after a long pause in the inquiry, once more eagerly to attempt some explanation of the ultimate constitution of matter.

This problem, a favourite one with many great men, has come to be looked upon by most persons as insoluble; nay, the attempt to solve it is sometimes treated as impious; but knowing that all the phenomena of light are explained by particular motions of a medium constituted according to simple laws, and so perfectly explained that the exact motions corresponding to all the colours of the spectrum, with their modifications due to reflection, refraction, and polarisation, can be defined in form,

[1] Review of Munro's *Lucretius*, second edition. From the *North British Review*, 1868.

speed, and magnitude—knowing this, we may reasonably expect that the other complex attributes of inorganic matter may be deduced from some simple theory, involving only as an assumption the existence of some original material possessing properties far less complex than those of the gross matter apparent to our senses. It is only in this sense that we can hope ever to understand the ultimate constitution of matter; but as the undulatory theory of light has both suggested the discovery of new facts, and has connected all known facts concerning light into one intelligible series of logical deductions, so any true theory of the constitution of matter would suggest new inquiries, and would group the apparently disjointed fragments of knowledge, now called the various branches of science, into one intelligible whole. To frame some such theory as this was the first aim of Greek philosophers, and to establish the true theory will be the greatest triumph of modern science. Of all the subtle guesses made by the Greeks at this enigma, one only, we think, has been fruitful, and that the one expounded by Lucretius, but learnt by him from Epicurus, who in his turn seems to have derived his most valuable conceptions from Democritus and Leucippus. As, however, we possess fragments only of these earlier writers, it is convenient to speak of the theory as that of Lucretius, though he seems to have been simply the eloquent and clear expounder of a doctrine wholly invented by others.

Before explaining how far the views of Lucretius are still held by naturalists, and how far they contain the germs of many modern theories, we must endeavour to give a clear account of what his views really were, in which attempt we shall be much aided by the admirable edition and translation of his works by Mr. Munro.

The principles of the atomic theory are all contained in the first two books; attention being generally called in the original to each new proposition by a '*nunc age*,' or some such expression. Lucretius begins by stating that 'nothing is ever begotten of nothing.' To this principle, which is assumed as true in all physical treatises of the present day, he unnecessarily adds, that this is not done even by divine power, about which he could know nothing. Lucretius felt little reverence for the pagan divinities, and states this principle so roundly as at first to shock our feelings; but if we limit the application of the principle to matter once created, and such as we can observe, his principle is true, and invariably acted upon. Not even by divine power is matter now created out of nothing—nor does any effect happen without what we call a natural cause. Lucretius. seizes the opportunity of stating that men think things are done by divine power because they do not understand how they happen, whereas he will show how all things are done without the hand of the gods—a bold proposition truly, but one which, translated into modern language, means simply that natural phenomena are subject to definite laws, and are not unintelligible miracles. Lucretius fails to perceive that definite physical laws are consistent with the work of God; and the difficulty of reconciling the two ideas, unreal as it seems to us, has been felt by able men even nowadays, when the conception of divine power is very different from any present to the mind of Lucretius. To most of us the very conception of a law suggests a lawgiver, while he, to prove the existence of laws, thought it necessary to deny the action of beings who could set those laws at nought. The demonstration which he gives of his first principle is loose, and goes rather to establish the fact that natural phenomena occur according to definite rules than to prove that no matter is created out of nothing, except in so far as this creation would, he thinks, disturb the order of nature. This first principle, as to the creation of matter, cannot indeed be otherwise than loosely stated by Lucretius, for no definition is given of what should measure the quantity of matter,[1] and until we have defined how this quantity is to be measured, we cannot experimentally determine whether matter is being created or not. But Lucretius meant his proposition to include the statement that nothing happens without a cause, and without a material cause, and his proof of this is precisely that which we should still adduce, being the perfect regularity with which in nature similar effects follow similar causes.

The next proposition is that ' nothing is ever annihilated, but simply dissolved into its first bodies,' or, as we should say,

---

[1] Afterwards, i. 360, the quantity of body is assumed as proportional to weight.

components. This statement is complementary to the first. Together the two propositions affirm that constancy in the total
quantity of matter which is a commonplace truth now, but
which to Lucretius must have been unsupported by any rigorous
proof. His own arguments in support of the law go no farther
than to show that we have no proof of the destruction of any
portion of matter. He shows that rain when it falls is not lost,
but produces leaves and trees, that ' by them in turn our race
and the race of wild beasts is fed; ' but he makes no effort to
measure accurately the quantity of matter apparently disappear-
ing, but reappearing in the new form, and without that measurement his proposition could not be rigorously proved ; moreover,
in the mind of Lucretius, the indestructibility referred to all
kinds of causes ; so that, to make our proposition co-extensive
with his, we must interpret it to mean that matter is indestructible, and that no cause fails to produce an equivalent effect,
though Lucretius probably did not conceive these two parts of
his proposition separate one from the other.

Occasion is taken at this point to state that the components
into which bodies are resolved, or out of which they are built,
may be invisible. The third distinct proposition states that ' all
things are not on all sides jammed together and kept in by
body : there is also void in things.' Lucretius thought that, in
order to explain the properties of matter, it was absolutely
necessary to admit the existence of vacuum, or empty space,
containing nothing whatever. If there were not void, he says,
things could not move at all ! And it does seem, at first sight,
that in a universe absolutely full, like a barrel full of herrings,
so shaped as to leave not a cranny between them, no motion
whatever would be possible ; but reflection shows us that what
is called re-entering motion is possible, even under those circumstances, provided we do not suppose our fish to stick to one
another; there may be an eddy in which the fish swim round
and round one after the other, without leaving any vacant space
between them or on either side, and yet without enlarging,
diminishing, or disturbing the barrel as they move.[1] Lucretius
either failed to perceive this, or declined to admit the possibility that all the movements of gross matter could be of this

[1] A homogeneous plenum may also be conceived as compressible.

class ; but he has another argument in favour of a vacuum : ' Why do we see one thing surpass another in weight, though not larger in size ? ' How can things be of various densities unless we admit empty pores in bodies ? His proof is insufficient ; but here again modern research has confirmed his conclusion, so far as it affects gross matter only, and Lucretius conceived no other. His explanation of varying density is that which is universally received and taught, and even the modern disbelievers in a vacuum do not deny that some space may be unoccupied by gross matter, but simply affirm, on grounds to be hereafter stated, that all space is full of something, though not of ponderable matter. In support of his proposition, Lucretius points to the pores found in all bodies, and uses the following ingenious though fallacious argument to prove a vacuum : ' If two broad bodies after contact quickly spring asunder, the air must surely fill all the void which is formed between the bodies. Well, however rapidly it stream together with swift circling currents, yet the whole space will not be able to be filled up in one moment ; for it must occupy first one spot, and then another, until the whole is taken up ; ' therefore, in the middle a void must have existed for a sensible time.

We are next informed by our author that matter exists, or, in the language of Lucretius, ' all nature, then, as it exists by itself, has been founded on two things : there are bodies, and there is void in which these bodies are placed, and through which they move about.' In his first and second propositions, Lucretius uses the word thing, *res*, which, as we have already explained, comprehended all kinds of things, such as matter, force, motion, thought, life, &c. He now states the existence of matter, and few will be disposed to contradict him ; indeed, he appeals to the general feeling of mankind in proof of his assumption. Unless you grant this, he says, ' there will be nothing to which we can appeal to prove anything by reasoning.'

Lucretius now affirms that nothing exists but matter and void, or, as put in Mr. Munro's translation, ' there is nothing which you can affirm to be at once separate from all body and quite distinct from void, which would, so to speak, count as the discovery of a third nature.' Here at last we reach debatable ground. Lucretius hardly adduces a single argument in support of this proposition, contenting himself with showing, first, that no tangible thing but matter exists—a mere begging the question ; and, secondly, that properties and accidents are not entities distinct from matter—which is true, but little to the point. As examples of properties he gives weight, heat, fluidity; as examples of accidents, poverty, riches, liberty, &c. Time, he says, exists not by itself, but simply from the things which happen ; actions do not exist by themselves, but may be fairly called accidents of matter, and of the space in which they severally go on. Even if all this be granted, we shall not necessarily concede that matter and void have alone a separate existence ; but we must not complain that Lucretius does not support his proposition more strongly at this point, for indeed his six books form one long argument in support of his proposition. Lucretius undertakes to show that every fact in the world can be explained by the properties of matter, and that matter itself may be conceived as possessed of but a very few simple properties, from the construction of which the complex facts we see may follow. Of course, he fails to do this ; but if the proposition be restricted to what are called physical phenomena, it becomes, if not certainly true, nevertheless an hypothesis well worthy of consideration, and not yet proved false. Lucretius admits no subtle ethers, no variety of elements with fiery, watery, light, heavy principles ; he does not suppose light to be one thing, fire another—electricity a fluid, magnetism a vital principle—but treats all phenomena as mere properties or accidents of simple matter, and produced in simple ways; but to understand what he meant by matter, or ' bodies,' we must pass on.

The next proposition of Lucretius describes the composition of matter as we perceive it. Bodies are either atoms, or compounded of atoms and void, or, more at length, they ' are partly first beginnings of things, partly those which are formed of a union of first beginnings.' The words which Mr. Munro here translates as ' first beginnings of things' describe the Lucretian atoms ; Lucretius does not use the word *atoms*, but calls these ' primordia,' or ' semina rerum.' These atoms are necessarily solid, or they could not mark off void space from full. They cannot be broken, because they have no void within them to admit a cutting body, or wet, or cold, or fire, therefore they must be everlasting and indestructible. Lucretius, too, is so persuaded of the great wear and tear that is going on, that he remarks, if atoms had not been indestructible, everything would have been destroyed by this time. The constancy of all phenomena is a very good argument in favour of the indivisible atom, for unless the component parts of a machine are unchanged, how can the results produced be constant? unless there be really something indestructible and indivisible in sodium, how can it happen that every little fragment shall retain every physical property of sodium, so that, for instance, when glowing with heat, it shall continually, as it were, ring out the same notes of light, imparting such vibrations to our eye as paint the well-known double yellow line? If we could divide the little bodies which, vibrating at those special speeds, prove sodium to be glowing in the flame, they would no more vibrate at those speeds than a cut violin-string would give out the true note to which it had been tuned. By such division sodium would be destroyed; whatever might be the result, the body named sodium would exist no longer ; but as yet no man has been able thus to divide the sodium atom, and no one expects that bodies will ever be decomposed into elements simpler than such as would ring out a single note, a single line in the spectrum. In other words, all men of science believe—consciously or not—in atoms indivisible and imperishable. Lucretius certainly knew nothing of spectrum analysis, nor of the law owing to which chemical compounds have forced an atomic theory into daily language ; but the arguments drawn from these sources are simply special applications of his general theorem ; if matter really obeys definite unchangeable laws, the ultimate materials employed to make matter must themselves be definite and unchangeable. Newton's exposition of this argument, quoted by Mr. Munro to illustrate our author, is admirably clear :

While the particles continue entire they may compose bodies of one and the same nature and texture in all ages ; but should they wear away or break in pieces, the nature of things depending on them would be changed. Water and earth composed of old worn-out particles would not be of the same nature and texture now with water and earth composed of entire particles in the beginning. And, therefore, that nature may be lasting, the changes of corporeal things are to be placed only in various separations and new associations and motions of these permanent particles, compound bodies being apt to break, not in the midst of solid particles, but where those particles are laid together and only touch in a few points.

We confess that these arguments seem to us unanswerable, as proving the existence of some unalterable basis of matter. Lucretius described his atoms as small, but not infinitely small, nay, having parts, yet 'strong in everlasting singleness,' impenetrably hard, indivisible, unalterable, eternal.

Having reached his atom, before proceeding with the consequences of his assumption, Lucretius pauses to demolish rival theorists ; but though he does this very well, we prefer to follow out his own propositions in their natural order, remarking, however, that the next proposition occurs incidentally, as it were, while refuting his antagonists, and is to the effect that the differences between all bodies ·may be accounted for by the different arrangement of the atoms, and the different way in which they move ; or, more literally, 'the motions which they mutually impart and receive.' Lucretius conceived matter as formed by atoms in continual motion, rebounding, as it were, from one another. His conception is most remarkable, as being very far removed from the impression produced by inert matter on our own senses, and yet almost indisputably true. Arguments drawn from the laws of the elasticity of gases and from the diffusion of fluids go far to prove the proposition. The former laws may be deduced from the assumption of atoms rebounding in a void : and it is hard to conceive why different fluids or liquids should mix with extraordinary rapidity whenever placed in contact one with its neighbour, unless molecules were continually fluttering as it were at the limits of each fluid, restrained only from continuing their course by the opposition of other atoms. If these arguments seem insufficient, we may refer to the conception of heat as a mode of motion. If heat be a mode of motion of gross matter, then, as all bodies are more or less hot, the molecules of all bodies will be moving with more or less speed—precisely what Lucretius taught. Lucretius was led to his conception by considerations very analogous to those which lead us to consider heat and other forms of energy as modes of motion. Probably the reason why he does not state this seventh proposition as a dogma by itself is, that the proof could not as yet be given; but in discussing rival doctrines he is led to anticipate his own views.

He proceeds to assert that there is no limit to space, nor yet to the total quantity of matter; but these are rather metaphysical than physical questions, although he seems to think that, unless infinite space were full of matter, the universe could not hold together, for he will not hear of gravitation, by which ' all things press to the centre of the sum.'   He is almost comically unfortunate in denouncing the idea, that heavy bodies which are beneath the earth shall press upwards, or that living things walk head downwards, and that when these see the sun we behold the stars of night; but although it is very interesting to observe that these doctrines were then held, we will examine only the propositions strictly necessary for his theory of matter, passing over also his assertion that atoms were not arranged by design, until we examine how he himself conceived that they were arranged.   This explanation is given in the Second Book, containing what we should term the Kinetic branch of his theory, or, to use his own language, he next explains ' by what *motion* the begetting bodies of matter do beget different things, and, after they are begotten, again break them up, and by what they are compelled so to do.'   The book opens with the proposition that matter does not ' cohere inseparably massed together;' it does not stick together as a mere inert mass. Lucretius infers this from the continual change which we perceive, and by which all things wax and wane, although the sum remains constant.

A modern physical treatise would attribute these changes to chemical affinity, heat, gravitation, &c., or possibly, in more general terms, to the various forms of what we term Energy. Lucretius can only suppose this energy to be represented by atoms in motion; and if this be not universally true, it is probably true for many cases.   This perpetual motion of the atoms is next reasserted as a distinct proposition.   ' No atom,' he says, ' can ever stop, giving up its motion to its neighbour.' At first sight, nothing can be more contrary to our ideas of the laws of motion.   We repeatedly see a ball strike another, and set it in motion, remaining itself apparently quiescent after the blow; but nevertheless it is quite impossible that the relative motion of two perfectly hard elastic bodies, such as Lucretius imagined, can ever be altered by knocking one against the other. Motion is essentially relative; we only know that a body moves by observing that it changes its position relatively to another. When, therefore, treating of two isolated bodies only, we need only speak of their relative velocity. The motion of the centre of gravity of any system of bodies remains quite unaffected by their collision one with another, and, in considering our two isolated atoms, we may as well, for simplicity's sake, assume the motion of their joint centre of gravity to be *nil*, though this is not necessary to our argument. Moreover, it is found that a certain quantity, sometimes called *vis viva*, sometimes the kinetic energy of the system, is also constant after and before any collision. This quantity is proportional for each body to the mass of the body, and to the square of its velocity. It must be remembered that we are now speaking of two simple bodies which have only the properties of hardness and elasticity, not being compressible, hot, or susceptible of vibration, so that the transformation of energy due to motion into other forms of energy such as heat is excluded by hypothesis.

Now, in the case of two such bodies striking one another, since their mass will not change, it is impossible that this quantity should remain constant unless each body kept its own velocity. The one cannot hand over a part of its velocity to the other, for in that case the centre of gravity of the system would acquire motion. The velocity of the two cannot increase or decrease simultaneously, or the *vis viva* of the system would alter, so the bodies have no choice but to bound back or to glance aside with their original velocity. In the latter case a spinning motion might represent the *vis viva*, but this would not be rest. If it be asked how it is that we do see the relative motion of bodies alter after striking one another, we answer that heat and other forms of energy have been found equivalent to *vis viva*, which may therefore pass into these forms, and so allow a change in the relative velocities of bodies. Had Lucretius known this he would have answered, that heat can only be equivalent to *vis viva* inasmuch as it substitutes the motion of small parts for the motion of the whole;—this being.

the very answer given by Leibnitz to the above objection, urged as fatal to the doctrine of *vis viva* which he had enounced.

It may be seen that our two bodies need not continue to move in straight lines after striking; they may glance off, so as to spin round. The *vis viva*, or energy, will be perfectly represented by the velocity of the rotating masses, and the centre of gravity may remain undisturbed. When two actual bodies strike and come to rest, it is probable that their atoms do acquire some periodic motion, such as spinning, which motion produces the appearance of heat, but is on so small a scale as to be otherwise invisible to our senses. When we consider the collision of a multitude of bodies, innumerable changes may take place in their relative velocities without violating the two principles, that the motion of the centre of gravity and the energy of the system shall both remain unchanged. Among these combinations some will admit of one or more parts of the system coming wholly to rest, contrary to Lucretius's views, but the following consideration shows that it is difficult to see how this would be brought about if we adhered strictly to his assumption, that the motion of a hard mass is the sole form of energy. He almost unconsciously, and certainly without any express statement, assumes elasticity as a property of his atoms, which he describes as rebounding one from another; but, reverting to our two hard bodies, if they do strike and rebound they must gradually slacken speed, stop for an inconceivably short time, and then gradually resume their pace in an opposite direction, so that, if they rebound, they must stop and pass through all speeds intermediate between zero and their original velocity; so that if we admit no form of energy but a hard mass in motion, we must conclude that no two bodies ever could strike one another, and yet, as neither we nor Lucretius have assumed anything to keep them apart, we find ourselves in a droll dilemma, which seems to prove the impossibility of the existence of a universe containing simple hard atoms in motion. We moderns jump out of the difficulty at once by saying that the hard bodies are elastic, and elasticity is a form of energy, so that the energy or *vis viva* which at one time was represented by the body in motion, is at another time represented by the potential energy of elasticity. Lucretius would have shaken

his head at this explanation, and would have much preferred
the theory just started by Sir William Thomson, and long since
vaguely suggested by Hobbes, that the elasticity of atoms may
be due to the motion of their parts—a proposition exemplified
by one smoke-ring bounding away from another in virtue of the
relative motions of their parts, these not being necessarily themselves elastic. The energy of the molecule at that point where
it strikes its neighbour and changes velocity is on this theory
transferred to another part of the molecule which moves faster
as the first part moves more slowly. If the molecules of gross
matter are made up of atoms in rapid motion, as Lucretius
believed, or of a portion of whirling fluid, as Sir William Thomson
suggests, and if elasticity itself be only a secondary property,
not possessed by the *primordia rerum* at all, then the proposition
that a molecule never can come to rest is undoubtedly true;—
such rest would be equivalent to the destruction of matter.
Lucretius could not have proved this, nor even have understood
the proof. He did not know the laws of motion even of two
elastic bodies, but it is singular to find modern science returning to the never-ending motion of the old Greek atom.

The next proposition of our author explains the varying density of bodies. He says that the greater or less density of bodies
depends on the smaller or greater distance to which the atoms
in each continue to rebound after striking one another. They
never stop striking and rebounding; they are in perpetual
motion, tossed about by blows. Mr. Munro's translation fails,
it seems to us, to convey this view, reading as though the atoms
struck, rebounded and remained quiet afterwards, hooked as it
were together; but Lucretius in many passages describes the
never-ending restlessness of his atoms, tossed like motes in a
sunbeam, which he describes to illustrate the motion of the
atoms in void. This explanation of the varying density of
matter is still commonly received, and will be found in all popular text-books; the density of the ultimate particles of gravi-
tating matter is very generally assumed to be the same, the
greater or less density of gross matter being supposed due to
empty pores, of greater or smaller magnitude, separating the
molecules. At first sight it is very difficult to see how any
other explanation of varying density can be given, since we

find that by compression we actually can increase the density of bodies without altering their weight or mass in any way. Now, unless there were a void space separating the molecules, where can these go to when squeezed ? Most men[1] will find a difficulty in conceiving that space absolutely full of matter, soft or hard, can be made to hold more; but the same space does hold sometimes more and sometimes less gross matter, so that in the latter case it cannot be quite full, or, in other words, the body it contains is composed in part of empty pores. The proof is incomplete, and, if molecules be formed by the motion of a fluid, greater density may possibly be due to a modification in the motion of molecules, and not only to the greater frequency of the eddying molecules in a given space.

Lucretius next points out that his atoms must move very rapidly. In vacuum atoms travel faster than light. His proof of this is extremely vague. He says the light and the heat of the sun (which he calls ' vapours ') are forced to travel slowly, cleaving the waves of air, and several minute bodies of the heat (vapour) are entangled together and impede one another, but atoms of solid singleness can go ahead wholly unimpeded in a vacuum—not a very satisfactory proof. The idea running in the mind of the writer seems to have been that any matter moving in a medium would be impeded by friction, and therefore necessarily move more slowly than a free atom moving in a void; he may also have felt that, if all the power of the universe depended on the motion of exceedingly small particles, it was necessary to suppose them endowed with great velocity; but we do not find this argument used, although it has led the modern believers in atoms to the conviction that if their motion does represent energy, their velocity must be enormous. Lucretius would be glad to know that Herapath, Joule, Krönig, Clausius, and Clerk Maxwell have been able to calculate it; $\frac{1}{400000}$ inch is the distance named by Maxwell.

The nature of the original motion of atoms is next defined. Atoms which have not struck one another move in straight parallel lines, sheer downwards; gravitation is the evidence of this. An infinite number of atoms eternally pour from infinite

---

[1] Sir William Thomson and Professor Tait find no difficulty in this conception.

space above to infinite space below with enormous velocity. This velocity is conceived as the explanation of the power or energy of the universe. Gravitation thus understood was a property of all matter. The apparent exceptions are correctly explained by Lucretius. The idea of his eternal infinite rain of atoms is enough to turn one giddy; it can be best discussed after we have stated the next most singular proposition. The atoms, at quite uncertain times and uncertain places, swerve a very little from the straight line, then they strike, and from their clashing, matter and all natural phenomena are produced. As Mr. Munro translates it, ' When bodies are borne downwards sheer through void at quite uncertain times and uncertain points of space, they swerve a little from their equal poise, you just and only just can call it a change of inclination. If they were not used to swerve, they would all fall down like drops of rain through the deep void, and no clashing would have been begotten nor blow produced among the first beginnings; thus nature never would have produced aught.'

Most people will think Nature would not have produced much had she started in this way, and they are probably right; this is the head and front of our philosopher's offending, and, indeed, there is not much to be said in his defence. Let us, nevertheless, in spite of the ridicule which from Cicero's time downwards has been heaped on this unhappy doctrine of the ' Declination of Atoms,' try to enter into the mind of Lucretius, and to understand what he sought for and thought he had found. As already said, he sought for power in the velocity of the atoms—power which, deflected hither and thither by obstacles of all kinds, should be the origin of every motion, every force observed on earth. Gravitation in its apparent action seemed to show a universal tendency in one direction; this, then, he claimed as an inherent property of his atoms—a claim no broader than the claim made by Newton, that every atom of matter should attract all other atoms at whatever distance they might be—and at first sight much more conceivable; at first sight only, for, indeed, atoms pouring onward, as imagined by our author, could be no source of power. Motion in mechanics has no meaning except as denoting a change of relative position; all atoms moving, as Lucretius fancied, at one speed, and in parallel lines, would relatively to one another have been in perfect rest. A bag of marbles in a railway train could not be employed as a source of energy in the train; they lie at rest; and it is only when brought into collision with something moving at a different pace from the train that they can develop any power, which may then be considerable. But more than this: How are we to conceive direction in space except relatively to something?—what is up and what is down in space? If it be answered, The place atoms come from is up above us, we answer, How, when all atoms are all one relatively to one another in a perfectly similar position, are we poor atoms to know that they are coming from anywhere? So far as we can see, an absolute motion in space is devoid of all meaning. We must conceive a shape or position for space before we can conceive of motion relatively to space, and as we are at perfect liberty to conceive any shape or position, or none at all, it follows that absolute motion in space is anything you please, that is to say, a mere fancy. Lucretius unconsciously assumed the world as his basis by which to measure direction and velocity. The direction in which things fall on the earth was sheer down in void; but really his assumption was meaningless, or, at least, explained in no way the power or force which he wished to explain. Not so, by the way, the older conception of Democritus, who thought atoms moved in all directions freely and indifferently;—a universe so constituted originally might at least contain all the energy we require. One atom would then exert its force on another, but the Lucretian atoms would have remained in profound stillness, except for that occasional swerve at quite uncertain times and places, the cause of which he leaves wholly unaccounted for. This swerving seems but a silly fancy, and yet consider this: It is a principle of mechanics that a force acting at right angles to the direction in which a body is moving does no work, although it may continually and continuously alter the direction in which the body moves. No power, no energy, is required to deflect a bullet from its path, provided the deflecting force acts always at right angles to that path—an apparent paradox, which is, nevertheless, quite true and familiar to the engineer. It is clear to us that Epicurus, when he devised his doctrine of a little swerving from the straight path of an atom, had an imperfect perception of this mechanical doctrine; a little swerving would bring his atoms into contact, and a modern mechanician would tell him you require no power to make them swerve. With what triumph Epicurus, and Lucretius his scholar, would have hailed the demonstration; but, alas! their triumph would have been short-lived; they would soon have perceived that their atoms were described as in deadly stillness—a death from which no life could spring, a rest from which they could never swerve until inspired with power from a source of life. Still we can see that their conception was not stupid, it was simply false, as all physical explanations of the origin of energy and matter must be. There is little to be said for the further conception that matter with its present properties would result from the mere accidental clashing of atoms; this one doctrine of Lucretius is so well known and so little valued, that we will waste no further time on it, merely pointing out that the worthlessness of these ideas as an explanation of the origin of things does not impair the value of the conception of moving atoms as the constituent parts of gross matter as it exists.

The motive for devising the curious doctrine that atoms might swerve now and then from the straight path without being acted upon by other atoms, was, as Mr. Munro observes, undoubtedly the desire to devise an explanation of Free-will. Lucretius believed in free-will. If you believe in free-will and in atoms, you have two courses open to you. The first alternative may be put as follows: Something which is not atoms must be allowed an existence, and must be supposed capable of acting on the atoms. The atoms may, as Democritus believed, build up a huge mechanical structure, each wheel of which drives its neighbour in one long inevitable sequence of causation; but you may assume that beyond this ever-grinding wheelwork there exists a power not subject to but partly master of the machine; you may believe that man possesses such a power, and if so, no better conception of the manner of its action could be devised than the idea of its deflecting the atoms in their onward path to the right or left of that line in which they would naturally move. The will, if it so acted, would add nothing sensible to nor take anything sensible from the energy of the universe. The modern believer in free-will will probably adopt this view, which is certainly consistent with observation, although not proved by it. Such a power of moulding circumstances, of turning the torrent to the right, where it shall fertilise, or to the left, where it shall overwhelm, but in nowise of arresting the torrent, adding nothing to it, taking nothing from it,—such is precisely the apparent action of man's will; and though we must allow that possibly the deflecting action does but result from some smaller subtler stream of circumstance, yet if we may trust to our direct perception of free-will, the above theory, involving a power in man beyond that of atoms, would probably be our choice. Lucretius chose the second alternative as an exit from the difficulty: Atoms with strict causation did exist, and free-will too. We will then grant free-will to atoms, one and all, not in perpetual exercise, but at quite uncertain times. The idea is startling, but not illogical, and the form in which atoms are supposed to exercise their free-will is quite unexceptionable. We cannot but admire the audacity of the man who, called upon to grant free-will as a *tertium quid*, either to man or to atoms, chooses the atoms without a qualm. We do not agree with him, because observation has detected no such action on the part of atoms, or the constituents of matter.

We cannot hope that natural science will ever lend the least assistance towards answering the Free-will and Necessity question. The doctrines of the indestructibility of matter and of the conservation of energy seem at first sight to help the Necessitarians, for they might argue that if free-will acts it must add something to or take something from the physical universe, and if experiment shows that nothing of the kind occurs, away goes free-will; but this argument is worthless, for if mind or will simply deflects matter as it moves, it may produce all the consequences claimed by the Wilful school, and yet it will neither add energy nor matter to the universe. Lucretius thought atoms acted thus; we do not, because we observe no action of the kind in matter, but, on the contrary, strict causation or sequence of phenomena. Whether what we call mind act so or not must also be a matter of observation, but as people have not been able to agree as to the results of observation about

free-will made during a great many centuries, we fear the path
of observation will lead us no further than we have already come.

We beg pardon for this little digression, which was really
necessary to the understanding of our author's physical theory.
Lucretius proceeds to state that atoms have always moved and
always will move with the same velocity, or, as translated by
Mr. Munro, 'The bodies of the first beginnings in time gone by
moved in the same way in which they now move, and will ever
hereafter be borne along in like manner, and the things which
have been wont to be begotten, will be begotten after the same
law,' for there is nothing 'extra,' nothing outside and beyond
the atoms which can either. add to or take away from what we
should call the energy of the universe. This proposition fore-
shadows the doctrine of conservation of energy. It is coupled
with the assertion that the sum of matter was never denser or
rarer than it now is, a proposition which we may admit, in the
sense that the mean density of the universe is constant, but
the connection of this proposition with what may be called the
constancy of the total amount of motion in the universe escapes
us. But it is clear, in all his work, that Lucretius conceived
two things as quite constant : atoms were neither created nor
destroyed, and their motion could neither be created nor
destroyed. He believed that each atom kept its velocity unaltered. The modern doctrine is that the total energy of the
universe is constant, but may be variously distributed, and is
possibly due to motion alone ultimately, though this last point
has not been yet proved. Many a fierce battle has been waged
over the question, whether what was called the 'quantity of
motion' in the universe was constant. Newton, with perfect
accuracy, declared that it was not, defining the quantity of
motion in a body as the product of mass and velocity. Leibnitz
declared that it was constant, defining the quantity of motion
as the product of the mass and the square of its velocity, but
observing that when apparently the quantity of motion diminished, it was simply transferred to the molecules of the body, so
as to escape our observation as motion. Davy and Joule have
proved him right in some cases, and shown that our senses still
detect the motion as heat. It is conceivable, but not yet proved,
that Leibnitz may be right in all cases, and that what we call

the potential energy of gravitation, elasticity, &c., may really be due to the motion either of the atoms of gross matter, or of their constituent parts. If matter in motion be conceived as the sole ultimate form of energy, Leibnitz's proposition is absolutely true, and Lucretius must be allowed great merit in having taught that the motion of matter was as indestructible as its material existence, although he knew neither the laws of momentum nor of *vis viva*. If energy, as he believed, be due solely to motion, then his doctrine is true.

It is unnecessary further to state our author's theory in distinct propositions. He proceeds to explain the necessary properties of atoms. It is not odd, he says, that though they are in continual motion, their sum (i.e. gross matter) seems to rest in supreme repose. Atoms are too minute to be perceived; their forms, he says, are various, but the number of these forms is finite. This doctrine corresponds to the modern idea of simple or elementary chemical substances, each with its special atom, but limited in number. There are, he thinks, an infinite number of similar atoms. Infinite or not, the chemical theory requires that there shall be a great many similar atoms, but nothing, thought Lucretius, is formed of simple atoms; all bodies, however minute, are compounds. Atoms have no colour, nor are they hot or cold in themselves; they have neither sound, scent, nor moisture as properties. All these properties Lucretius believed to be dependent on the shape, motion, and relative position of his atoms, but he makes only the most feeble attempt to explain how these various properties can be thus conferred, nor could this be done with the slightest hope of success until the laws of these properties had been established by long series of experiments. Something may now be done in this direction, but it remains to be done, with one exception. The motions producing the phenomenon of light are known, but we do not know what moves.

Of course, Lucretius believed organic bodies to be made of atoms, and atoms only. Sentient beings, he thought, did not require to be built up of sentient materials; but we need not discuss this conclusion, which follows of course from his assumption that nothing but atoms and void exists, a mere assumption, until the manner how atoms can build sentient beings be discovered. He determines in favour of a plurality of worlds, for what has chanced to happen here must certainly have chanced to happen elsewhere.

The Second Book concludes by a contrast between the miserable inefficiency of the gods, who pass a calm time in tranquil peace, and the mighty power of the infinite sum of clashing atoms, now building up new worlds, now slowly but inevitably crumbling heaven and earth to dust by the unceasing aggression of their never-ending flood.

He thinks Memmius his friend ought to be very glad when this conclusion is reached, and if fine poetry could please Memmius he probably enjoyed the peroration ; otherwise it is doubtful how far looking upon himself as a curious and complicated result of the accidental collision of little bits of hard stuff is calculated to make a man cheerful.

We do not propose to follow Lucretius further. The applications which he makes of his theory are no doubt curious and amusing, but they contain little that is true, while any criticism of them would lead us to consider the whole field of physical research ; nor do they add much to the clearness of his doctrine as to the constitution of matter. Let us rather reconsider what that doctrine was, and what merit it can claim. We shall find that almost all the propositions which refer simply to the constitution of matter are worthy of the highest admiration, as either certainly true, or as foreshadowing in a remarkable way doctrines since held by most eminent naturalists. Confine the following statements to matter as we can observe it, to physical science in fact, and they form a basis which even now would require but little modification to be acceptable to a modern student of physics.

Nothing is made out of nothing, nor can anything perish ; both matter and vacuum have a real existence, and gross matter, such as we perceive, contains absolutely solid particles separated by empty spaces. The absolutely solid particles are atoms. These are impenetrable, hard, indivisible, indestructible. These atoms are in continual motion, and the difference between various bodies consists, first, in the difference of the shapes of original atoms, and, secondly, in their arrangement and their motion. The velocity with which atoms move is exceedingly great, and their motion is indestructible ; it can neither increase nor diminish. This motion escapes our senses only because atoms are very small. But they are not infinitely small. Atoms have no colour, nor are they of themselves hot, cold, noisy, moist, coloured, or scented. These properties are given by motion, shape, and arrangement. We shall better understand the extraordinary merit and good sense of these propositions after considering some rival theories.

Where Lucretius breaks down is in the attempt to account for the origin of the power found in the universe, and for the various regulated motions required to explain what we observe and for the apparent anomaly between the strict causation required and perceived in inanimate nature, and the free-will of which he was conscious. Here he fails entirely, and many others have failed too. Although *he* would have cared little for our commendation of his physics, coupled with a rejection of his proud claim to have set free mankind from grovelling superstition, by explaining the mystery of the existence of matter and man's mind, *we* may derive sincere pleasure in recognising the early germs of discoveries which have required two thousand years to reach their present development. Let us not be too indignant at his scornful rejection of divine agency. Divinity to him meant either the old pagan gods or the pale abstract idea of a First Cause, which explained nothing, being but one form of statement that something was left to be explained. What wonder that he rejected both ? We may admire those old philosophers who could clothe divinity with noble attributes, and find in their own hearts the motive for their faith, but we need not therefore despise those who, smitten with the great truth that nature's laws are constant, fancied that in this constancy they saw the proof that nature's laws are self-existent. But we are diverging from our subject.

We will not compare our author's views with other ancient theories at any great length ; these at first sight seem greatly inferior to the atomic doctrine. Of the idea that the universe is composed of four elements, earth, fire, air, and water, no trace remains except in language, but careful investigation might show that the believers in these elements, or in some one or more of them, as the material of the universe, meant something

very different and much more sensible than the vulgar inter--
pretation of their doctrine. Lucretius abuses these philosophers,
some because they denied a vacuum, a denial which he thought
inconsistent with motion, some because their material wanted
the character of indestructibility which he thought essential,
some because he quite failed to perceive how all things could
be made out of the element chosen—fire, for instance ; but we
must not take Lucretius's account of rival theories as fair ; we
may with the exercise of a good deal of fancy see in the doctrine of homœomeria, which taught that all things contained the
materials of everything else in a latent state, a foreshadowing
of the chemical theory which proves that our bodies are made
of the same chemical materials as peas, cabbages, &c., but it
requires an elastic imagination to link the old and new creed
together. Any explanation of the metaphysical conceptions of
matter would also be out of place here. To Aristotle the existence
of an atom with any properties at all, and the nature of motion,
were mysteries demanding, as he says, speculation of a far
deeper kind than Democritus and the atomic school attempted.
This is true enough, but we think Aristotle and his followers
got entangled in the ' snares of words,' to use Hobbes's language,
and their teaching led to little or no progress in what we call
science. Let us then pass on some two thousand years, and see
at the revival of philosophy what some modern great men have
taught and written on the possible constitution of matter. We
need choose no smaller men than Leibnitz and Descartes to
serve as foils to our author.

Descartes, after a hypocritical flourish to the effect that he
knew the complete fallacy of all he was going to say, since it
did not agree with the orthodox theory of creation, but still that
it would be interesting to consider how God might have created
the world if He had been of Descartes's mind as to the simplest
way of proceeding, propounds the following plan :

The universe at first was quite full of something ; it was all
alike, and there was no void anywhere. This universal plenum
by-and-by was broken up into pieces. The pieces of plenum
rubbed against one another till they became quite round : the
dust rubbed off their angles filled up the interstices—for of
course no void could possibly occur once the universe was quite

full. The dust and round balls he calls the first and second kind of materials of which the universe, as we know it, is composed; but besides the dust and balls there is a third material; all the edges of the first fragments of plenum did not get ground into dust; a fair number were merely rubbed into a kind of snake-shape of triangular section—such a shape as would slip through the interstices in a pile of cannon-balls. These snake-shaped pieces sometimes got entangled, and when so entangled they composed the solid matter which is apparent to our senses. The balls and dust fill all space, the dust forms the great vortices which carry the planets round the sun, the balls are light and go flying about, so do the snakes, which, getting entangled, form gross matter. It is far more interesting to endeavour to understand the views of great men, however removed they may be from our own, than to look merely on the ludicrous side which their ideas may happen to present; but we are unable in all Descartes's theory of matter to perceive anything beyond the most childish fancy. It does not seem to have occurred to him that there would be any difficulty in breaking up an absolute plenum; what would be the nature of the separation between the fragments, what could define the boundary, he nowhere says; he sends his balls, dust, and snakes flying about in any direction he may think convenient; the balls and dust are imponderable, the knotted snakes, made of the same stuff, and intermediate between the two other kinds, are ponderable. Why three kinds —balls, dust, and snakes? Why not rather fragments of infinite variety of shape and size, from big bits of plenum to dust? No answer to all this, but long dissertation on the knotting of snakes to form spots on the sun. His laws of motion are false, and he knew it, but says we must not judge from our experience of gross matter; and yet, this man insisted on clear conceptions as the very test of truth.

Leibnitz about the same time declared against atoms, against a vacuum, and against Descartes. He will have it to be inconsistent with the perfection of God, that a vacuum can exist. It is out of the question that God should leave any part of space unemployed. John Bernoulli, in whose correspondence with Leibnitz these questions are treated with much dexterity, very properly replies that vacuum may be useful, since it may be a condition without which matter would not have its present properties ; if so, the void could not properly be called unemployed. Still, Bernoulli admitting that a void is not necessary to the theory of matter, gives it up. We must of course remember that these men did not mean by void the absence of gross matter—the Torricellian vacuum was then known—they meant absolute emptiness. This argument about what God could or could not do, because it was derogatory to His dignity or wisdom, was at this time pulled in upon all occasions, and led to the strangest paradoxes about His free-will and omnipotence. We do not use the argument now in support of the laws of mechanics ; we do not speak of circles as more perfect than other figures, and therefore more consistent with Divine wisdom, but in morals a claim of the kind is still frequently made, and Darwin applies this argument to stripes on horses' legs, which he thinks God would not have stooped to create. We are far from saying that an appeal of the kind is without meaning. The argument may be turned thus, when it will no longer seem altogether foolish : We observe great regularity and very perfect adaptation of means to ends throughout creation, so that what we do understand seems to be perfectly done, and we infer that the contrivances we do not understand are equally perfect. Any contrivance which we can show to be bad or imperfect will therefore by that very fact be proved impossible as a part of creation. The main proposition will very generally be granted; the difficulty lies in applying the minor premiss. When a man says that a vacuum is an imperfect contrivance, he only means that he dislikes it; and the application of the argument to moral questions is generally open to like criticism. Bernoulli asked Leibnitz how he accounted for the existence of moral evil as part of a perfect universe. Leibnitz returned Bernoulli's own argument about a vacuum. Evil may be necessary to allow of good, just as Bernoulli thought a vacuum might be necessary to allow of matter.

Leibnitz, though he protested against atoms, himself devised what must be called an atomic theory, though his atoms were not separated by a vacuum. They were a kind of bubble (*bulla*) with a glassy shell containing ether. They were of various composition, containing more or less fire, earth, air, or water ;

not the gross things known by that name, but essences of some kind. Leibnitz does not think his bubbles existed from all eternity, but gives the strangest account of their formation in his 'Theoria motus concreti.' He sets the sun and earth spinning in the midst of a universal ether. Molecules of the sun's mass, too, had a special motion of their own, which impelled some thing or some action, we are not sure which, along the ether, producing light; this light striking the earthy, airy, watery globe of the earth, sets the whole in fermentation; the dense parts formed in hollow bubbles containing ether; these spun round and so acquired consistency. (This idea of giving consistency by motion, taken by Leibnitz from Hobbes, was in opposition to Descartes, who derived consistency from rest.) Leibnitz explains his meaning by a metaphor : In a glass-blower's, glasses of a simple artificial form result from the straight motion of breath, combined with the circular motion of fire, and so ' bullae ' were produced from the straight motion of light and the circular motion of the earth. These bubbles are the seeds of things—Lucretius's own' phrase—the origin of various kinds of things, the receptacles of ether, the basis of bodies, the cause of the force we admire in motions.

The bubbles varied in ' contents through density ; ' in ' contents through size ; ' in emptiness, or perfect fulness, and in more or less emptiness and fulness. He explains how bubbles for the animal, vegetable, and mineral reigns, of sterile or productive qualities; salt, sulphurous, mercurial bubbles, &c. &c., are formed, and gives the special combination of qualities wanted for each. Thus, one of his bubbles is empty-extraordinary-alkaline-colourable-feminine, another full-extraordinary-acid-coloured-masculine—these two kinds of seeds differ in their way of acting. This seems like idiocy to persons not familiar with the scholastic habit of bracketing off qualities and categories, distinguishing and dividing things into a kind of verbal Chinese pattern. We have not made out the constitution of Leibnitz's ether, or his earthy, watery, airy globe, out of which he blew his bubbles, but we have found enough to show a very unfavourable contrast with Lucretius, even omitting monads, pre-established harmony, and many other interesting ideas, proposed by the man who claimed to have run a race with Newton in inventing the higher calculus of mathematics, and who enounced the doctrine of *vis viva*.

Adhesion, he thought, was obtained by motion, but how we fail to understand. His explanation runs somewhat thus—that two bodies in motion, one after the other, are both trying to be in the same place at once, and, as they cannot accomplish this, stick together. Even Bernoulli, familiar with the views and terms of the day, found Leibnitz's theory extremely difficult to understand; as found in his 'Hypothesis physica nova,' it is contained in a series of short dogmatic sentences with very little elucidation; we may therefore be unjust to him in our ignorance, but his criticism contained in his correspondence with Bernoulli seems to us much more valuable than this blowing of little complex bubbles. Thus he would not hear of the usual explanation of solidity, by the supposition that particles were hooked together or entangled by their shape, as taught both by Lucretius and Descartes. What, he asks, is to keep the hook together? and he got no answer. He refused to admit Lucretius's postulate of infinitely hard bodies and infinitely elastic bodies; indeed, the two properties do seem incompatible. The elasticity which we observe is given by a change of position of the parts of the body, and if the parts never change position it is hard to see by what the energy required for elasticity can be represented. He further objected to the assumption that atoms were indivisible, since, however small we conceive a particle to be, we can invariably think of its parts. Leibnitz was not to be satisfied with the idea which Lucretius seems to hold, that a thing may exist just big enough to have parts too small in themselves for independent existence. John Bernoulli, however, did not quite abandon atoms in consequence of this attack; like a sensible man he does not like assumptions of infinite hardness and infinite elasticity, but he replies to Leibnitz that atoms may be so constituted that they may be really indivisible by any process to which they can be subjected by other atoms, although they may have an infinity of parts such as the mind can conceive.

We will now endeavour to trace the development of the school which, discarding the hard solid elastic atoms of Lucretius, attempts to deduce the properties of matter from the

motion of an all-pervading fluid endowed with comparatively
simple qualities. This conception of matter probably differs
little from the tenets of those ancient philosophers who held that
the universe was built of some one element, such as air, fire, or
water. Descartes, who has at least the merit of reviving the
idea, in opposition to Gassendi and others who followed Lucretius, could devise no rational hypothesis from this assumption ;
but Hobbes, contemporary with Descartes, held views which
bear a striking resemblance to those recently broached by Sir
William Thomson. Hobbes thought that a moist fluid ether
fills the universe, so that it left no empty space at all. He
understood by fluidity, that which is made such by nature
equally in every part of the fluid body—not as dust is fluid, for
so a house which is falling in pieces may be called fluid—but in
such manner as water seems fluid ; he defines ' a hard body to
be that whereof no part can be sensibly moved unless the whole
be moved ; ' and in explanation how a fluid can compose a hard
body, he says, ' Whatsoever, therefore, is soft or fluid can never
be made hard, but by such motion as makes many of the parts
together stop the motion of some one part by resisting the
same ;' an admirable explanation of a recent discovery due to
Helmholtz, described below, contrasting most favourably with
Leibnitz's subsequent mere verbal quibble on the same point.
More than this, Hobbes perceived that elasticity need not be a
primary quality of matter, but might be conferred by motion.
' If the cause of this restitution (elasticity) be asked, I say it
may be in this manner, namely, that the particles of the bended
body, whilst it is held bent, do nevertheless retain their motion,
and by this motion they restore it as soon as the force is removed by which it is bent.' These are most remarkable propositions, and, should Thomson's ideas be established, will
entitle Hobbes to a very high position as the precursor of the
true theory. Unfortunately, Hobbes did not compose an harmonious system out of the above ideas. He missed the conception of vortices of ether as atoms, and introduced particles of
gross matter, distinct from ether, which may after all be true.
He also could not get free from the old nomenclature of elements, and even devised those same glassy bubbles full of ether,
which now serve chiefly to prove that Leibnitz took (without

acknowledgment which we can find) the best of Hobbes's ideas, without being able to leave the dross behind. Hobbes has a kind of undulating theory of light, which he thought was produced by the motion of an ether; Leibnitz took that too; but Galileo might perhaps claim this, as well as the notion that it was the action of this ether which was meant by the spirit brooding on the waters at creation. Leibnitz took that too, and altogether he seems to have been a great hand at appropriation.

Malebranche, who followed Descartes in most things, gave up to a great extent the balls and dust and snakes, and broached the idea that gross matter was made up of molecules, each of which was an eddy or vortex of the primeval fluid. Here we reach an intelligible conception, greatly in advance of the crude and somewhat confused views of Hobbes. The molecule is separated from the surrounding medium by the motion of its parts, it has a distinct existence, and may have very different properties from all the rest of the medium or fluid. If the parts of this fluid do not cohere in any way, but move frictionless, our little vortex-atom may have quite a sharp boundary, and if inertia be granted as an original property in our fluid, the little vortex may go on spinning for ever. Moreover, if it goes at a very great rate it may contain almost infinitely more energy or power than other parts of the medium, even when these are displaced by the motion of the vortex-atom, or a congeries of these, through the medium, which must of course then form a comparatively slow eddy coming in behind our vortex-atoms as fast as it is shoved away in front. The vortex plays the part of the Lucretian atom, the medium of the Lucretian void. A few vortices in a given space constitute a rare body; a dense body contains many vortices in the same space. The idea is one of remarkable merit, and has received several recent developments. Malebranche conceives the medium itself as full of vortices, almost infinitely small as compared with those constituting gross matter. He thought that cohesion was the result of pressure from this elastic medium against gross matter, as the two halves of a Magdeburg sphere were pressed together by the elastic air outside when the air inside is removed. Here we have a fresh explanation of hardness, as due to the motion of a fluid—an idea adopted in an unintelligible form by Leibnitz from Hobbes, and also by John Bernoulli, who further argues that this property may be given by re-entering motion.

This very idea, first due, we think, to Hobbes, and now proved possible by rigid mathematics, is perhaps the latest contribution to our subject. Helmholtz has proved that in a perfect fluid one vortex or whirlpool cannot destroy another, cannot cut through it or divide in any way from the outside— so that a ring-shaped vortex, for instance, would be quite indestructible by other vortices; by a ring-vortex we do not mean one in which the fluid moves round in a simple circle, but a ring built up of a series of such little circles side by side ; each little circle placed as a circlet of thread tied on a marriage ring would be. Such a ring-vortex as this, once set going in a perfect fluid, in which no friction occurs, would go on for ever, if we suppose our fluid endowed with inertia. Our ring-vortex might be stretched, squeezed, even knotted by other similar vortices, but it could never be pierced by them, never destroyed, and would, in all its metamorphoses, retain some of its original characteristics, depending on the velocity of its particles and its magnitude. Sir William Thomson at once pounced on this indestructible vortex as possibly fulfilling the conditions required for a practical atom. Each vortex would be indestructible, since we could never bring to bear on it anything but other like vortices. It would be elastic, in virtue of the motion of its parts only, without any assumption of elasticity in its materials— an idea this hard to grasp, but to be practically felt by anyone who tries to upset a good heavy top. He will find that, as he pushes it over, it resists, and will come upright again, exerting what we may call a kind of elasticity due to motion only. Moreover, Thomson shows that these very vortices have necessary modes of vibration, which may correspond to the special waves of light which the chemical atom of each elementary substance is capable of exciting or receiving ; knotted, or even knitted, they would explain cohesion and chemical properties without any supposition of attraction or repulsion between atoms. By their impact they may explain the elasticity of gases in the manner proposed by a later Bernoulli ; by other motions, such as those treated of by Thomson himself and Clerk Maxwell, they may cause magnetism and electricity. Nor is more required for the explanation of heat; and although it cannot be said that we yet know with any certainty what motions are required for the explanation of these phenomena, we do begin to know some of the relations which must exist between the several motions; nor need we despair even of explaining light and gravitation with the same machinery. Having traced the theory of a continuous fluid to its development in the hands of Thomson, we find that this school too has arrived at indestructible elastic atoms as the secondary constituents of gross matter, though they reject the crude atoms of Lucretius as a primary material.

Bacon was very cautious about atomic theories, but on the whole believed in atoms. He devised the idea of groups or knots of atoms, saying, in reference to the argument of Democritus, that if only one kind of atom existed, all things could be made out of all things; 'there is no doubt but that the seeds of things though equal, as soon as they have thrown themselves into certain groups or knots, completely assume the nature of dissimilar bodies till those groups or knots are dissolved.'

Newton, while approving of some form of the atomic theory, was very guarded in expressing his opinions; but his discovery of the laws of gravitation exercised great influence on most subsequent hypotheses as to the constitution of matter. The conception of atoms having the property of exerting various forces across a void space, followed as a matter of course from the idea of universal gravitation. A school arose which taught that atoms might have the property of exerting force at a distance, and that this property might be inherent in the atoms, just as Lucretius taught that hardness and elasticity were original indefeasible properties of the seeds of things. Force came to be considered as having a real existence apart from matter; but this idea, though very popular now, was not established without a hard struggle, and may yet have to be abandoned.

This view is in direct contradiction with the old axiom that matter could not act where it was not, or, as Hobbes put it, 'there can be no cause of motion except a body contiguous and moved'—no unnatural idea, but, on the contrary, universally or almost universally believed till Newton's time. We do not think that the fact of gravitation justifies the assumption that

atoms can exert a force upon one another across a void, but
Newton spoke of gravitation as an action between two distant
bodies, and since then we have got quite accustomed to the idea
of finite molecules of matter acting everywhere in the universe,
and that, too, without any material medium of communication.
This to Leibnitz was either miraculous or absurd. But, in fact,
Newton did not teach this; he stated a fact, he did not devise
hypotheses; he found that from the law of gravitation the vast
mass of facts observed about falling bodies and planetary
motions could be logically deduced. The one statement comprehended all the others; his great discovery was the short
statement with its proof; he invented no explanation of how
the law of gravitation could be brought about, and neither
asserted nor denied that some medium of communication must
exist. Leibnitz and other doubters said, How can this be, this
attraction at a distance? We cannot see how it can be done, so
we will not believe it; it is miraculous or absurd. Newton
could only reply it was a fact, and we have been so satisfied with
the answer as to be somewhat in danger of forgetting that the
question, 'How can it be?' deserves consideration; that the
statement of the law of gravitation, though a wonderful discovery, does not set a bound to further inquiry.

The law of gravitation considered as a result is beautifully
simple; in a few words it expresses a fact from which most
numerous and complex results may be deduced by mere reasoning, results found invariably to agree with the records of observation; but this same law of gravitation looked upon as an
axiom or first principle is so astoundingly far removed from all
ordinary experience as to be almost incredible. What! every
particle in the whole universe is actively attracting every other
particle through void without the aid of any communication by
means of matter or otherwise—each particle unchecked by
distance, unimpeded by obstacles, throws this miraculous influence to infinite distance without the employment of any
means! No particle interferes with its neighbour, but all
these wonderful influences are co-existent in every point of
space! The result is apparent at each particle, but the condition of this intermediate space is exactly the same as though
no such influence were being transmitted across it! Earth

attracts Sirius across space, and yet the space between is as if neither Earth nor Sirius existed! Can these things be? We think not; and Newton himself did not affirm this; his work was to prove a fact, and he neither affirmed nor denied the possibility of a medium of communication. That was a secondary question then, but now that the fact of the attraction is established the secondary question has risen to the first rank, and we must consider whether the intermediate space really contains nothing which plays a part in gravitation.

Analogy is against such a supposition. The influence exerted at a distance by electricity, magnetism, heat, and light, is effected by the substances filling intermediate space. For every one of these influences we suppose some intermediate material, and the existence of this material, often called an ether, is almost demonstrated. Faraday, by proving the influence of the intermediate material in the case of electrical action, by his discovery of magneto-optic rotation, and by showing how lines of force arose in media, rudely shook the theory of attraction and repulsion, exerted at a distance across a perfect void. Light gives us a very perfect analogy to illustrate our assertion that the law of gravitation is not an hypothesis, but a result capable of and requiring further explanation. Gravitation is not perceived directly by the senses, except in the case of the attraction of the earth. We have a special sense for the perception of light, yet many phenomena of radiation are not detected by the eye. Similarly, some of the phenomena of gravitation may escape our observation. Newton detected some of these. Suppose we had all been blind, Newton, instead of discovering universal gravitation, might have discovered light and its laws. From observations on the growth of vegetation, the sensation of heat, chemical decomposition, and other facts perceptible to blind creatures, he with vast genius might have discovered that a body existed at a great distance from the earth, from which a peculiar influence was periodically rained upon the earth; that this influence could also be produced by fire and in other ways by men living on the earth, and was in a given medium inversely proportional to the square of the distance from the source of light, as we call it. He might have discovered the transparency and opacity of bodies, and the simpler laws of refraction and reflection. To any one of his blind compeers who objected that such a supposition as an influence starting from an amazing distance, occupying no sensible time in the traject, transmitted, reflected, and refracted without the interference of one ray with another, was either miraculous or absurd, and wholly unworthy of consideration as a physical hypothesis, he would have answered : Light exists for all that, and its laws I can prove to you by mathematical reasoning from experiment. He would have been perfectly right, as he was about gravitation, but that need not have prevented subsequent philosophers from devising the undulatory theory of light if they had been clever enough ; quite similarly, the fact that gravitation as discovered by Newton does exist need not prevent our trying to devise a scheme which shall explain its action, starting from much simpler postulates than that of a universal influence of each atom on all others at a distance.

The action of a body on its neighbour can be explained without the idea of a force acting even across a small void, by the simple assumption that two bodies cannot be in the same place at the same time, an assumption only tacitly made by Lucretius, and generally received as undeniable, though it admits of rational doubt, for experiment is by no means conclusive as to its certainty. Still, most people will be and have been unable to doubt it. With this assumption, motion and influence of all kinds can be transmitted either through a fluid constituting a plenum, or from one atom to another, as they clash in a vacuum. By successive blows or extended currents action can produce results at a great distance from its origin upon either of these hypotheses, without the assumption that matter can act where it is not. Some explanation of gravity may be found requiring only the above assumption coupled with the other dogma, that matter once in motion will continue to move till stopped, and no atomic theory can be received as complete which does not explain gravitation as one of its consequences.

Lesage, a Genevese, undertook to deduce the laws of gravitation as a necessary consequence of the atomic theory, reverting, however, to the chaotic motion of atoms in all directions taught by Democritus, instead of the rectilinear parallel motions of Lucretius. Lesage asked you to conceive two solid bodies in space, say the earth and sun, and atoms coming to assail them equally in all directions; but one side of the earth would be partially screened by the sun, and the corresponding side of the sun would be partially screened by the earth, so what we would call the front faces of the earth and sun, which looked towards one another, would be less bombarded by the atoms than all the other faces. The atoms hitting at the back of the two bodies would push them together. The atoms hitting the sides would of course balance one another. The idea is ingenious, but requires some strong assumptions. The attraction of gravitation is not as the surface of the bodies, but as their mass. Lesage had therefore to suppose his solid bodies not solid but excessively porous, built up of molecules like cages, so that an infinite number of atoms went through and through them, allowing the last layer of the sun or earth to be struck by just as many atoms as the first, otherwise clearly the back part of the sun and earth would gravitate more strongly than the front or nearer sides, which would be struck only by the siftings of the previous layers of matter. This notion involves a prodigious quantity of material in the shape of flying atoms, where we perceive no gross matter, but very little material in solid bodies where we do find gross matter, and it further requires that the accumulation of atoms which strike the solid bodies perpetually should be insensible, and on these grounds, independently of dynamical imperfections, we must reject the theory in its crude form, though it may prove fruitful some day. Meanwhile it serves to show that the school which denies action at a distance need not have recourse to an absolute plenum.

Three distinct atomic theories have now been discussed: we have found believers in atoms of ' solid singleness,' in atoms due to the motion of a continuous fluid, and in atoms having the property of exerting force at a distance. Naturally the three elementary conceptions have been compounded in a variety of ways. Leibnitz mentions with great disapproval a certain Hartsoeker who supposed that atoms moved in an ambient fluid, though the idea is not unlike his own. It is difficult to trace the origin of the hypothesis, but Galileo and Hobbes both speak of a subtle ether. The conception of an all-pervading imponderable fluid of this kind has formed part of many theories, and ether came to be very generally adopted as a favourite name for the fluid, but caloric was also much thought of as a medium. We even find half-a-dozen imponderable co-existent fluids regarded with favour—one called heat, another electricity, another phlogiston, another light, and what not, with little hard atoms swimming about, each endowed with forces of repulsion and attraction of all sorts, as was thought desirable. This idea of the constitution of matter was perhaps the worst of all. These imponderable fluids were mere names, and these forces were suppositions, representing no observed facts. No attempt was made to show how or why the forces acted, but gravitation being taken as due to a mere 'force,' speculators thought themselves at liberty to imagine any number of forces, attractive or repulsive, or alternating, varying as the distance, or the square, cube, fifth power of the distance, &c: At last Boscovich got rid of atoms altogether, by supposing them to be the mere centres of forces exerted by a position or point only, where nothing existed but the power of exerting a force. A medium composed of molecules flying in all directions has been shown by Maxwell to have certain properties in which it resembles a solid body rather than a fluid. The less the molecules interfere with each other's motion the more decided do these properties become, till in the ultimate case in which they do not interfere at all, Maxwell states that the elastic properties of the medium are precisely those deduced by French mathematicians from the hypothesis of centres of force at rest acting on one another at a distance. Thus the most opposite hypotheses sometimes conduct to the same result. Dalton, assuming that the idea of an atom with an ambient ether was generally believed in, gave an immense support to the atomic theory by his discovery of the simple relations in which substances combine chemically. Since then it has been heretical to doubt atoms, until Sir Benjamin Brodie the other day broached ideas which seemed independent if not subversivè of the simple atomic faith.

Reviewing the various doctrines, we find that the problem of the constitution of matter is yet unsolved ; but at least it can now be fairly stated. We know with much accuracy the conditions to be fulfilled by any hypothesis, and we possess a

mathematical machinery by which we can test how completely
any hypothesis does fulfil those conditions. The materials for
the work are not wanting, though the architect has not appeared. Inertia and motion seem the most indispensable elements in the conception of the *materia prima* extended in space.
Once in motion, it must continue in motion till stopped; when
at rest, it must not move without a cause; when in motion, it
represents energy, or power, and can exert force. How? The
simplest but not the only mode conceivable, is by displacement,
in virtue of the property that two parts of it cannot occupy one
and the same part of space. The believers in displacement may
assume that space is quite full, or that in parts it is wholly
empty; that it contains one, two, or more kinds of primary ingredients capable of displacing one another, or each its own
products merely.

The most plausible suggestion yet made by this school is,
that a single omnipresent fluid, ether, fills the universe; that
by various motions, of the nature of eddies, the qualities of cohesion, elasticity, hardness, weight, mass, or other universal properties of matter are given to small portions of the fluid which
constitute the chemical atoms; that these, by modifications in
their combination, form and motion, produce all the accidental
phenomena of gross matter; that the primary fluid, by other
motions, transmits light, radiant heat, magnetism, and gravitation; that in certain ways the portions of the fluid transmuted
into gross matter can be acted upon by the primary fluid which
remains imponderable or very light; but that these ways differ
very much from those in which one part of gross matter acts
upon another; that the transmutation of the primary fluid into
gross matter, or of gross matter into primary fluid, is a creative
action wholly denied to us, the sum of each remaining constant.

Gross matter, on this view, would be merely an assemblage
of parts of the medium moving in a particular way; groups of
ring-vortices, for instance. There appears to be some difficulty
in determining the fundamental properties to be assumed for our
medium. We must grant it inertia or it would not continue in
motion.

The believers in hard atoms can hardly restrict themselves to
the combination and motion of atoms of gross matter; these will

not explain light, gravitation, and analogous phenomena, for which a second kind of very subtle matter is required; but this may be supposed to consist of almost infinitely finer atoms. If the molecules of gross matter, be supposed constructed from these finer atoms moving in certain special ways, this doctrine would be in accordance with that of Lucretius, and would differ little from the fluid theory, except that it would admit a void. Thus far the displacement school.

Those who believe in force exerted at a distance without a means of communication have more elbow-room. They may assume attractive and repelling forces, perhaps oblique and tangential forces; they may assume that these forces vary according to laws, simple or very complex; they may wholly deny the existence of anything but force, and grant extension and inertia to a field of force regulated in a special fashion. This little field of force, or a combination of such fields, may build their chemical atom, and the motions of these atoms, in their turn as above, produce some of the properties or accidents of gross matter; they may believe in a plenum or a partial vacuum, and in one or more kinds of matter, precisely as the other school may do; and, indeed, it is impossible to set a limit to their conjectures; because when once the mind admits this conception of an abstract force, such as that of gravitation as popularly understood, it will not refuse to entertain the idea of any other kind of force varying according to infinitely different laws, nor is there any mental limit to the possible set of co-existent forces.

Let each party try. Mathematics provide a sure test of success, though impotent to suggest a theory. The existence of the chemical atom, already quite a complex little world, seems very probable, and the description of the Lucretian atom is wonderfully applicable to it. We are not wholly without hope that the real weight of each such atom may some day be known, not merely the relative weight of the several atoms, but the number in a given volume of any material; that the form and motion of the parts of each atom, and the distance by which they are separated, may be calculated; that the motions by which they produce heat, electricity, and light may be illustrated by exact geometrical diagrams; and that the fundamental properties of the intermediate and possibly constituent medium may be arrived at. Then the motion of planets and music of the spheres will be neglected for a while in admiration of the maze in which the tiny atoms turn. Those who doubt the possibility of this achievement should read the writings of Thomson, Clausius, Rankine, and Clerk Maxwell. They will there gain some insight into what is meant by an explanation of such things as heat, electricity, and magnetism, as caused by the motion of matter, ponderable or imponderable. They will also perceive the vast difference between the old hazy speculations and the endeavours of modern science. Yet when we have found a mechanical theory by which the phenomena of inorganic matter can be mathematically deduced from the motion of materials endowed with a few simple properties, we must not forget that Democritus, Leucippus, and Epicurus began the work, and we may even now recognise their merits, and acknowledge Lucretius not only as a great poet, but as the clear expositor of a very remarkable theory of the constitution of matter, though we must admit that he failed in his bolder attempts to abolish the gods, and dispense with creation, or even to reconcile universal causation with free-will.

DARWIN AND THE ORIGIN OF SPECIES.[1]

THE theory proposed by Mr. Darwin as sufficient to account for the origin of species has been received as probable, and even as certainly true, by many who from their knowledge of physiology, natural history, and geology, are competent to form an intelligent opinion. The facts, they think, are consistent with the theory. Small differences are observed between animals and their offspring. Greater differences are observed between varieties known to be sprung from a common stock. The differences between what have been termed species are sometimes hardly greater in appearance than those between varieties owning a common origin. Even when species differ more widely, the difference, they say, is one of degree only, not of kind. They can see no clear, definite distinction by which to decide in all cases whether two animals have sprung from a common ancestor or not. They feel warranted in concluding, that for aught the structure of animals shows to the contrary, they may be descended from a few ancestors only— nay, even from a single pair.

The most marked differences between varieties known to have sprung from one source have been obtained by artificial breeding. Men have selected, during many generations, those individuals possessing the desired attributes in the highest degree. They have thus been able to add, as it were, small successive differences, till they have at last produced marked varieties. Darwin shows that by a process, which he calls natural selection, animals more favourably constituted than their fellows will survive in the struggle for life, will produce descendants resembling themselves, of which the strong will

[1] Review of Darwin's *Origin of Species*: from the *North British Review*, June 1867.

live, the weak die; and so, generation after generation, nature, by a metaphor, may be said to choose certain animals, even as man does when he desires to raise a special breed. The device of nature is based on the attributes most useful to the animal; the device of man on the attributes useful to man, or admired by him. All must agree that the process termed natural selection is in universal operation. The followers of Darwin believe that by that process differences might be added, even as they are added by man's selection, though more slowly, and that this addition might in time be carried to so great an extent as to produce every known species of animal from one or two pairs, perhaps from organisms of the lowest known type.

A very long time would be required to produce in this way the great differences observed between existing beings. Geologists say their science shows no ground for doubting that the habitable world has existed for countless ages. Drift and inundation, proceeding at the rate we now observe, would require cycles of ages to distribute the materials of the surface of the globe in their present form and order; and they add, for aught we know, countless ages of rest may at many places have intervened between the ages of action.

But if all beings are thus descended from a common ancestry, a complete historical record would show an unbroken chain of creatures, reaching from each one now known back to the first type, with each link differing from its neighbour by no more than the several offspring of a single pair of animals now differ. We have no such record; but geology can produce vestiges which may be looked upon as a few out of the innumerable links of the whole conceivable chain, and what, say the followers of Darwin, is more certain than that the record of geology must necessarily be imperfect? The records we have show a certain family likeness between the beings living at each epoch, and this is at least consistent with our views.

There are minor arguments in favour of the Darwinian hypothesis, but the main course of the argument has, we hope, been fairly stated. It bases large conclusions as to what has happened upon the observation of comparatively small facts now to be seen. The cardinal facts are the production of varieties by man, and the similarity of all existing animals. About the truth and extent of those facts none but men possessing a special knowledge of physiology and natural history have any right to an opinion; but the superstructure based on those facts enters the region of pure reason, and may be discussed apart from all doubt as to the fundamental facts.

Can natural selection choose special qualities, and so breed special varieties as man does? Does it appear that man has the power indefinitely to magnify the peculiarities which distinguish his breeds from the original stock? Is there no other evidence than that of geology as to the age of the habitable earth? and what is the value of the geological evidence? How far, in the absence of other knowledge, does the mere difficulty in classifying organised beings justify us in expecting that they have had a common ancestor? And finally, what value is to be attached to certain minor facts supposed to corroborate the new theory? These are the main questions to be debated in the present essay, written with a belief that some of them have been unduly overlooked. The opponents of Darwin have been chiefly men having special knowledge similar to his own, and they have therefore naturally directed their attention to the cardinal facts of his theory. They have asserted that animals are not so similar but that specific differences can be detected, and that man can produce no varieties differing from the parent stock, as one species differs from another. They naturally neglect the deductions drawn from facts which they deny. If your facts were true, they say, perhaps nature would select varieties, and in endless time all you claim might happen; but we deny the facts. You produce no direct evidence that your selection took place, claiming only that your hypothesis is not inconsistent with the teaching of geology. Perhaps not, but you only claim a 'may be,' and we attack the direct evidence you think you possess.

To an impartial looker-on the Darwinians seem rather to have had the best of the argument on this ground, and it is at any rate worth while to consider the question from the other point of view; admit the facts, and examine the reasoning. This we now propose to do, and for clearness will divide the subject into heads corresponding to the questions asked above, as to the extent of variability, the efficiency of natural selection,

the lapse of time, the difficulty of classification, and the value
of minor facts adduced in support of Darwin.

Some persons seem to have thought his theory dangerous to
religion, morality, and what not. Others have tried to laugh it
out of court. We can share neither the fears of the former nor
the merriment of the latter ; and, on the contrary, own to feeling
the greatest admiration both for the ingenuity of the doctrine
and for the temper in which it was broached, although, from a
consideration of the following arguments, our opinion is adverse
to its truth.

Variability.—Darwin's theory requires that there shall be no
limit to the possible difference between descendants and their
progenitors, or, at least, that if there be limits they shall be at
so great a distance as to comprehend the utmost differences
between any known forms of life. The variability required, if
not infinite, is indefinite. Experience with domestic animals
and cultivated plants shows that great variability exists. Darwin calls special attention to the differences between the various
fancy pigeons, which, he says, are descended from one stock ;
between various breeds of cattle and horses, and some other
domestic animals. He states that these differences are greater
than those which induce some naturalists to class many specimens as distinct species. These differences are infinitely small
as compared with the range required by his theory, but he assumes that by accumulation of successive differences any degree
of variation may be produced ; he says little in proof of the
possibility of such an accumulation, seeming rather to take for
granted that if Sir John Seabright could with pigeons produce
in six years a certain head and beak of say half the bulk possessed by the original stock, then in twelve years this bulk could
be reduced to a quarter, in twenty-four to an eighth, and so
farther. Darwin probably never believed or intended to teach
so extravagant a proposition, yet by substituting a few myriads
of years for that poor period of six years, we obtain a proposition
fundamental in his theory. That theory rests on the assumption that natural selection can do slowly what man's selection
does quickly ; it is by showing how much man can do, that
Darwin hopes to prove how much can be done without him.

But if man's selection cannot double, treble, quadruple, centuple, any special divergence from a parent stock, why should we imagine that natural selection should have that power? When we have granted that the 'struggle for life' might produce the pouter or the fantail, or any divergence man can produce, we need not feel one whit the more disposed to grant that it can produce divergences beyond man's power. The difference between six years and six myriads, blinding by a confused sense of immensity, leads men to say hastily that if six or sixty years can make a pouter out of a common pigeon, six myriads may change a pigeon to something like a thrush; but this seems no more accurate than to conclude that because we observe that a cannon-ball has traversed a mile in a minute, therefore in an hour it will be sixty miles off, and in the course of ages that it will reach the fixed stars. This really might be the conclusion drawn by a savage seeing a cannon-ball shot off by a power the nature of which was wholly unknown to him, and traversing a vast distance with a velocity confusing his brain, and removing the case from the category of stones and arrows, which he well knows will not go far, though they start fast. Even so do the myriads of years confuse our speculations, and seem to remove natural selection from man's selection; yet Darwin would be the first to allow that the same laws probably or possibly govern the variation, whether the selection be slow or rapid. If the intelligent savage were told, that though the cannon-ball started very fast, it went slower and slower every instant, he would probably conclude that it would not reach the stars, but presently come to rest like his stone and arrow. Let us examine whether there be not a true analogy between this case and the variation of domestic animals.

We all believe that a breeder, starting business with a considerable stock of average horses, could, by selection, in a very few generations, obtain horses able to run much faster than any of their sires or dams; in time perhaps he would obtain descendants running twice as fast as their ancestors, and possibly equal to our race-horses. But would not the difference in speed between each successive generation be less and less? Hundreds of skilful men are yearly breeding thousands of racers. Wealth and honour await the man who can breed one horse to run one part in five thousand faster than his fellows. As a matter of experience, have our racers improved in speed by one part in a thousand during the last twenty generations? Could we not double the speed of a cart-horse in twenty generations? Here is the analogy with our cannon-ball; the rate of variation in a given direction is not constant, is not erratic; it is a constantly diminishing rate, tending therefore to a limit.

It may be urged that the limit in the above case is not fixed by the laws of variation, but by the laws of matter; that bone and sinew cannot make a beast of the racer size and build go faster. This would ,be an objection rather to the form than to the essence of the argument. The existence of a limit, as proved by the gradual cessation of improvement, is the point which we aim at establishing. Possibly in every case the limit depends on some physical difficulty, sometimes apparent, more often concealed; moreover, no one can à *priori* calculate what bone and sinew may be capable of doing, or how far they can be improved; but it is unnecessary further to combat this objection, for whatever be the peculiarity aimed at by fancy breeders, the same fact recurs. Small terriers are valuable, and the limit below which a terrier of good shape would be worth its weight in silver, perhaps in gold, is nearly as well fixed as the possible speed of a race-horse. The points of all prize cattle, of all prize flowers, indicate limits. A rose called ' Senateur Vaisse ' weighs 300 grains, a wild rose weighs 30 grains. A gardener, with a good stock of wild roses, would soon raise seedlings with flowers of double, treble, the weight of his first briar flowers. He or his grandson would very slowly approach the ' Cloth of Gold ' or ' Senateur Vaisse,' and if the gradual rate of increase in weight were systematically noted, it would point with mathematical accuracy to the weight which could not be surpassed.

We are thus led to believe that whatever new point in the variable beast, bird, or flower, be chosen as desirable by a fancier, this point can be rapidly approached at first, but that the rate of approach quickly diminishes, tending to a limit never to be attained. Darwin says that our oldest cultivated plants still yield new varieties. Granted; but the new variations are not successive variations in one direction. Horses could be produced with very long or with very short ears, very long or short hair, with large or small hoofs, with peculiar colour, eyes, teeth, perhaps. In short, whatever variation we perceive of ordinary occurrence might by selection be carried to an extravagant excess. If a large annual prize were offered for any of these novel peculiarities, probably the variation in the first few years would be remarkable, but in twenty years' time the judges would be much puzzled to which breeder the prize should fall, and the maximum excellence would be known and expressed in figures, so that an eighth of an inch more or less would determine success or failure.

A given animal or plant appears to be contained, as it were, within a sphere of variation; one individual lies near one portion of the surface, another individual, of the same species, near another part of the surface; the average animal at the centre. Any, individual may produce descendants varying in any direction, but is more likely to produce descendants varying towards the centre of the sphere, and the variations in that direction will be greater in amount than the variations towards the surface. Thus, a set of racers of equal merit indiscriminately breeding will produce more colts and foals of inferior than of superior speed, and the falling off of the degenerate will be greater than the improvement of the select. A set of Clydesdale prize horses would produce more colts and foals of inferior than superior strength. More seedlings of ' Senateur Vaisse ' will be inferior to him in size and colour than superior. The tendency to revert, admitted by Darwin, is generalised in the simile of the sphere here suggested. On the other hand, Darwin insists very sufficiently on the rapidity with which new peculiarities are produced; and this rapidity is quite as essential to the argument now urged as subsequent slowness.

We hope this argument is now plain. However slow the rate of variation might be, even though it were only one part in a thousand per twenty or two thousand generations, yet if it were constant or erratic we might believe that, in untold time, it would lead to untold distance; but if in every case we find that deviation from an average individual can be rapidly effected at first, and that the rate of deviation steadily diminishes till it reaches an almost imperceptible amount, then we are as much entitled to assume a limit to the possible deviation as we are to

the progress of a cannon-ball from a knowledge of the law of
diminution in its speed. This limit to the variation of species
seems to be established for all cases of man's selection. What
argument does Darwin offer showing that the law of variation
will be different when the variation occurs slowly, not rapidly?
The law may be different, but is there any experimental ground
for believing that it *is* different? Darwin says (p. 153): 'The
struggle between natural selection on the one hand, and the
tendency to reversion and variability on the other hand, will in
the course of time cease, and that the most abnormally developed
organs may be made constant, I can see no reason to doubt.'
But what reason have we to believe this? Darwin says the
variability will disappear by the continued rejection of the
individuals tending to revert to a former condition; but is there
any experimental ground for believing that the variability *will*
disappear; and, secondly, if the variety can become fixed, that
it will in time become ready to vary still more in the original
direction, passing that limit which we think has just been
shown to exist in the case of man's selection? It is peculiarly
difficult to see how natural selection could reject individuals
having a tendency to produce offspring reverting to an original
stock. The tendency to produce offspring more like their
superior parents than their inferior grandfathers can surely be
of no advantage to any individual in the struggle for life. On
the contrary, most individuals would be benefited by producing
imperfect offspring, competing with them at a disadvantage;
thus it would appear that natural selection, if it select anything,
must select the most perfect individuals, having a tendency to
produce the fewest and least perfect competitors; but it may be
urged that though the tendency to produce good offspring is
injurious to the parents, the improved offspring would live and
receive by inheritance the fatal tendency of producing in their
turn parricidal descendants. Yet this is contending that in the
struggle for life natural selection can gradually endow a race
with a quality injurious to every individual which possesses it.
It really seems certain that natural selection cannot tend to
obliterate the tendency to revert; but the theory advanced
appears rather to be that, if owing to some other qualities a race
is maintained for a very long time different from the average or

original race (near the surface of our sphere), then it will in time spontaneously lose the tendency to relapse, and acquire a tendency to vary outside the sphere. What is to produce this change? Time simply, apparently. The race is to be kept constant, to all appearance, for a very long while, but some subtle change due to time is to take place; so that, of two individuals just alike in every feature, but one born a few thousand years after the other, the first shall tend to produce relapsing offspring, the second shall not. This seems rather like the idea that keeping a bar of iron hot or cold for a very long time would leave it permanently hot or cold at the end of the period when the heating or cooling agent was withdrawn. This strikes us as absurd now, but Bacon believed it possibly true. So many things may happen in a very long time, that time comes to be looked on as an agent capable of doing great and unknown things. Natural selection, as we contend, could hardly select an individual because it bred true. Man does. He chooses for sires those horses which he sees not only run fast themselves, but produce fine foals. He never gets rid of the tendency to revert. Darwin says species of pigeons have bred true for centuries. Does he believe that it would not be easier by selection to diminish the peculiarities of the pouter pigeon than to increase them? and what does this mean, but that the tendency to revert exists? It is possible that by man's selection this tendency may be diminished as any other quality may be somewhat increased or diminished, but, like all other qualities, this seems rapidly to approach a limit which there is no obvious reason to suppose ' time ' will alter.

But not only do we require for Darwin's theory that time shall first permanently fix the variety near the outside of the assumed sphere of variation, we require that it shall give the power of varying beyond that sphere. It may be urged that man's rapid selection does away with this power; that if each little improvement were allowed to take root during a few hundred generations, there would be no symptom of a decrease of the rate of variation, no symptom that a limit was approached. If this be so, breeders of race-horses and prize flowers had better change their tactics; instead of selecting the fastest colts and finest flowers to start with, they ought to begin with very ordinary beasts and species. They should select the descendants which might be rather better in the first generation, and then should carefully abstain from all attempts at improvement for twenty, thirty, or one hundred generations. Then they might take a little step forward, and in this way in time they or their children's children would obtain breeds far surpassing those produced by their over-hasty competitors, who would be brought to a stand by limits which would never be felt or perceived by the followers of the maxim, *Festina lente*. If we are told that the time during which a breeder or his descendants could afford to wait bears no proportion to the time used by natural selection, we may answer that we do not expect the enormous variability supposed to be given by natural selection, but that we do expect to observe some step in that direction, to find that by carefully approaching our limit by slow degrees, that limit would be removed a little farther off. Does anyone think this would be the case?

There is indeed one view upon which it would seem natural to believe that the tendency to revert may diminish. If the peculiarities of an animal's structure are simply determined by inheritance, and not by any law of growth, and if the child is more likely to resemble its father than its grandfather, its grandfather than its great-grandfather, &c., then the chances that an animal will revert to the likeness of an ancestor a thousand generations back will be slender. This is perhaps Darwin's view. It depends on the assumption that there is no typical or average animal, no sphere of variation, with centre and limits, and cannot be made use of to prove that assumption. The opposing view is that of a race maintained by a continual force in an abnormal condition, and returning to that condition so soon as the force is removed; returning not suddenly, but by similar steps with those by which it first left the average state, restrained by the tendency to resemble its immediate progenitors. *A priori*, perhaps, one view is as probable as the other; or in other words, as we are ignorant of the reasons why atoms fashion themselves into bears and squirrels, one fancy is as likely to meet with approval as another. Experiments conducted in a limited time point, as already said, to a limit, with a tendency to revert. And while admitting that the tendency to revert may be diminished though not extinguished, we are unaware of any reason for supposing that pouters, after a thousand generations of true breeding, have acquired a fresh power of doubling their crops, or that the oldest breed of Arabs are likely to produce 'sports' vastly surpassing their ancestors in speed. Experiments conducted during the longest time at our disposal show no probability of surpassing the limits of the sphere of variation, and why should we concede that a simple extension of time will reverse the rule?

The argument may be thus resumed.

Although many domestic animals and plants are highly variable, there appears to be a limit to their variation in any one direction. This limit is shown by the fact that new points are at first rapidly gained, but afterwards more slowly, while finally no further perceptible change can be effected. Great, therefore, as the variability is, we are not free to assume that successive variations of the same kind can be accumulated. There is no experimental reason for believing that the limit would be removed to a greater distance, or passed, simply because it was approached by very slow degrees, instead of by more rapid steps. There is no reason to believe that a fresh variability is acquired by long selection of one form; on the contrary, we know that with the oldest breeds it is easier to bring about a diminution than an increase in the points of excellence. The sphere of variation is a simile embodying this view;—each point of the sphere corresponding to a different individual of the same race, the centre to the average animal, the surface to the limit in various directions. The individual near the centre may have offspring varying in all directions with nearly equal rapidity. A variety near the surface may be made to approach it still nearer, but has a greater tendency to vary in every other direction. The sphere may be conceived as large for some species and small for others.

Efficiency of Natural Selection.—Those individuals of any species which are most adapted to the life they lead, live on an average longer than those which are less adapted to the circumstances in which the species is placed. The individuals which live the longest will have the most numerous offspring, and as

the offspring on the whole resemble their parents, the descendants from any given generation will on the whole resemble the
more favoured rather than the less favoured individuals of the
species. So much of the theory of natural selection will hardly
be denied ; but it will be worth while to consider how far this
process can tend to cause a variation in some one direction. It
is clear that it will frequently, and indeed generally, tend to
prevent any deviation from the common type. The mere existence of a species is a proof that it is tolerably well adapted
to the life it must lead ; many of the variations which may
occur will be variations for the worse, and natural selection will
assuredly stamp these out. A white grouse in the heather, or a
white hare on a fallow, would be sooner detected by its enemies
than one of the usual plumage or colour. Even so, any favourable deviation must, according to the very terms of the statement, give its fortunate possessor a better chance of life ; but
this conclusion differs widely from the supposed consequence
that a whole species may or will gradually acquire some one
new quality, or wholly change in one direction, and in the same
manner. In arguing this point, two distinct kinds of possible
variation must be separately considered : *first*, that kind of
common variation which must be conceived as not only possible,
but inevitable, in each individual of the species, such as longer
and shorter legs, better or worse hearing, &c. ; and, *secondly*,
that kind of variation which only occurs rarely, and may be
called a sport of nature, or more briefly a ' sport,' as when a
child is born with six fingers on each hand. The common
variation is not limited to one part of any animal, but occurs in
all ; and when we say that on the whole the stronger live longer
than the weaker, we mean that in some cases long life will have
been due to good lungs, in others to good ears, in others to good
legs. There are few cases in which one faculty is pre-eminently
useful to an animal beyond all other faculties, and where that is
not so, the effect of natural selection will simply be to kill the
weakly, and insure a sound, healthy, well-developed breed.
If we could admit the principle of a gradual accumulation or
improvements, natural selection would gradually improve the
breed of everything, making the hare of the present generation
run faster, hear better, digest better, than his ancestors ; his

enemies, the weasels, greyhounds, &c., would have 'improved likewise, so that perhaps the hare would not be really better off; but at any rate the direction of the change would be from a war of pigmies to a war of Titans. Opinions may differ as to the evidence of this gradual perfectibility of all things, but it is beside the question to argue this point, as the origin of species requires not the gradual improvement of animals retaining the same habits and structure, but such modification of those habits and structure as will actually lead to the appearance of new organs. We freely admit, that if an accumulation of slight improvements be possible, natural selection might improve hares as hares, and weasels as weasels, that is to say, it might produce animals having every useful faculty and every useful organ of their ancestors developed to a higher degree; more than this, it may obliterate some once useful organs when circumstances have so changed that they are no longer useful, for since that organ will weigh for nothing in the struggle of life, the average animal must be calculated as though it did not exist.

We will even go further: if, owing to a change of circumstances, some organ becomes pre-eminently useful, natural selection will undoubtedly produce a gradual improvement in that organ, precisely as man's selection can improve a special organ. In all cases the animals above the average live longer, those below the average die sooner, but in estimating the chance of life of a particular animal, one special organ may count much higher or lower according to circumstances, and will accordingly be improved or degraded. Thus, it must apparently be conceded that natural selection is a true cause or agency whereby in some cases variations of special organs may be perpetuated and accumulated, but the importance of this admission is much limited by a consideration of the cases to which it applies: first of all we have required that it should apply to variations which must occur in every individual, so that enormous numbers of individuals will exist, all having a little improvement in the same direction; as, for instance, each generation of hares will include an enormous number which have longer legs than the average of their parents, although there may be an equally enormous number who have shorter legs; secondly, we require that the variation shall occur in an organ already useful owing to the

habits of the animal. Such a process of improvement as is
described could certainly never give organs of sight, smell, or
hearing to organisms which had never possessed them. It could
not add a few legs to a hare, or produce a new organ, or even
cultivate any rudimentary organ which was not immediately
useful to an enormous majority of hares. No doubt half the
hares which are born have longer tails than the average of their
ancestors; but as no large number of hares hang by their tails,
it is inconceivable that any change of circumstances should breed
hares with prehensile tails; or, to take an instance less shocking
in its absurdity, half the hares which are born may be presumed to
be more like their cousins the rabbits in their burrowing organs
than the average hare ancestor was; but this peculiarity cannot be improved by natural selection as described above, until
a considerable number of hares begin to burrow, which we have
as yet seen no likelihood of their doing. Admitting, therefore,
that natural selection may improve organs already useful to
great numbers of a species, does not imply an admission that it
can create or develop new organs, and so originate species.

But it may be urged, although many hares do not burrow,
one may, or at least may hide in a hole, and a little scratching
may just turn the balance in his favour in the struggle for life.
So it may, and this brings us straight to the consideration of
'sports,' the second kind of variation above alluded to. A hare
which saved its life by burrowing would come under this head;
let us here consider whether a few hares in a century saving
themselves by this process could, in some indefinite time, make
a burrowing species of hare. It is very difficult to see how this
can be accomplished, even when the sport is very eminently
favourable indeed; and still more difficult when the advantage
gained is very slight, as must generally be the case. The
advantage, whatever it may be, is utterly outbalanced by
numerical inferiority. A million creatures are born; ten thousand survive to produce offspring. One of the million has
twice as good a chance as any other of surviving; but the
chances are fifty to one against the gifted individuals being one
of the hundred survivors. No doubt, the chances are twice as
great against any one other individual, but this does not prevent
their being enormously in favour of *some* average individual.

However slight the advantage may be, if it is shared by half the individuals produced, it will probably be present in at least fifty-one of the survivors, and in a larger proportion of their offspring ; but the chances are against the preservation of any one 'sport' in a numerous tribe. The vague use of an imperfectly understood doctrine of chance has led Darwinian supporters, first, to confuse the two cases above distinguished ; and, secondly, to imagine that a very slight balance in favour of some individual sport must lead to its perpetuation. All that can be said, is that in the above example the favoured sport would be preserved once in fifty times. Let us consider what will be its influence on the main stock when preserved. It will breed and have a progeny of say 100 ; now this progeny will, on the whole, be intermediate between the average individual and the sport. The odds in favour of one of this generation of the new breed will be, say $1\frac{1}{2}$ to 1, as compared with the average individual ; the odds in their favour will therefore be less than that of their parent ; but owing to their greater number, the chances are that about $1\frac{1}{2}$ of them would survive. Unless these breed together, a most improbable event, their progeny would again approach the average individual ; there would be 150 of them, and their superiority would be say in the ratio of $1\frac{1}{4}$ to 1 ; the probability would now be that nearly two of them would survive, and have 200 children, with an eighth superiority. Rather more than two of these would survive ; but the superiority would again dwindle, until after a few generations it would no longer be observed, and would count for no more in the struggle for life than any of the hundred trifling advantages which occur in the ordinary organs. An illustration will bring this conception home. Suppose a white man to have been wrecked on an island inhabited by negroes, and to have established himself in friendly relations with a powerful tribe, whose customs he has learnt. Suppose him to possess the physical strength, energy, and ability of a dominant white race, and let the food and climate of the island suit his constitution ; grant him every advantage which we can conceive a white to possess over the native ; concede that in the struggle for existence his chance of a long life will be much superior to that of the native chiefs ; yet from all these admissions there does not follow the conclusion that, after a limited or unlimited number of generations, the inhabitants of the island will be white. Our shipwrecked hero would probably become king; he would kill a great many blacks in the struggle for existence; he would have a great many wives and children, while many of his subjects would live and die as bachelors; an insurance company would accept his life at perhaps one-tenth of the premium which they would exact from the most favoured of the negroes. Our white's qualities would certainly tend very much to preserve him to a good old age, and yet he would not suffice in any number of generations to turn his subjects' descendants white. It may be said that the white colour is not the cause of the superiority. True, but it may be used simply to bring before the senses the way in which qualities belonging to one individual in a large number must be gradually obliterated. In the first generation there will be some dozens of intelligent young mulattoes, much superior in average intelligence to the negroes. We might expect the throne for some generations to be occupied by a more or less yellow king; but can anyone believe that the whole island will gradually acquire a white or even a yellow population, or that the islanders would acquire the energy, courage, ingenuity, patience, self-control, endurance, in virtue of which qualities our hero killed so many of their ancestors, and begot so many children; those qualities, in fact, which the struggle for existence would select, if it could select anything?

Here is a case in which a variety was introduced, with far greater advantages than any sport ever heard of, advantages tending to its preservation, and yet powerless to perpetuate the new variety.

Darwin says that in the struggle for life a grain may turn the balance in favour of a given structure, which will then be preserved. But one of the weights in the scale of nature is due to the number of a given tribe. Let there be 7,000 A's and 7,000 B's, representing two varieties of a given animal, and let all the B's, in virtue of a slight difference of structure, have the better chance of life by $\frac{1}{7000}$th part. We must allow that there is a slight probability that the descendants of B will supplant the descendants of A; but let there be only 7,001 A's against 7,000 B's at first, and the chances are once more equal, while if there be 7,002 A's to start, the odds would be laid on the A's. True, they stand a greater chance of being killed; but then they can better afford to be killed. The grain will only turn the scales when these are very nicely balanced, and an advantage in numbers counts for weight, even as an advantage in structure. As the numbers of the favoured variety diminish, so must its relative advantage increase, if the chance of its existence is to surpass the chance of its extinction, until hardly any conceivable advantage would enable the descendants of a single pair to exterminate the descendants of many thousands if they and their descendants are supposed to breed freely with the inferior variety, and so gradually lose their ascendency. If it is impossible that any sport or accidental variation in a single individual, however favourable to life, should be preserved and transmitted by natural selection, still less can slight and imperceptible variations, occurring in single individuals, be garnered up and transmitted to continually increasing numbers; for if a very highly favoured white cannot blanch a nation of negroes, it will hardly be contended that a comparatively very dull mulatto has a good chance of producing a tawny tribe; the idea, which seems almost absurd when presented in connection with a practical case, rests on a fallacy of exceedingly common occurrence in mechanics and physics generally. When a man shows that a tendency to produce a given effect exists, he often thinks he has proved that the effect must follow. He does not take into account the opposing tendencies, much less does he measure the various forces, with a view to calculate the result. For instance, there is a tendency on the part of a submarine cable to assume a catenary curve, and very high authorities once said it would; but, in fact, forces neglected by them utterly alter the curve from the catenary. There is a tendency on the part of the same cables, as usually made, to untwist entirely; luckily there are opposing forces, and they untwist very little. These cases will hardly seem obvious; but what should we say to a man who assented that the centrifugal tendency of the earth must send it off in a tangent? One tendency is balanced or outbalanced by others; the advantage of structure possessed by an isolated specimen is enormously outbalanced by the advantage of numbers possessed by the others.

A Darwinian may grant all that has been said, but contend that the offspring of ' sports' is not intermediate between the new sport and the old species; he may say that a great number of the offspring will retain in full vigour the peculiarity constituting the favourable sport. Darwin seems with hesitation to make some such claim as this, and though it seems contrary to ordinary experience, it will be only fair to consider this hypothesis. Let an animal be born with some useful peculiarity, and let all his descendants retain his peculiarity in an eminent degree, however little of the first· ancestor's blood be in them; then it follows, from mere mathematics, that the descendants of our gifted beast will probably exterminate the descendants of his inferior brethren. If the animals breed rapidly the work of substitution would proceed with wonderful rapidity, although it is a stiff mathematical problem to calculate the number of generations required in any given case. To put this case clearly beside the former, we may say that if in a tribe of a given number of individuals there appears one super-eminently gifted, and if the advantage accruing to the descendants bears some kind of proportion to the amount of the ancestor's blood in their veins, the chances are considerable that for the first few generations he will have many descendants; but by degrees this advantage wanes, and after many generations the chances are so far from being favourable to his breed covering the ground exclusively, that they are actually much against his having any descendants at all alive, for though he has a rather better chance of this than any of his neighbours, yet the chances are greatly against any one of them. It is infinitely improbable that the descendants of any one should wholly supplant the others. If, on the contrary, the advantage given by the sport is retained by all descendants, independently of what in common speech might be called the proportion of blood in their veins directly derived from the first sport, then these descendants will shortly supplant the old species entirely, after the manner required by Darwin.

But this theory of the origin of species is surely not the Darwinian theory; it simply amounts to the hypothesis that, from time to time, an animal is born differing appreciably from its progenitors, and possessing the power of transmitting the difference to its descendants. What is this but stating that, from time to time, a new species is created? It does not, indeed, imply that the new specimen suddenly appears in full vigour, made out of nothing; but it offers no explanation of the cause of the divergence from the progenitors, and still less of the mysterious faculty by which the divergence is transmitted unimpaired to countless descendants. It is clear that every divergence is not thus transmitted, for otherwise one and the same animal might have to be big to suit its father and little to suit its mother, might require a long nose in virtue of its grandfather and a short one in virtue of its grandmother, in a word, would have to resume in itself the countless contradictory peculiarities of its ancestors, all in full bloom, and unmodified one by the other, which seems as impossible as at one time to be and not to be. The appearance of a new specimen capable of perpetuating its peculiarity is precisely what might be termed a creation, the word being used to express our ignorance of how the thing happened. The substitution of the new specimens, descendants from the old species, would then be simply an example of a strong race supplanting a weak one, by a process known long before the term 'natural selection' was invented. Perhaps this is the way in which new species are introduced, but it does not express the Darwinian theory of the gradual accumulation of infinitely minute differences of every-day occurrence, and apparently fortuitous in their character.

Another argument against the efficiency of natural selection is, that animals possess many peculiarities the special advantage of which it is almost impossible to conceive; such, for instance, as the colour of plumage never displayed; and the argument may be extended by pointing how impossible it is to conceive that the wonderful minutiæ of, say a peacock's tail, with every little frond of every feather differently barred, could have been elaborated by the minute and careful inspection of rival gallants or admiring wives; but although arguments of this kind are probably correct, they admit of less absolute demonstration than the points already put. A true believer can always reply, 'You do not know how closely Mrs. Peahen inspects her husband's toilet, or you cannot be absolutely certain that under some unknown circumstances that insignificant feather was really unimportant;' or finally, he may take refuge in the word correlation, and say, other parts were useful, which by the law of correlation could not exist without these parts; and although he may have not one single reason to allege in favour of any of these statements, he may safely defy us to prove the negative, that they are not true. The very same difficulty arises when a disbeliever tries to point out the difficulty of believing that some odd habit or complicated organ can have been useful before fully developed. The believer who is at liberty to invent any imaginary circumstances, will very generally be able to conceive some series of transmutations answering his wants.

He can invent trains of ancestors of whose existence there is no evidence; he can marshal hosts of equally imaginary foes; he can call up continents, floods, and peculiar atmospheres; he can dry up oceans, split islands, and parcel out eternity at will; surely with these advantages he must be a dull fellow if he cannot scheme some series of animals and circumstances explaining our assumed difficulty quite naturally. Feeling the difficulty of dealing with adversaries who command so huge a domain of fancy, we will abandon these arguments, and trust to those which at least cannot be assailed by mere efforts of imagination. Our arguments as to the efficiency of natural selection may be summed up as follows:

We must distinguish several kinds of conceivable variation in individuals.

First. We have the ordinary variations peculiar to each individual. The effect of the struggle for life will be to keep the stock in full vigour by selecting the animals which in the main are strongest. When circumstances alter, one special organ may become eminently advantageous, and then natural selection will improve that organ. But this efficiency is limited to the cases in which the same variation occurs in enormous numbers of individuals, and in which the organ improved is already used by the mass of the species. This case does not apply to the appearance of new organs or habits.

Secondly. We have abnormal variations called sports, which may be supposed to introduce new organs or habits in rare individuals. This case must be again subdivided: we may suppose the offspring of the sports to be intermediate between

their ancestor and the original tribe. In this case the sport
will be swamped by numbers, and after a few generations its
peculiarity will be obliterated. Or, we may suppose the offspring
of the sport faithfully to reproduce the advantageous peculiarity
undiminished. In this case the new variety will supplant the
old species; but this theory implies a succession of phenomena
so different from those of the ordinary variations which we see
daily, that it might be termed a theory of successive creations; it
does not express the Darwinian theory, and is no more dependent
on the theory of natural selection than the universally admitted
fact that a new strong race, not intermarrying with an old
weak race, will surely supplant it. So much may be conceded.

Lapse of Time.—Darwin says with candour that he ' who
does not admit how incomprehensibly vast have been the past
periods of time,' may at once close his volume, admitting thereby that an indefinite, if not infinite, time is required by his
theory. Few will on this point be inclined to differ from the
ingenious author. We are fairly certain that a thousand years
has made no very great change in plants or animals living in
a state of nature. The mind cannot conceive a multiplier vast
enough to convert this trifling change by accumulation into
differences commensurate with those between a butterfly and
an elephant, or even between a horse and a hippopotamus. A
believer in Darwin can only say to himself, Some little change
does take place every thousand years; these changes accumu-
late, and if there be no limit to the continuance of the process,
I must admit that in course of time any conceivable differences
may be produced. He cannot think that a thousandfold the
difference produced in a thousand years would suffice, according to our present observation, to breed even a dog from a cat.
He may perhaps think that by careful selection, continued for
this million years, man might do quite as much as this; but he
will readily admit that natural selection does take a much
longer time, and that a million years must by the true believer
be looked upon as a minute. Geology lends her aid to convince
him that countless ages have elapsed, each bearing countless
generations of beings, and each differing in its physical conditions very little from the age we are personally acquainted

with. This view of past time is, we believe, wholly erroneous. So far as this world is concerned, past ages are far from countless; the ages to come are numbered; no one age has resembled its predecessor, nor will any future time repeat the past. The estimates of geologists must yield before more accurate methods of computation, and these show that our world cannot have been habitable for more than an infinitely insufficient period for the execution of the Darwinian transmutation.

Before the grounds of these assertions are explained, let us shortly consider the geological evidence. It is clear that denudation and deposition of vast masses of matter have occurred while the globe was habitable. The present rate of deposit and denudation is very imperfectly known, but it is nevertheless sufficiently considerable to account for all the effects we know of, provided sufficient time be granted. Any estimate of the time occupied in depositing or denuding a thousand feet of any given formation, even on this hypothesis of constancy of action, must be very vague. Darwin makes the denudation of the Weald occupy 300,000,000 years, by supposing that a cliff 500 feet high was taken away one inch per century. Many people will admit that a strong current washing the base of such a cliff as this, might get on at least a hundredfold faster, perhaps a thousandfold; and on the other hand, we may admit that, for aught geology can show, the denudation of the Weald may have occupied a few million times more years than the number Darwin arrives at. The whole calculation savours a good deal of that known among engineers as 'guess at the half and multiply by two.'

But again, what are the reasons for assuming uniformity of action, for believing that currents were no stronger, storms no more violent, alternations of temperature no more severe, in past ages than at present? These reasons, stated shortly, are that the simple continuance of actions we are acquainted with would produce all the known results, that we are not justified in assuming any alteration in the rate or violence of those actions without direct evidence, that the presence of fossils and the fineness of the ancient deposits show directly that things of old went on much as now. This last reason, apparently the strongest, is really the weakest; the deposits would assuredly take place in still waters, and we may fairly believe that still waters then resembled still waters now. The sufficiency of present actions is an excellent argument in the absence of all proof of change, but falls to utter worthlessness in presence of the direct evidence of change. We will try to explain the nature of the evidence, which does prove not only that the violence of all natural changes has decreased, but also that it is decreasing, and must continue to decrease.

Perpetual motion is popularly recognised as a delusion ; yet perpetual *motion* is no mechanical absurdity, but in given conditions is a mechanical necessity. Set a mass in motion and it must continue to move for ever, unless stopped by something else. This something else takes up the motion in some other form, and continues it till the whole or part is again transmitted to other matter ; in this sense perpetual motion is inevitable. But this is not the popular meaning of ' perpetual motion,' which represents a vague idea that a watch will not go unless it is wound up. Put into more accurate form, it means that no finite construction of physical materials can continue to *do work* for an infinite time ; or in other words, one part of the construction cannot continue to part with its energy and another part to receive it for ever, nor can the action be perpetually reversed. All motion we can produce in this world is accompanied by the performance of a certain amount of work in the form of overcoming friction, and this involves a redistribution of energy. No continual motion can therefore be produced by any finite chemical, mechanical, or other physical construction. In this case, what is true on a small scale is equally true on a large scale. Looking on the sun and planets as a certain complex physical combination, differing in degree but not in kind from those we can produce in the workshop by using similar materials subject to the same laws, we at once admit that if there be no resistance the planets may continue to revolve round the sun for ever, and may have done so from infinite time. Under these circumstances, neither the sun nor planets gain or lose a particle of energy in the process. Perpetual motion is, therefore, in this case quite conceivable. But when we find the sun raising huge masses of water daily from the sea to the skies, lifting yearly endless vegetation from the earth, setting breeze and hurricane in motion, dragging the huge tidal wave round and round our earth ; performing, in fine, the great bulk of the endless labour of this world and of other worlds, so that the energy of the sun is continually being given away ; then we may say this continual work cannot go on for ever. This would be precisely the perpetual motion we are for ever ridiculing as an exploded delusion, and yet how many persons will read these lines, to whom it has occurred that the physical work done in the world requires a motive power, that no physical motive power is infinite or indefinite, that the heat of the sun, and the sum of all chemical and other physical affinities in the world, is just as surely limited in its power of doing work as a given number of tons of coal in the furnace of a steam engine. Most readers will allow that the power man can extract from a ton of coals is limited, but perhaps not one reader in a thousand will at first admit that the power of the sun and that of the chemical affinities of bodies on the earth is equally limited.

There is a loose idea that our perpetual motions are impossible because we cannot avoid friction, and that friction entails somehow a loss of power, but that Nature either works without friction, or that in the general system friction entails no loss, and so her perpetual motions are possible ; but Nature no more works without friction than we can, and friction entails a loss of available power in all cases. When the rain falls, it feels the friction as much as drops from Hero's fountain ; when the tide rolls round the world it rubs upon the sea-floor, even as a ball of mercury rubs on the artificial inclined planes used by ingenious inventors of impossibilities ; when the breeze plays among the leaves, friction occurs according to the same laws as when artificial fans are driven through the air. Every chemical action in nature is as finite as the combustion of oxygen and carbon. The stone which, loosened by the rain, falls down a mountain-side, will no more raise itself to its first height, than the most ingeniously devised counterpoise of mechanism will raise an equal weight an equal distance. How comes it, then, that the finite nature of natural actions has not been as generally recognised as the finite nature of the so-called artificial combinations ? Simply because, till very lately, it was impossible to follow the complete cycle of natural operations in the same manner as the complete cycle of any mechanical operations could be followed. All the pressures and resistances of the machine were calculable; we knew not so much as if there were analogous pressures and resistances in nature's mechanism. The establishment of the doctrine of conservation of energy, showing a numerical equivalence between the various forms of physical energy exhibited by *vis viva*, heat, chemical affinity, electricity, light, elasticity, and gravitation, has enabled us to examine the complete series of any given actions in nature, even as the successive actions of a train of wheels in a mill can be studied. There is no missing link; there is no unseen gearing, by which, in our ignorance, we might assume that the last wheel of the set somehow managed to drive the first. We have experimentally proved one law—that the total quantity of energy in the universe is constant, meaning by energy something perfectly intelligible and mensurable, equivalent in all cases to the product of a mass into the square of a velocity, sometimes latent, that is to say, producing or undergoing no change; at other times in action, that is to say, in the act of producing or undergoing change, not a change in amount, but a change of distribution. First, the hand about to throw a ball, next the ball in motion, lastly the heated wall struck by the ball, contain the greater part of the energy of the construction; but, from first to last, the sum of the energies contained by the hand, the ball, and the wall, is constant. At first sight, this constancy, in virtue of which no energy is ever lost, but simply transferred from mass to mass, might seem to favour the notion of a possible eternity of change, in which the earlier and later states of the universe would differ in no essential feature. It is to Professor Sir W. Thomson of Glasgow that we owe the demonstration of the fallacy of this conception, and the establishment of the contrary doctrine of a continual dissipation of energy, by which the available power to produce change in any finite quantity of matter diminishes at every change of the distribution of energy. A simple illustration of the meaning of this doctrine is afforded by an unequally heated bar of iron. Let one end be hot and the other cold. The total quantity of heat (representing one form of energy) contained by the bar is mensurable and finite, and the bar contains within itself the elements of change—the heated end may become cooler, and the cold end warmer. So long as any two parts differ in temperature, change may occur; but so soon as all parts of the bar are at one temperature, the bar *quoad* heat can produce no change in itself, and yet if we conceive radiation or conduction from the surface to have been prevented, the bar will contain the same total energy as before. In the first condition, it had the power of doing work, and if it had not been a simple bar, but a more complex arrangement of materials of which the two parts had been at different temperatures, this difference might have been used to set wheels going, or to produce a thermo-electric current; but gradually the wheels would have been stopped by friction producing heat once more, the thermo-electric current would have died out, producing heat in its turn, and the final quantity of heat in the system would have been the same as before. Its distribution only, as in the simple case, would have been different. At first, great differences in the distribution existed; at last, the distribution was absolutely uniform; and in that condition the system could suffer no alteration until affected by some other body in a different condition, outside itself. Every change in the distribution of energy depends on a difference between bodies, and every change tends, on the whole, to diminish this difference, and so render the total future possible change less in amount. Heat is the great agent in this gradual decay. No sooner does energy take this form than it is rapidly dissipated, i.e. distributed among a large number of bodies, which assume a nearly equal temperature; once energy has undergone this transformation, it is practically lost. The equivalent of the energy is there; but it can produce no change until some fresh body, at a very different temperature, is presented to it. Thus it is that friction is looked upon as the grand enemy of so-called perpetual motion; it is the commonest mode by which *vis viva* is converted into heat; and we all practically know, that once the energy of our coal, boiling water, steam, piston, fly-wheel, rolling mills, gets into this form, it is simply conducted away, and is lost to us for ever; just so, when the chemical or other energies of nature, contained, say, in our planetary system, once assume the form of heat, they are in a fair way to be lost for all available purposes. They will produce a greater or less amount of change according to circumstances. The greater the difference of the temperature produced between the surrounding objects, the greater the physical changes they will effect, but the degradation is in all cases inevitable. Finally, the sun's rays take the form of heat, whether they raise water or vegetation, or do any other work, and in this form the energy quits the earth radiated into distant space. Nor would this gradual degradation be altered if space were bounded and the planets inclosed in a perfect non-conducting sphere. Everything inside that sphere would gradually become equally hot, and when this consummation was reached no further change would be possible. We might say (only we should not be alive) that the total energy of the system was the same as before, but practically the universe would contain mere changeless death, and to this condition the material universe tends, for the conclusion is not altered even by an unlimited extension of space. Moreover, the rate at which the planetary system is thus dying is perfectly mensurable, if not yet perfectly measured. An estimate of the total loss of heat from the sun is an estimate of the rate at which he is approaching the condition of surrounding space, after reaching which he will radiate no more. We intercept a few of his rays, and can measure the rate of his radiation very accurately: we know that his mass contains many of the materials our earth is formed of, and we know the capacity for heat and other forms of energy which those materials are capable of, and so can estimate the total possible energy contained in the sun's mass. Knowing thus approximately, how much he has, and how fast he is losing it, we can, or Professor Thomson can, calculate how long it will be before he will cool down to any given temperature. Nor is it possible to assume that, *per contra*, he is receiving energy to an unlimited extent in other ways. He may be supplied with heat and fuel by absorbing certain planetary bodies, but the supply is limited, and the limit is known and taken into account in the calculation, and we are assured that the sun will be too cold for our or Darwin's purposes before many millions of years—a long time, but far enough from countless ages; quite similarly past countless ages are inconceivable, inasmuch as the heat required by the sun to have allowed him to cool from time immemorial, would be such as to turn him into mere vapour, which would extend over the whole planetary system, and evaporate us entirely. It has been thought necessary to give the foregoing sketch of the inevitable gradual running down of the heavenly mechanism, to show that this reasoning concerning the sun's heat does not depend on any one special fact, or sets of facts, about heat, but is the mere accidental form of decay, which in some shape is inevitable, and the very essential condition of action. There is a kind of vague idea, when the sun is said to be limited in its heating powers, that somehow chemistry or electricity, &c., may reverse all that; but it has been explained that every one of these agencies is subject to the same law; they can never twice produce the same change in its entirety. Every change is a decay, meaning by change a change in the distribution of energy.

Another method by which the rate of decay of our planetary system can be measured, is afforded by the distribution of heat in the earth. If a man were to find a hot ball of iron suspended in the air, and were carefully to ascertain the distribution in the ball, he would be able to determine whether the ball was being heated or cooled at the time. If he found the outside hotter than the inside, he would conclude that in some way the ball was receiving heat from outside; if he found the inside hotter than the outside, he would conclude that the ball was cooling, and had therefore been hotter before he found it than when he found it. So far mere common-sense would guide him, but with the aid of mathematics and some physical knowledge of the properties of iron and air he would go much farther, and be able to calculate how hot the ball must have been at any given moment, if it had not been interfered with. Thus he would be able to say, the ball must have been hung up less than say five hours ago, for at that time the heat of the ball would have been such, if left in its present position, that the metal would be fused, and so could not hang where he saw it. Precisely analogous reasoning holds with respect to the earth; it is such a ball; it is hotter inside than outside. The distribution of the heat near its surface is approximately known. The properties of the matter of which it is composed are approximately known, and hence an approximate calculation can be made of the period of time within which it must have been hot enough to fuse the materials of which it is composed, provided it has occupied its present position, or a similar position, in space. The data for this calculation are still very imperfect, but the result of analogous calculation applied to the sun, as worked out by Professor Sir W. Thomson, is five hundred million years, and the results derived from the observed temperatures of the earth are of the same order of magnitude. This calculation is a mere approximation. A better knowledge of the distribution of heat in the interior of the globe may modify materially our estimates. A better knowledge of the conducting powers of rocks, &c., for heat, and their distribution in the earth, may modify it to a less degree, but unless our information be wholly erroneous as to the gradual increase of temperature as we descend towards the centre of the earth, the main result of the calculation, that the centre is gradually cooling, and if uninterfered with must, within a limited time, have been in a state of complete fusion, cannot be overthrown. Not only is the time limited, but it is limited to periods utterly inadequate for the production of species according to Darwin's views. We have seen a lecture-room full of people titter when told that the world would not, without supernatural interference, remain habitable for more than one hundred million years. This period was to those people ridiculously beyond anything in which they could take an interest. Yet a thousand years is an historical period well within our grasp—as a Darwinian or geological unit it is almost uselessly small. Darwin would probably admit that more than a thousand times this period, or a million years, would be no long time to ask for the production of a species differing only slightly from the parent stock. We doubt whether a thousand times more change than we have any reason to believe has taken place in wild animals in historic times, would produce a cat from a dog, or either from a common ancestor. If this be so, how preposterously inadequate are a few hundred times this unit for the action of the Darwinian theory!

But it may be said they are equally inadequate for the geological formations which we know of, and therefore your calculations are wrong. Let us see what conclusions the application of the general theory of the gradual dissipation of energy would lead to, as regards these geological formations. We may per-

haps find the solution of the difficulty in reconciling the results
of the calculation of the rate of secular cooling with the results
deduced from the denudation or deposition of strata in the
following consideration. If there have been a gradual and continual dissipation of energy, there will on the whole have been
a gradual decrease in the violence or rapidity of all physical
changes. When the gunpowder in a gun is just lighted, the
energy applied in a small mass produces rapid and violent
changes; as the ball rushes through the air it gradually loses
speed; when it strikes rapid changes again occur, but not so
rapid as at starting. Part of the energy is slowly being diffused
through the air; part is being slowly conducted as heat from
the interior to the exterior of the gun, only a residue shatters
the rampart, and that residue, soon changing into heat, is finally
diffused at a gradually decreasing rate into surrounding matter.
Follow any self-contained change, and a similar gradual diminution on the whole will be observed. There are periods of greater
and less activity, but the activity on the whole diminishes.
Even so must it have been, and so will it be, with our earth.
Extremes tend to diminish; high places become lower, low
places higher, by denudation. Conduction is continually endeavouring to reduce extremes of heat and cold; as the sun's
heat diminishes, so will the violence of storms; as inequalities
of surface diminish, so will the variations of climate. As the
external crust consolidates, so will the effect of internal fire
diminish. As internal stores of fuel are consumed, or other
stores of chemical energy used up, the convulsions or gradual
changes they can produce must diminish; on every side, and
from whatever cause changes are due, we see the tendency to
their gradual diminution of intensity or rapidity. To say that
things must or can always have gone on at the present rate is a
sheer absurdity, exactly equivalent to saying that a boiler fire
once lighted will keep a steam engine going for ever at a
constant rate. To say that changes which have occurred, or will
occur, since creation, have been due to the same causes as those
now in action; and further, that such causes have not varied in
intensity according to any other laws than those which now
prevail, is, we believe, a correct scientific statement; but then
we contend that those causes must and do hourly diminish in

intensity, and have since the beginning diminished in intensity and will diminish, till further sensible change ceases, and a dead monotony is the final physical result of the mechanical laws which matter obeys.

Once this is granted, the calculations as to the length of geological periods, from the present rates of denudation and deposit, are blown to the winds. They are rough, very rough, at best. The present assumed rates are little better than guesses; but even were these really known, they could by no means be simply made use of in a rule-of-three sum, as has generally been done. The rates of denudation and deposition have been gradually, on the whole, slower and slower, as the time of fusion has become more and more remote. There has been no age of cataclysm, in one sense, no time, when the physical laws were other than they now are, but the results were as different as the rates of a steam engine driven with a boiler first heated to 1,500 degrees Fahrenheit, and gradually cooling to 200.

A counter argument is used, to the effect that our argument cannot be correct, since plants grew quietly, and fine deposits were formed in the earliest geological times. But, in truth, this fact in no way invalidates our argument. Plants grow just as quietly on the slope of Vesuvius, with a few feet between them and molten lava, as they do in a Kentish lane; but they occasionally experience the difference of the situation. The law according to which a melted mass cools would allow vegetation to exist, and animals to walk unharmed over an incredibly thin crust. There would be occasional disturbances; but we see that a few feet of soil are a sufficient barrier between molten lava and the roots of the vine; each tendril grows not the less slowly and delicately because it is liable in a year or two to be swallowed up by the stream of lava. Yet no one will advance the proposition that changes on the surface of a volcano are going on at the same rate as elsewhere. Even so in the primeval world, barely crusted over, with great extremes of climate, violent storms, earthquakes, and a general rapid tendency to change, tender plants may have grown, and deep oceans may have covered depths of perfect stillness, interrupted occasionally by huge disturbances. Violent currents or storms in some regions do not preclude temperate climates in others, and after all the evidence of tranquillity is very slight. There are coarse deposits as well as fine ones; now a varying current sifts a deposit better than a thousand sieves, the large stones fall first in a rapid torrent, then the gravel in a rapid stream, then the coarse sand, and finally the fine silt cannot get deposited till it meets with still water. And still water might assuredly exist at the bottom of oceans, the surface of which was traversed by storms and waves of an intensity unknown to us. The soundings in deep seas invariably produce samples of almost intangible ooze. All coarser materials are deposited before they reach regions of such deathlike stillness, and this would always be so. As to the plants, they may have grown within a yard of red-hot gneiss.

Another class of objections to the line of argument pursued consists in the suggestion that it is impossible to prove that since the creation things always have been as they are. Thus, one man says: 'Ah, but the world and planetary system may have passed through a warm region of space, and then your deductions from the radiation of heat into space go for nothing; or, a fresh supply of heat and fuel may have been supplied by regular arrivals of comets or other fourgons; or the sun and centre of the earth may be composed of materials utterly dissimilar to any we are acquainted with, capable of evolving heat from a limited space at a rate which we have no example of, leaving coal or gunpowder at an infinite distance behind them. Or it may please the Creator to continue creating energy in the form of heat at the centre of the sun and earth; or the mathematical laws of cooling and radiation, and conservation of energy and dissipation of energy, may be actually erroneous, since man is, after all, fallible.' Well, we suppose all these things *may* be true, but we decline to allow them the slightest weight in the argument until some reason can be shown for believing that any one of them *is* true.

To resume the arguments in this chapter: Darwin's theory requires countless ages, during which the earth shall have been habitable, and he claims geological evidence as showing an inconceivably great lapse of time, and as not being in contradiction with inconceivably greater periods than are even geologically indicated — periods of rest between formations, and periods anterior to our so-called first formations, during which the rudimentary organs of the early fossils became degraded from their primeval uses. In answer, it is shown that a general physical law obtains, irreconcilable with the persistence of active change at a constant rate ; in any portion of the universe, however large, only a certain capacity for change exists, so that every change which occurs renders the possibility of future change less, and, on the whole, the rapidity or violence of changes tends to diminish. Not only would this law gradually entail in the future the death of all beings and cessation of all change in the planetary system, and in the past point to a state of previous violence equally inconsistent with life, if no energy were lost by the system, but this gradual decay from a previous state of violence is rendered far more rapid by the continual loss of energy going on by means of radiation. From this general conception pointing either to a beginning, or to the equally inconceivable idea of infinite energy in finite materials, we pass to the practical application of the law to the sun and earth, showing that their present state proves that they cannot remain for ever adapted to living beings, and that living beings can have existed on the earth only for a definite time, since in distant periods the earth must have been in fusion, and the sun must have been mere hot gas, or a group of distant meteors, so as to have been incapable of fulfilling its present functions as the comparatively small centre of the system. From the earth we have no very safe calculation of past time, but the sun gives five hundred million years as the time separating us from a condition inconsistent with life. We next argue that the time occupied in the arrangement of the geological formations need not have been longer than is fully consistent with this view, since the gradual dissipation of energy must have resulted in a gradual diminution of violence of all kinds, so that calculations of the time occupied by denudations or deposits based on the simple division of the total mass of a deposit, or denudation by the annual action now observed, are fallacious, and that even as the early geologists erred in attempting to compress all action into six thousand years, so later geologists have outstepped all bounds in their figures, by assuming that the world has always gone on much as it now does, and that the planetary system contains an inexhaustible motive power, by which the vast labour of the system has been and can be maintained for ever. We have endeavoured to meet the main objections to these views, and conclude that countless ages cannot be granted to the expounder of any theory of living beings, but that the age of the inhabited world is proved to have been limited to a period wholly inconsistent with Darwin's views.

Difficulty of Classification.—It appears that it is difficult to classify animals or plants, arranging them in groups as genera, species, and varieties; that the line of demarcation is by no means clear between species and sub-species, between sub-species and well-marked varieties, or between lesser varieties and individual differences; that these lines of demarcation, as drawn by different naturalists, vary much, being sometimes made to depend on this, sometimes on that organ, rather arbitrarily. This difficulty chiefly seems to have led men to devise theories of transmutation of species, and is the very starting-point of Darwin's theory, which depicts the differences between various individuals of any one species as identical in nature with the differences between individuals of various species, and supposes all these differences, varying in degree only, to have been produced by the same causes; so that the subdivision into groups is, in this view, to a great extent arbitrary, but may be considered rational if the words variations, varieties, sub-species, species, and genera, be used to signify or be considered to express that the individuals included in these smaller or greater groups have had a common ancestor very lately, some time since, within the later geological ages, or before the primary rocks. The common terms, explained by Darwin's principles, signify, in fact, the more or less close blood-relationship of the individuals. This, if it could be established, would undoubtedly afford a less arbitrary principle of classification than pitching on some one organ and dragging into a given class all creatures that had this organ in any degree similar. The application of the new doctrine might offer some difficulty, as it does not clearly appear what would be regarded as the sign of more or less immediate descent from a common ancestor, and perhaps each classifier would have pet marks by which to decide the question, in which case the new principle would not be of much practical use; yet if the theory were really true, in time the marks of common ancestry would probably come to be known with some accuracy, and meanwhile the theory would give an aim and meaning to classification, which otherwise might be looked upon as simply a convenient form of catalogue.

If the arguments already urged are true, these descents from common ancestors are wholly imaginary. ' How, then,' say the supporters of transmutation, ' do you account for our difficulty in distinguishing, *à priori*, varieties from species ? The first, we know by experience, have descended from a common ancestor : the second you declare have not, and yet neither outward inspection nor dissection will enable us to distinguish a variety from what you call a species. Is not this strange, if there be an essential difference ? '

No, it is not strange. There is nothing either wonderful or peculiar to organised beings in the difficulty experienced in classification, and we have no reason to expect that the differences between beings which have had no common ancestor should be obviously greater than those occurring in the descendants of a given stock. Whatever origin species may have had, whether due to separate creation or some yet undiscovered process, we ought to expect a close approximation between these species and difficulty in arranging them as groups. We find this difficulty in all classification, and the difficulty increases as the number of objects to be classified increases. Thus the chemist began by separating metals from metalloids, and found no difficulty in placing copper and iron in one category, and sulphur and phosphorus in the other. Nowadays, there is or has been a doubt whether hydrogen gas be a metal or no. It probably ought to be so classed. Some physical properties of tellurium would lead to its classification as a metal; its chemical properties are those of a metalloid. Acids and bases were once very intelligible headings to large groups of substances. Nowadays there are just as finely drawn distinctions as to what is an acid and what a base, as eager discussions which substance in a compound plays the part of acid or base, as there can possibly be about the line of demarcation between animal or vegetable life, and any of the characteristics used to determine the group that shall claim a given shell or plant. Nay, some chemists are just as eager to abandon the old terms altogether, as Darwin to abolish species. His most advanced disciple will hardly contend that metals and metalloids are the descendants of organic beings, which, in the struggle for life, have gradually lost all their organs : yet is it less strange that inorganic substances should be hard to class, than that organic beings, with their infinitely greater complexity, should be difficult to arrange in neat, well-defined groups ? In the early days of chemistry, a theory might well have been started, perhaps was started, that all metals were alloys of a couple of unknown substances. Each newly discovered metal would have appeared to occupy an intermediate place between old metals. Alloys similarly occupied an intermediate place between the metals composing them ; why might not all metals be simply sets of alloys, of which the elements were not yet discovered ? An alloy can no more be distinguished by its outward appearance than a hybrid can. Alloys differ as much from one another, and from metals, as metals do one from another, and a whole set of Darwinian arguments might be used to prove all metals alloys. It is only of late, by a knowledge of complicated electrical and other properties, that we could feel a certainty that metals were not alloys.

Other examples may be given, and will hereafter be given, of analogous difficulties of classification; but let us at once examine what expectations we might naturally form, à priori, as to the probable ease or difficulty in classifying plants and animals, however these may have originated. Are not animals and plants combinations, more or less complex, of a limited number of elementary parts? The number of possible combinations of a given number of elements is limited, however numerous these elements may be. The limits to the possible number of combinations become more and more restricted, as we burden these combinations with laws more and more complicated—insisting, for instance, that the elements shall only be combined in groups of threes or fives, or in triple groups of five each, or in n groups, consisting respectively of a, b, c, d . . . n elements arranged each in a given order. But what conceiv-

able complexity of algebraic arrangement can approach the
complexity of the laws which regulate the construction of an
organic being out of inorganic elements ? Let the chemist tell
us the laws of combination of each substance found in an organised being. Let us next attempt to conceive the complexity
of the conditions required to arrange these combinations in
a given order, so as to constitute an eating, breathing, moving,
feeling, self-reproducing thing. When our mind has recoiled
baffled, let us consider whether it is not probable, nay certain, that there should be a limit to the possible number of combinations, called animals, or vegetables, produced out of a few
simple elements, and grouped under the above inconceivably
complex laws. Next, we may ask whether, as in the mathematical permutations, combinations, and arrangements, the complete set of possible organised beings will not necessarily form a
continuous series of combinations each resembling its neighbour,
even as the letters of the alphabet, grouped say in all possible
sets of five each, might be arranged so as to form a continuous
series of groups, or sets of series, according as one kind of
resemblance or another be chosen to guide us in the arrangement. It is clear that the number of combinations or animals
will be immeasurably greater when these combinations are
allowed to resemble each other very closely, than when a condition is introduced, that given marked differences shall exist
between them. Thus, there are upwards of 7,890,000 words or
combinations of five letters in the English alphabet. These are
reduced to 26 when we insert a condition that no two combinations shall begin with the same letter, and to 5 when we
stipulate that no two shall contain a single letter alike. Thus
we may expect, if the analogy be admitted, to find varieties of a
given species, apparently though not really, infinite in number, since the difference between these varieties is very small,
whereas we may expect that the number of well-marked possible species will be limited, and only subject to increase by the
insertion of fresh terms or combinations, intermediate between
those already existing. Viewed in this light, a species is the
expression of one class of combination ; the individuals express
the varieties of which that class is capable.

It may be objected that the number of elements in an organised being is so great as practically to render the number of possible combinations infinite; but unless infinite divisibility of matter be assumed, this objection will not hold, inasmuch as the number of elements or parts in the germ or seed of a given animal or plant appears far from infinite. Yet it is certain that differences between one species and another, one variety and another, one individual and another, exist in these minute bodies, containing very simple and uniform substances if analysed chemically. Probably, even fettered by these conditions, the number of possible animals or plants is inconceivably greater than the number which exist or have existed; but the greater the number, the more they must necessarily resemble one another.

It may perhaps be thought irreverent to hold an opinion that the Creator could not create animals of any shape and fashion whatever; undoubtedly we may conceive all rules and all laws as entirely self-imposed by Him, as possibly quite different or non-existent elsewhere; but what we mean is this, that just as with the existing chemical laws of the world the number of possible chemical combinations of a particular kind is limited, and not even the Creator could make more without altering the laws He has Himself imposed, even so, if we imagine animals created or existing under some definite law, the number of species, and of possible varieties of one species, will be limited; and these varieties and species, being definite arrangements of organic compounds, will as certainly be capable of arrangement in series as inorganic chemical compounds are. These views no more imply a limit to the power of God than the statement that the three angles of a triangle are necessarily equal to two right angles.

It is assumed that all existing substances or beings of which we have any scientific knowledge exist under definite laws. Under any laws there will be a limit to the possible number of combinations of a limited number of elements. The limit will apply to size, strength, length of life, and every other quality. Between any extremes the number of combinations called animals or species can only be increased by filling in gaps which exist between previously existing animals, or between these and the possible limits, and therefore whatever the general laws of organisation may be, they must produce results similar to those which we observe, and which lead to difficulty in classification, and to the similarity between one species or variety and another. Turning the argument, we might say that the observed facts simply prove that organisms exist and were created under definite laws, and surely no one will be disposed to deny this. Darwin assumes one law, namely, that every being is descended from a common ancestor (which, by the way, implies that every being shall be capable of producing a descendant like any other being), and he seems to think this the only law which would account for the close similarity of species, whereas any law may be expected to produce the same results. We observe that animals eat, breathe, move, have senses, are born, and die, and yet we are expected to feel surprise that combinations, which are all contrived to perform the same functions, resemble one another. It is the apparent variety that is astounding, not the similarity. Some will perhaps think it absurd to say that the number of combinations is limited. They will state that no two men ever were or will be exactly alike, no two leaves in any past or future forest; it is not clear how they could find this out, or how they could prove it. But, as already explained, we quite admit that by allowing closer and closer similarity, the number of combinations of a fixed number of elements may be enormously increased. We may fairly doubt the identity of any two of the higher animals, remembering the large number of elements of which they consist, but perhaps two identical foraminiferæ have existed. As an idle speculation suggested by the above views, we might consider whether it would be possible that two parts of any two animals should be identical, without their being wholly identical, looking on each animal as one possible combination, in which no part could vary without altering all the others. It would be difficult to ascertain this by experiment.

It is very curious to see how man's contrivances, intended to fulfil some common purpose, fall into series, presenting the difficulty complained of by naturalists in classifying birds and beasts, or chemists in arranging compounds. It is this difficulty which produces litigation under the Patent Laws. Is or is not this machine comprised among those forming the subject of the

patent ? At first sight nothing can be more different than the
drawing in the patent and the machine produced in court, and
yet counsel and witnesses shall prove to the satisfaction of
judge, jury, and one party to the suit, that the essential part, the
important organ, is the same in both cases. The case will often
hinge on the question, What is the important organ ? Just the
question which Darwin asks; and quite as difficult to answer
about a patented machine as about an organic being.

This difficulty results from the action of man's mind con-
triving machines to produce a common result according to definite laws, the laws of mechanics. An instance of this is afforded
by the various forms of bridge. Nothing would appear more
distinct than the three forms of suspension bridge, girder, and
arch; the types of which are furnished by a suspended rope, a
balk of wood, and a stone arch; yet if we substitute an iron-
plate girder of approved form for the wooden balk, and then a
framed or lattice girder for the plate-iron girder, we shall see
that the girder occupies an intermediate place between the two
extremes, combining both the characteristics of the suspension
and arched rib—the upper plates and a set of diagonal strutts
being compressed like the stones of an arch, the lower plates
and a set of diagonal ties being extended like a suspended rope.
Curve the top plates, as is often done, and the resemblance to
an arch increases, yet every member of the girder remains.
Weaken the bracing, leaving top and bottom plates as before,
the bridge is now an arched bridge with the abutments tied
together. Weaken the ties gradually, and you gradually approach nearer and nearer to the common arch with the usual
abutments. Quite similarly the girder can be transformed into
a suspension bridge by gradual steps, so that none can say
when the girder ends and the suspension bridge begins. Nay,
take the common framed or lattice girder, do not alter its shape
in any way, but support it, first, on flat stones like a girder, then
wedge it between sloping abutments like an arch, and lastly,
hang it up between short sloping links like those of a suspension
bridge, attached to the upper corners at the end—you will so
alter the strains in the three cases that in order to bear the
same load, the relative parts of the framework must be altered
in their proportions in three distinct ways, resembling in the

arrangement of the strongest parts, first a girder, next an arch, and finally a suspension bridge. Yet the outline might remain the same, and not a single member be removed.

Thus we see, that though in three distinct and extreme cases it is easy to give distinctive names with clear characteristics, it is very difficult as the varieties multiply to draw distinct lines between them. Shall the distribution of strains be the important point? Then one and the same piece of framework will have to be included under each of three heads, according to the manner in which it is suspended or supported. Shall form be the important point? We may construct a ribbed arch of string, of a form exactly similar to many compressed arches, we may support this from below, and yet the whole arch shall be in tension, and bear a considerable load. Shall the mode of support be the important point? It would be an odd conclusion to arrive at, that any stiff beam hung up in a particular way was a suspension bridge. Nor is this difficulty simply a sophistical one invented for the occasion; the illustration was suggested by a practical difficulty met with in drawing up a patent; and in ordinary engineering practice one man will call a certain bridge a stiffened arch, while another calls it a girder of peculiar form; a third man calls a bridge a strengthened girder, which a fourth says differs in no practical way from a suspension bridge. Here, as in the case of animals or vegetables, when the varieties are few, classification is comparatively easy; as they are multiplied it becomes difficult; and when all the conceivable combinations are inserted it becomes impossible. Nor must it be supposed that this is due to the suggestion of one form by another in a way somewhat analogous to descent by animal reproduction. The facts would be the same however the bridges were designed. There are only certain ways in which a stream can be bridged; the extreme cases are easily perceived, and ingenuity can then only fill in an indefinite number of intermediate varieties. The possible varieties are not created by man, they are found out, laid bare. Which are laid bare will frequently depend on suggestion or association of ideas, so that groups of closely analogous forms are discovered about the same time; but we may *à priori* assert that whatever is discovered will lie between the known extremes and will render the task of classification, if attempted, more and more difficult.

Legal difficulties furnish another illustration. Does a particular case fall within a particular statute? is it ruled by this or that precedent? The number of statutes or groups is limited; the number of possible combinations of events almost unlimited. Hence, as before, the uncertainty which group a special combination shall be classed within. Yet new combinations being doubtful cases, are so precisely because they are intermediate between others already known.

It might almost be urged that all the difficulties of reasoning and all differences of opinion might be reduced to difficulties of classification, that is to say, of determining whether a given minor is really included in a certain major proposition and of discovering the major proposition or genus we are in want of. As trivial instances, take the docketing of letters or making catalogues of books. How difficult it is to devise headings, and how difficult afterwards to know under what head to place your book. The most arbitrary rule is the only one which has a chance of being carried out with absolute certainty.

Yet while these difficulties meet us wherever we turn, in chemistry, in mechanics, law, or mere catalogues of heterogeneous objects, we are asked to feel surprise that we cannot docket off creation into neat rectangular pigeon-holes, and we are offered a special theory of transmutation, limited to organic beings, to account for a fact of almost universal occurrence.

To resume this argument: Attention has been drawn to the fact that when a complete set of combinations of certain elements is formed according to a given law, they will necessarily be limited in number, and form a certain sequence, passing from one extreme to the other by successive steps.

Organised beings may be regarded as combinations, either of the elementary substances used to compose them, or of the parts recurring in many beings; for instance, of breathing organs, apparatus for causing blood to circulate, organs of sense, reproduction, &c., in animals. The conclusion is drawn that we can feel no reasonable surprise at finding that species should form a graduated series which it is difficult to group as genera, or that varieties should be hard to group into various distinct species.

Nor is it surprising that newly discovered species and varieties should almost invariably occupy an intermediate position between some already known, since the number of varieties of one species, or the number of possible species, can only be indefinitely increased by admitting varieties or species possessing indefinitely small differences one from another.

We observe that these peculiarities require no theory of transmutation, but only that the combination of the parts, however effected, should have been made in accordance with some law, as we have every reason to expect they would be.

In illustration of this conclusion, cases of difficult classification are pointed out containing nothing analogous to reproduction, and where no struggle for life occurs.

Observed Facts supposed to support Darwin's Views.—The chief arguments used to establish the theory rest on conjecture. Beasts may have varied; variations may have accumulated; they may have become permanent; continents may have arisen or sunk, and seas and winds been so arranged as to dispose of animals just as we find them, now spreading a race wildly, now confining it to one Galapagos island. There may be records of infinitely more animals than we know of in geological formations yet unexplored. Myriads of species differing little from those we know to have been preserved, may actually not have been preserved at all. There may have been an inhabited world for ages before the earliest known geological strata. The world may indeed have been inhabited for an indefinite time; even the geological observations may perhaps give a most insufficient idea of the enormous times which separated one formation from another: the peculiarities of hybrids may result from accidental differences between the parents, not from what have been called specific differences.

We are asked to believe all these maybe's happening on an enormous scale, in order that we may believe the final Darwinian 'maybe,' as to the origin of species. The general form of his argument is as follows :—All these things may have been, therefore my theory is possible, and since my theory is a possible one, all those hypotheses which it requires are rendered probable. There is little direct evidence that any of these may-be's actually *have been*.

In this essay an attempt has been made to show that many of these assumed possibilities are actually impossibilities, or at the best have not occurred in this world, although it is prover-bially somewhat difficult to prove a negative.

Let us now consider what direct evidence Darwin brings for-ward to prove that animals really are descended from a common ancestor. As direct evidence we may admit the possession of webbed feet by unplumed birds; the stripes observed on some kinds of horses and hybrids of horses, resembling not their parents, but other species of the genus; the generative vari-ability of abnormal organs; the greater tendency to vary of widely diffused and widely ranging species; certain peculiarities of distribution. All these facts are consistent with Darwin's theory, and if it could be shown that they could not possibly have occurred except in consequence of natural selection, they would prove the truth of this theory. It would, however, clearly be impossible to prove that in no other way could these phenomena have been produced, and Darwin makes no attempt to prove this. He only says he cannot imagine why unplumed birds should have webbed feet, unless in consequence of their direct descent from web-footed ancestors who lived in the water: that he thinks it would in some way be derogatory to the Creator to let hybrids have stripes on their legs, unless some ancestor of theirs had stripes on his leg. He cannot imagine why abnormal organs and widely diffused genera should vary more than others, unless his views be true; and he says he cannot account for the peculiarities of distribution in any way but one. It is perhaps hardly necessary to combat these arguments, and to show that our inability to account for certain phenomena, in any way but one, is no proof of the truth of the explanation given, but simply is a confession of our ignorance. When a man says a glowworm must be on fire, and in answer to our doubts challenges us to say how it can give out light unless it be on fire, we do not admit his challenge as any proof of his assertion, and indeed we allow it no weight whatever as against positive proof we have that the glowworm is not on fire. We conceive Darwin's theory to be in exactly the same case;

its untruth can, as we think, be proved, and his or our own inability to explain a few isolated facts consistent with his views
would simply prove his and our ignorance of the true explanation. But although unable to give any certainly true explanation of the above phenomena, it is possible to suggest explanations perhaps as plausible as the Darwinian theory, and though
the fresh suggestions may very probably not be correct, they
may serve to show that at least more than one conceivable explanation may be given.

It is a familiar fact that certain complexions go with certain
temperaments, that roughly something of a man's character may
be told from the shape of his head, his nose, or perhaps from
most parts of his body. We find certain colours almost always
accompanying certain forms and tempers of horses. There is a
connexion between the shape of the hand and the foot, and so
forth. No horse has the head of a cart-horse and the hind-
quarters of a racer; so that, in general, if we know the shape of
most parts of a man or horse, we can make a good guess at the
probable shape of the remainder. All this shows that there is
a certain correlation of parts, leading us to expect that when the
heads of two birds are very much alike, their feet will not be
very different. From the assumption of a limited number of
possible combinations or animals, it would naturally follow that
the combination of elements producing a bird having a head
very similar to that of a goose, could not fail to produce a foot
also somewhat similar. According to this view, we might expect
most animals to have a good many superfluities of a minor kind,
resulting necessarily from the combination required to produce
the essential or important organs. Surely, then, it is not very
strange that an animal intermediate by birth between a horse
and ass should resemble a quagga, which results from a combination intermediate between the horse and ass combination. The
quagga is in general appearance intermediate between the horse
and ass, therefore, *à priori*, we may expect that in general
appearance a hybrid between the horse and the ass will resemble
the quagga, and if in general appearance it does resemble a
quagga, we may expect that owing to the correlation of parts
it will resemble the quagga in some special particulars. It is
difficult to suppose that every stripe on a zebra or quagga, or

cross down a donkey's back, is useful to it. It seems possible, even probable, that these things are the unavoidable consequences of the elementary combination which will produce the quagga, or a beast like it. Darwin himself appears to admit that correlation will or may produce results which are not themselves useful to the animal; thus how can we suppose that the beauty of feathers which are either never uncovered, or very rarely so, can be of any advantage to a bird ? Nevertheless those concealed parts are often very beautiful, and the beauty of the markings on these parts must be supposed due to correlation. The exposed end of a peacock's feather could not be so gloriously coloured without beautiful colours even in the unexposed parts. According to the view already explained, the combination producing the one was impossible unless it included the other. The same idea may perhaps furnish the clue to the variability of abnormal organs and widely diffused species, the abnormal organ may with some plausibility be looked upon as the rare combination difficult to effect, and only possible under very special circumstances. There is little difficulty in believing that it would more probably vary with varying circumstances than a simple and ordinary combination. It is easy to produce two common wine-glasses which differ in no apparent manner; two Venice goblets could hardly be blown alike. It is not meant here to predicate ease or difficulty of the action of omnipotence; but just as mechanical laws allow one form to be reproduced with certainty, so the occult laws of reproduction may allow certain simpler combinations to be produced with much greater certainty than the more complex combinations. The variability of widely diffused species might be explained in a similar way. These may be looked on as the simple combinations of which many may exist similar one to the other, whereas the complex combinations may only be possible within comparatively narrow limits, inside which one organ may indeed be variable, though the main combination is the only possible one of its kind.

We by no means wish to assert that we know the above suggestions to be the true explanations of the facts. We merely wish to show that other explanations than those given by Darwin are conceivable, although this is indeed not required by our argument, since, if his main assumptions can be proved false, his theory will derive no benefit from the few facts which may be allowed to be consistent with its truth.

The peculiarities of geographical distribution seem very difficult of explanation on any theory. Darwin calls in alternately winds, tides, birds, beasts, all animated nature, as the diffusers of species, and then a good many of the same agencies as impenetrable barriers. There are some impenetrable barriers between the Galapagos Islands, but not between New Zealand and South America. Continents are created to join Australia and the Cape of Good Hope, while a sea as broad as the British Channel is elsewhere a valid line of demarcation. With these facilities of hypothesis there seems to be no particular reason why many theories should not be true. However an animal may have been produced, it must have been produced somewhere, and it must either have spread very widely, or not have spread, and Darwin can give good reasons for both results. If produced according to any law at all, it would seem probable that groups of similar animals would be produced in given places. Or we might suppose that all animals having been created anywhere or everywhere, those have been extinguished which were not suited to such climate; nor would it be an answer to say that the climate, for instance, of Australia, is less suitable now to marsupials than to other animals introduced from Europe, because we may suppose that this was not so when the race began; but in truth it is hard to believe any of the suppositions, nor can we just now invent any better; and this peculiarity of distribution, namely, that all the products of a given continent have a kind of family resemblance, is the sole argument brought forward by Darwin which seems to us to lend any countenance to the theory of a common origin and the transmutation of species.

Our main arguments are now completed. Something might be said as to the alleged imperfection of the geological records. It is certain that, when compared with the total number of animals which have lived, they must be very imperfect; but still we observe that of many species of beings thousands and even millions of specimens have been preserved. If Darwin's theory be true, the number of varieties differing one from another a very little must have been indefinitely great, so great indeed as probably far to exceed the number of individuals which have existed of any one variety. If this be true, it would be more probable that no two specimens preserved as fossils should be of one variety than that we should find a great many specimens collected from a very few varieties, provided, of course, the chances of preservation are equal for all individuals. But this assumption may be denied, and some may think it probable that the conditions favourable to preservation only recur rarely, at remote periods, and never last long enough to show a gradual unbroken change. It would rather seem probable that fragments, at least, of perfect series would be preserved of those beings which lead similar lives favourable to their preservation as fossils. Have any fragments of these Darwinian series been found where the individuals merge from one variety insensibly to another?

It is really strange that vast numbers of perfectly similar specimens should be found, the chances against their perpetuation as fossils are so great; but it is also very strange that the specimens should be so exactly alike as they are, if, in fact, they came and vanished by a gradual change. It is, however, not worth while to insist much on this argument, which by suitable hypotheses might be answered, as by saying, that the changes were often quick, taking only a few myriad ages, and that then a species was permanent for a vastly longer time, and that if we have not anywhere a gradual change clearly recorded, the steps from variety to variety are gradually being diminished as more specimens are discovered. These answers do not seem to us sufficient, but the point is hardly worth contesting, when other arguments directly disproving the possibility of the assumed change have been advanced.

These arguments are cumulative. If it be true that no species can vary beyond defined limits, it matters little whether natural selection would be efficient in producing definite variations. If natural selection, though it does select the stronger average animals and under peculiar circumstances may develop special organs already useful, can never select new imperfect organs such as are produced in sports, then, even though eternity were granted, and no limit assigned to the possible changes of animals, Darwin's cannot be the true explanation of the manner in which change has been brought about. Lastly,

even if no limit be drawn to the possible difference between offspring and their progenitors, and if natural selection were
admitted to be an efficient cause capable of building up even
new senses, even then, unless time, vast time, be granted, the
changes which might have been produced by the gradual selection of peculiar offspring have not really been so produced.
Any one of the main pleas of our argument, if established, is
fatal to Darwin's theory. What then shall we say if we believe
that experiment has shown a sharp limit to the variation of
every species, that natural selection is powerless to perpetuate
new organs even should they appear, that countless ages of a
habitable globe are rigidly proven impossible by the physical
laws which forbid the assumption of infinite power in a finite
mass? What can we believe but that Darwin's theory is an
ingenious and plausible speculation, to which future physiolo-
gists will look back with the kind of admiration we bestow on
the atoms of Lucretius, or the crystal spheres of Eudoxus, containing like these some faint half-truths, marking at once the
ignorance of the age and the ability of the philosopher. Surely
the time is past when a theory unsupported by evidence is
received as probable, because in our ignorance we know not why
it should be false, though we cannot show it to be true. Yet
we have heard grave men gravely urge, that because Darwin's
theory was the most plausible known, it should be believed.
Others seriously allege that it is more consonant with a lofty
idea of the Creator's action to suppose that he produced beings
by natural selection, rather than by the finikin process of making
each separate little race by the exercise of Almighty power.
The argument, such as it is, means simply that the user of it
thinks that this is how he personally would act if possessed of
almighty power and knowledge, but his speculations as to his
probable feelings and actions, after such a great change of circumstances, are not worth much. If we are told that our experience shows that God works by law, then we answer, ' Why the
special Darwinian law?' A plausible theory should not be
accepted while unproven; and if the arguments of this essay be
admitted, Darwin's theory of the origin of species is not only
without sufficient support from evidence, but is proved false by
accumulative proof.

A FRAGMENT ON TRUTH.[1]

I.

It is admitted on all hands that in all matters, whether of faith, knowledge or perception, we should endeavour to attain truth; to believe truly, know truly, feel truly. But, inasmuch as in all these respects all men differ one from another, a strong desire has ever been felt for some criterion or touchstone by which a man might, even in the simplest case, discern truth from false-hood. No such touchstone has or ever will be found, and in-deed the desire is rather akin to the vague longing for magic powers than to any healthy appetite.

The word truth is in this matter strangely abused. The word, truly used, signifies a concordance either between some verbal expression and an external fact or between some mental impression and an external fact. These two meanings shall be separately considered. In regard to the first, let us first re-member that no word has any definite or absolutely circum-scribed meaning, much less, any congeries of words such as form a sentence. No one word absolutely corresponds to any one thing, fitting it so as absolutely to express all its qualities and no others. To suppose that a sentence built of these imperfect bricks, and apprehended in the shadowy impression which words convey, shall correspond absolutely with even the simplest natural fact in all its relations is manifestly absurd.

Agreement between many minds as to any statement be-comes more and more probable as the statement is more and more restricted to the simplest class of facts. In this and in this alone lies the supposed superiority of mathematics, and science generally, in regard to truths. The truths of exact science mean very little; hence we agree fairly well as to the truth of the

[1] From an unfinished MS., dated May, 1885.

language by which they are expressed. The moment we apply these so-called truths to express the complex facts of any real combination, we are met by the discrepancy between theory and practice ; the statements and the facts accord imperfectly.

Using the word truth in this first sense, the mechanic well knows that absolute truth is unattainable, but he nevertheless values highly the approximation to truth which allows him approximately to ascertain the result of given simple material combinations. The knowledge that perfect agreement between the statement and the corresponding fact is impossible by no means leads him to avoid the calculation of the breaking weight of a bridge, or the power required to pump water for the supply of a town. The calculations are imperfect; the words, or symbols, and the facts do not agree ; it is inconceivable that they should agree ; at most he hopes for the approximate agreement between the conception indicated by the words in his mind and one small part of the resulting facts, but this approximate agreement is of infinite value to mankind, and the mechanic who rejects calculation because of its imperfect accord with fact is ignorant or foolish.

Coming now to the second conceivable meaning of truth, as signifying an accord between what seems and what is, between conception in our minds and the fact which gives rise to that conception, the very notion of absolute agreement or identity seems preposterous. The impression can not conceivably be identical with the impresser. A dint by a hammer can not be identical with the hammer which produces it. At most one may fit the other, and it is clear that in any such interaction the nature of the impressing or impressed matter greatly influences the perfection of fit. Our perception can never be the thing perceived, and can never correspond even roughly with that thing, except as regards some very limited quality ; as the dint may correspond with a part of the hammer head in respect of form, but in no wise as regards weight, colour, strength, magnetic properties, and so forth. Moreover, the impression even in regard to form will differ widely in different substances ; nay, no two pieces of money, all of one material, all struck from one die, shall be identical. How much less can we expect the impression on several minds, differing far more one from another

than one gold bar from another, to be identical? Not only is
the gold piece not the die, but no two coins are absolutely alike.
So our perceptions are not the things perceived, neither are the
perceptions of one thing the same in two men. But coin is that
which we can use, the die is no concern of ours—and the identity
between the coins is sufficient for our use. So natural facts are
important to us only as they impress us; with the things in
themselves we have no concern, and the impressions on various
men are so far alike as to render commerce possible.

Therefore we may say nothing is true, in either sense.
Words can express no truth wholly, and again no impression
corresponds wholly with the fact; and again we may say that
no man can honestly say that which is untrue, for if he say that
which he believes, his words express that which he perceives or
conceives. He may indeed use the words unskilfully, that is to
say, his words may convey to me some impression false to the
fact, yet to him the words may be true, as when a man awaking
says, I have not slept. He can but tell what is then in his mind,
and he truly tells this although he slept. He would be an untrue witness if having slept unawares he, not remembering it,
said, ' I slept.'

So again we may say that our senses never deceive us, for
given all the circumstances, the impression must be that which
it is. We might desire to leave out some of the circumstances,
as in trying to spear a fish we might desire to leave out the
refractive power of water, but our senses tell us all the fact.
They even tell us their own imperfection, the impression made
being partly the result of the fact and partly the result of our
own organism. In this sense every impression is true, being
the inevitable result of the whole complex circumstances. The
man who is colour-blind is not misled by his senses. His eye
tells him that to him a given surface is colourless, which is true,
although to another man this surface may be brilliantly coloured.
The one impression is no more true in this transcendental sense
than the other.

This matter of truth and untruth has been thus sifted merely
to expose the folly involved in any hope of a criterion of absolute
truth. Let there be a criterion—is it conceivable that we
should ever apply it with exactitude?

II.

When it is said that by scientific measurement things, not sensations, are compared, the word things must be interpreted as bearing a very wide meaning. Time, space, matter, energy, are included in the term. Measurements compare such things as position, velocity, tenacity, and so forth. Call these attributes, properties, abstractions, relations, conditions, actions, what you will, the fact remains that when the man of science measures these things, he compares the things themselves, not the sensations they produce in his mind. Neither the method nor the result of the comparison is affected by the individual mind, except so far as concerns the more or less complete accuracy of the measurement.

No doubt one individual will weigh with less error than another while using the same apparatus, and then the imperfection of our organs prevents absolute agreement about even measurable things; but the disagreement in these cases affects the mere fringe or margin of the thing, not the thing in bulk. One man comparing the weight of two masses, A and B, shall say that A is twice as heavy as B; another man shall say that this relation is really 2 to 1·0001; but no person can be found to measure so badly as to suppose that B is twice as heavy as A. A reversal of this kind, or indeed any wide disagreement between the measurements of fairly skilled observers is inconceivable, and all the measurers act successfully on the supposition that there is one real relation between the weights, and that this relation is independent of the peculiarities of their own minds. Those fields of knowledge in which this manner of comparison between the things themselves is possible cover the domain of exact knowledge rightly so called, although we may never hope for absolute exactitude in any measurement whatever. We know absolutely that in those fields relations exist and that our senses enable us to determine those relations with all the nicety which our wants require. In these regions dispute is settled by an appeal to things, not to minds; no criterion of truth is missed, and prolonged difference of opinion concerning the relations measured is impossible, except as regards a small and constantly lessening fringe or margin of the whole part.

III.

The mathematical relations between numbers and magnitudes require separate consideration.

In geometry they involve the comparison of dimensions, and dimensions are, in the sense explained above, things. But where mathematics deal with the relations between symbols, whether algebraic or numerical, they do not compare things. They essentially substitute a symbol for the thing itself, and use the symbols to facilitate not measurement but ratiocination concerning measurements. Number is not an object, a condition, a property, an attribute, or even an abstraction. Numbers express the result of measurement; they cannot themselves be measured. The relation between them is stated when they are stated. The very conception of number involves the assumption that two things can be identical, or may for a special purpose be regarded as identical. This is neither the result of abstraction nor reasoning; it is merely a short way of saying that any four things which we choose to regard as identical, may be counted in two ways, described as regards number in two different phrases————.

END OF THE FIRST VOLUME.

PRINTED BY
SPOTTISWOODE AND CO., NEW-STREET SQUARE
LONDON

Printed in the United States
By Bookmasters